Advanced Energetic Materials: Testing and Modeling

Advanced Energetic Materials: Testing and Modeling

Editors

Rui Liu
Yushi Wen
Weiqiang Pang

Basel • Beijing • Wuhan • Barcelona • Belgrade • Novi Sad • Cluj • Manchester

Editors

Rui Liu
Beijing Institute of
Technology
Beijing, China

Yushi Wen
China Academy of
Engineering Physics
Mianyang, China

Weiqiang Pang
Xi'an Modern Chemistry
Research Institute
Xi'an, China

Editorial Office
MDPI
St. Alban-Anlage 66
4052 Basel, Switzerland

This is a reprint of articles from the Special Issue published online in the open access journal *Crystals* (ISSN 2073-4352) (available at: https://www.mdpi.com/journal/crystals/special_issues/advanced_energetic_materials).

For citation purposes, cite each article independently as indicated on the article page online and as indicated below:

Lastname, A.A.; Lastname, B.B. Article Title. *Journal Name* **Year**, *Volume Number*, Page Range.

ISBN 978-3-0365-9456-9 (Hbk)
ISBN 978-3-0365-9457-6 (PDF)
doi.org/10.3390/books978-3-0365-9457-6

Contents

About the Editors

Rui Liu

Dr. Rui Liu, male, is currently an Associate Professor at the Beijing Institute of Technology. He mainly engaged in dynamic mechanical behavior and the safety of energetic composite materials. He has published more than 30 SCI international peer-reviewed papers. He served as a (young) Editorial Board member for several international journals, such as *Int J. Struct Integration.*, *J. Mech Behav Mater.*, *Def. Technol.*, *Journal of Ordnance Engineering*, *Energetic Materials*, and *Journal of Explosives & Propellants*.

Yushi Wen

Dr. Yushi Wen is an Associate Research Fellow at the Institute of Chemical Materials. His research focuses on the ignition mechanisms of energetic materials and molecular dynamics simulations of rapid chemical reactions. He has led various projects funded by the National Natural Science Foundation. He also supervises a research team and has successfully mentored a Ph.D. and a Master's student. He has published nearly 50 papers in notable SCI journals, earning over 600 citations. Dr. Wen also serves as a frequent reviewer for several scientific journals.

Weiqiang Pang

Prof. Dr. Weiqiang Pang is a visiting Scholar at Politecnico di Milano. For ten years, he has mostly been devoted to composite energy materials and their application in solid rocket propellants. He is the Editor-in-chief of *"The Journal of Ordnance Equipment Engineering"*, an Editorial Board member of *"The Chinese Journal of Explosives & Propellants"*, *"Journal of Solid Rocket Technology"*, etc. He has published 12 books nearly 200 scientific papers.

Preface

Energetic Materials (EMs) are a traditional branch of materials. They started more than 2000 years ago in China with black powder or something very close to it. In recent years, the demand for industrial and defense applications for energetic materials, including pyrotechnics, explosives, and propellants, has inspired new developments in this field. The occurrence of advanced energetic materials, in particular, offers a unique new opportunity to improve the performance of energetic formulations. DB propellants have been widely applied to solid rocket propulsion as homogeneous mixtures of NC (nitrocellulose) with NG (nitroglycerin). The composite modified double base (CMDB) propellant, often used in military missiles and space vehicles, properly combines specific features of the two previous kinds of propellant. They are smokeless and are suitable for free-standing motor grain configurations thanks to their high strength and elastic modulus. Composite propellants are a multi-phase mixture of oxidizing particles (such as AP) and metal fuel particles (such as Al) with high polymers as the matrix. For example, hydroxyl-terminated polybutadiene (HTPB) is a polymer widely used in propulsion for composite solid propellants and hybrid fuels. As a typical composite propellant, high-energy HTPB propellant has the advantages of excellent combustion performance and mechanical properties, low flame temperature, low molecular combustion products, and low infrared radiation. However, it also has a high probability of detonation and the risk of detonation. Today, an important challenge concerns solid oxidizers and insensitive compositions. On top of the energetic performance, large density, low cost, low sensitivity to multiple stimuli, low characteristic signature, slow aging, reliable safety, green features before and after burning easy disposal, and reuse technology are also of great interest to researchers and users of solid propellants. Innovative, energetic materials, such as DNTF, RDX@FOX composite, and so on, are often incorporated into propellant composition and require attention. When they were introduced to the energetic system, the performance of the system would be influenced significantly. It is critical to fully consider the properties of both the material and the composite system when selecting the cladding material to ensure that the propellant energy, ignition, density, and other characteristics are maintained while effectively improving the crystallization of DNTF and maintaining stable control of its crystallization amount. Recently, the emergence of high-energy density compounds, such as CL-20, 3,4-Bis(3-nitrofurazan-4-yl) furozan (DNTF), and Ni/Al energetic structural materials, allowed the formulation of new propellants with an increased energy density. Thus, high-energy, low vulnerability, and green solid propellants, laser-driven combustion, etc., are now hot topics worldwide. Moreover, with the development of computer science and technology, many theoretical simulation methods, such as molecular dynamics simulation, artificial neural networks (ANN), nonlinear dynamics software LS-DYNA, etc., can be used to investigate the performance of EMs. For example, the high-energy materials genome (HEMG) is based on the experimental data on the combustion and detonation characteristics of various high-energy materials (HEMs) under various conditions, being also based on the metadata on the quantum and physicochemical characteristics of HEMs components, as well as the thermodynamic characteristics of HEM as a whole.

Rui Liu, Yushi Wen, and Weiqiang Pang

Editors

Editorial

Advanced Energetic Materials: Testing and Modeling

Rui Liu [1], Yushi Wen [2] and Weiqiang Pang [3,*]

[1] State Key Laboratory of Explosion Science and Technology, Beijing Institute of Technology, Beijing 100081, China; liurui1985@bit.edu.cn

[2] Institute of Chemical Materials, China Academy of Engineering Physics, Mianyang 621999, China; wenys@caep.cn

[3] Xi'an Modern Chemistry Research Institute, Xi'an 710065, China

* Correspondence: nwpu_pwq@163.com

Citation: Liu, R.; Wen, Y.; Pang, W. Advanced Energetic Materials: Testing and Modeling. *Crystals* **2023**, *13*, 1100. https://doi.org/10.3390/cryst13071100

Received: 7 July 2023
Accepted: 13 July 2023
Published: 14 July 2023

Energetic Materials (EMs) are a traditional branch of materials. It started more than 2000 years ago in China with black powder, or something very close to it. In recent years, the demand for industrial and defense applications for energetic materials, including pyrotechnics, explosives, and propellants, inspired new developments in this field. The occurrence of advanced energetic materials in particular offers a unique new opportunity to improve the performance of energetic formulations. DB propellants, as homogeneous mixtures of NC (nitrocellulose) with NG (nitroglycerin), have been widely applied to solid rocket propulsion. The composite modified double base (CMDB) propellant, often used in military missiles and space vehicles, properly combines specific features of the two previous kinds of propellant. They are smokeless and are suitable for free standing motor grain configurations thanks to their high strength and elastic modulus. Composite propellants are a multi-phase mixture of oxidizing particles (such as AP) and metal fuel particles (such as Al) with high polymers as the matrix. For example, hydroxyl-terminated polybutadiene (HTPB) is a polymer widely used in propulsion both for composite solid propellants and hybrid fuels. As a typical composite propellant, high-energy HTPB propellant has the advantages of excellent combustion performance and mechanical properties, low flame temperature, low molecular combustion products, and low infrared radiation. However, it also has a high probability of detonation and the risk of detonation. Today, an important challenge concerns the solid oxidizers and insensitive compositions. On top of the energetic performance, large density, low-cost, low-sensitivity to multiple stimuli, low characteristic signature, slow aging, reliable safety, green features before and after burning, easy disposal, and reuse technology are also of great interest to researchers and users of solid propellants. Innovative energetic materials, such as DNTF, RDX@FOX composite, and so on, are often incorporated into propellant composition and require attention. When they were introduced to the energetic system, the performance of the system would be influenced significantly. It is critical to fully consider the properties of both the material and the composite system when selecting the cladding material to ensure that the propellant energy, ignition, density, and other characteristics are maintained while effectively improving the crystallization of DNTF and maintaining stable control of its crystallization amount. Recently, the emergence of high-energy density compounds, such as CL-20, 3,4-Bis(3-nitrofurazan-4-yl) furozan (DNTF), and Ni/Al energetic structural materials, allowed formulating new propellants with an increased energy density. Thus, high-energy, low vulnerability, and green solid propellants, laser-driven combustion, etc., are now hot topics worldwide. Moreover, with the development of computer science and technology, many theoretical simulation methods, such as molecular dynamics simulation, artificial neural networks (ANN), nonlinear dynamics software LS-DYNA, etc., can be used to investigate the performance of EMs. For example, the high-energy materials genome (HEMG) is based on the experimental data on the combustion and detonation characteristics of various high-energy materials (HEMs)

under various conditions, being based also on the metadata on the quantum and physico-chemical characteristics of HEMs components as well as the thermodynamic characteristics of HEM as a whole.

To accelerate the potential applications, various works focused on the physical and chemical characteristics through theory, experiments, and simulations. The aim of this issue is to collect comprehensive knowledge on materials synthesis, characterization, combustion, mechanical, detonation, and safety. This Special Issue *Advanced Energetic Materials: Testing and Modeling* explores innovative EMs and EMs ingredients as well as formulations test and models. It collected contributions covering recent progress and models of energetic materials in chemical propulsion. Attention was focused on the design, model, properties, and state of-the-art of this class of thermochemical propulsion devices. A total of 13 papers were selected for publication after a standard peer review process, which summarize the most recent achievements of famous research groups, participation of young authors with novel/innovative concepts was especially encouraged, of course, with the assistance of their supervisors.

To investigate the crystallization of DNTF in modified double-base propellants, gly-cidyl azide polymer (GAP) was used as the coating material for the in situ coating of DNTF, and the performance of the coating was investigated to inhibit the crystallization. Molecular dynamics was used to construct a bilayer interface model of GAP and DNTF with different growth crystal surfaces, and molecular dynamics' calculations of the binding energy and mechanical properties of the composite system were carried out by Qin, Y. [1]. It was found that GAP can form a white gel on the surface of DNTF crystals and has a good coating effect which can significantly reduce the impact sensitivity and friction sensitivity of DNTF. GAP could effectively improve the mechanical properties of DNTF. GAP can be referred to as a better cladding layer for DNTF, which is feasible for inhibiting the DNTF crystallization problem in propellants.

In order to study the ignition process and response characteristics of cast polymer-bonded explosives (PBXs) under the action of friction, HMX-based cast PBXs were used to carry out friction ignition experiments at a 90° swing angle and obtain the critical ignition loading pressure was 3.7 MPa. The friction temperature rise process was numerically simulated at the macro and micro scale, and the ignition characteristics were judged by Yuan, J. [2]. It was found that the maximum temperature rise was 55 °C, and the temperature rise of the whole tablet was not enough to ignite the explosive. HMX crystal particles can be ignited at a temperature of 619 K under 4 MPa hydraulic pressure loaded by friction sensitivity instrument. The external friction heat between cast PBX tablet and sliding column had little effect on ignition.

To study the engine safety against fragment in complex battlefield environments, the fragment impact safety simulation study of a high-energy four-component HTPB propellant solid engine was conducted. The equation of state parameters and reaction rate equation parameters of the detonation product of HTPB propellant were calibrated by using a 50 mm diameter cylinder test and Lagrange test combined with genetic algorithm. The nonlinear dynamics software LS-DYNA was used to build a finite element model of the fragment impact engine and simulate the mechanical response of the high-energy HTPB propellant under different operating conditions by Liu, Z. [3]. It was found that the critical detonation velocity decreased with the increase in the number of fragments. When the number of fragments was more than five, the influence of this factor on the critical detonation velocity was no longer obvious. Under the same loading strength conditions, the greater metal shell strength and the greater the shell wall thickness, the more difficult it was for the HTPB propellant to be detonated by the shock. This study can provide a reference for the design and optimization analysis of solid rocket engine fragment impact safety.

For the solid propellant burning rate prediction, high-energy materials genome (HEMG) is an analytical and calculation tool that contains relationships between vari-ables of the object, which allows researchers to calculate the values of one part of the variables through others, solve direct and inverse tasks, predict the characteristics of non-experimental objects, predict parameters to obtain an object with desired characteristics, and execute virtual experiments for conditions which cannot be organized or have difficultly being organized. The history and current status of the emergence of HEMG are presented herein. The fundamental basis of the artificial neural networks (ANN) as a methodological HEMG base, as well as some examples of HEMG conception used to create multifactor computational models (MCM) of solid rocket propellants (SRP) combustion, was presented by Pang, W. [4].

To study the role of complex composition of 2:17R-cell boundaries in the realization of magnetization reversal processes of (Sm, Zr)(Co, Cu, Fe)$_z$ alloys intended for high-energy permanent magnets, the micromagnetic simulation was performed using the modified sandwich model of a (Sm, Zr)(Co, Cu, Fe)$_z$ magnet, which includes additional domain-wall pinning barriers in the form of 2:7R or 5:19R phase layers by Zheleznyi, M. [5]. It was found that there was a possibility of reaching the increased coercivity at the expense of 180°-domain wall pinning at the additional barriers within cell boundaries. The phase and structural states of the as-cast $Sm_{1-x}Zr_x(Co_{0.702}Cu_{0.088}Fe_{0.210})_z$ alloy sample with $x = 0.13$ and $z = 6.4$ were studied, and the presence of the above phases in the vicinity of the 1:5H phase was demonstrated.

Research on energetic materials continuously develops energetic materials with higher detonation performance and energy density, taking it as an eternal quest. Due to the introduction of oxygen atoms, N-oxide energetic compounds have a unique oxygen balance, excellent detonation properties, and a high energy density, attracting the extensive attention of researchers all over the world. Synthetic strategies towards azine N-oxides and azole N-oxides of N-oxides were fully reviewed. Corresponding reaction mechanisms towards the aromatic N-oxide frameworks and examples that use the frameworks to create high-energy substances were discussed. Moreover, the energetic properties of N-oxide energetic compounds were compared and summarized by She, W. [6].

As we know, aluminum (Al) has been widely used in micro-electromechanical systems (MEMS), polymer-bonded explosives (PBXs), and solid propellants. Its typical core–shell structure (the inside active Al core and the external alumina (Al$_2$O$_3$) shell) determines its oxidation process, which is mainly influenced by oxidant diffusion, Al$_2$O$_3$ crystal transformation and melt-dispersion of the inside active Al. Metastable intermixed composites (MICs), flake Al, and nano Al can improve the properties of Al by increasing the diffusion efficiency of the oxidant. Fluorine, titanium carbide (TiC), and alloy can crack the Al$_2$O$_3$ shell to improve the properties of Al. Furthermore, those materials with good thermal conductivity can increase the heat transferred to the internal active Al, which can also improve the reactivity of Al. The integration of different modification methods was employed by Wang, D.to further improve the properties of Al [7]. With the ever-increasing demands on the performance of MEMS, PBXs, and solid propellants, Al-based composite materials with high stability during storage and transportation, and high reactivity for usage will become a new research focus in the future.

It is difficult for the reactor to achieve uniform quality of composite material, which affects its application performance. 1,3,5-trinitro-1,3,5-triazacyclohexane (RDX) and 1,1-diamino-2,2-dinitroethylene (FOX-7) are famous high-energy and insensitive explosives. The preparation of RDX@FOX-7 composites can meet the requirements with high-energy and low sensitivity. Based on the principle of solvent-anti-solvent, the recrystallization process was precisely controlled by microfluidic technology. The RDX@FOX-7 composites with different mass ratios were prepared by Yu, J. [8]. It was found that at the mass ratio of 10%, the RDX@FOX-7 composites were ellipsoid of about 15 μm with uniform distribution and quality. The advantages of microscale fabrication of composite materials were verified. With the increase in FOX-7 mass ratios, the melting temperature of RDX was advanced, the thermal decomposition peak of RDX changed to double peaks, and the activation energy of RDX@FOX-7 composite decreased. These changes were more pronounced between 3 and 10%, but not between 10 and 30%. The ignition delay time of RDX@FOX-7 was shorter than that of RDX and FOX-7. RDX@FOX-7 burned more completely than RDX indicating that FOX-7 can assist heat transfer and improve the combustion efficiency of RDX.

Ni/Al energetic structural materials attracted much attention due to their high energy release, but understanding their thermal reaction behavior and mechanism in order to guide their practical application is still a challenge. A novel understanding of the thermal reaction behavior and mechanism of Ni/Al energetic structural materials in the inert atmosphere were reported. The reaction kinetic model of Ni/Al energetic structural materials with Ni/Al molar ratios was obtained. The effect of the Ni/Al molar ratios on their thermal reactions was discussed based on the products of a Ni/Al thermal reaction by Wang, K. [9]. It was found that the liquid Al was adsorbed on the surface of Ni with high contact areas, leading in an aggravated thermal reaction of Ni/Al.

Taking the combustion tear gas mixture as the research object, the system formula was optimized by adding a different mass fraction of 5-amino-1H-tetrazole (5AT). TG-DSC, a thermocouple, and a laser smoke test system were used to characterize the combustion temperature and velocity, as well as the smoke concentration and particle size. Starink's method, the Flynn–Wall–Ozawa method, and the Coats–Redfern method were used to evaluate the pyrolysis kinetic parameters of the samples by Zhai, H. [10]. It was found that when the mass fraction of 5AT in the system was 10%, the maximum combustion temperature of the sample decreased by nearly 70 °C and the smoke concentration increased by 12.81%. Adding an appropriate amount of the combustible agent 5AT to the combustion tear gas mixture can improve its combustion performance and smoking performance, which provides an important, new idea for the development of a new generation of safe, efficient, and environmentally friendly tear gas mixtures.

To study the design method and pressure relief effect of the mitigation structure of a shell under the action of thermal stimulation, a systematic research method of theoretical calculation-simulation-experimental verification of the mitigation structure was established by Liang, J. [11]. The pressure relief effect of the mitigation structure was verified by the low-melting alloy plug with refined crystal structure for sealing the pressure relief hole and the cook-off test. It was found that the critical pressure relief area is when the ratio of the area of the pressure relief hole to the surface area of the charge is $A_V/S_B = 0.0189$. When the number of openings increased to 6, the required pressure relief coefficient decreased to $A_V/S_B = 0.0110$. When the length/diameter ratio was greater than 5, the opening at one end cannot satisfy the reliable pressure relief of the shell. The designed low-melting-point alloy mitigation structure can form an effective pressure relief channel.

To study the crystal mechanical properties of DNTF and hexanitrohexaazaisowurtzitane (CL-20) deeply, the crystals of DNTF and CL-20 were prepared by the solvent evaporation method. The crystal micromechanical loading procedure was characterized by the nanoindentation method. In addition, the crystal fracture behaviors were investigated with scanning probe microscopy (SPM) by Nan, H. [12]. It was found that the hardness for DNTF and CL-20 was 0.57 GPa and 0.84 GPa, and the elastic modulus was 10.34 GPa and 20.30 GPa, respectively. CL-20 obviously exhibited a higher hardness, elastic modulus, and local energy-dissipation, and a smaller elastic recovery ability of crystals than those of DNTF. CL-20 crystals are more prone to cracking and have a lower fracture toughness value than DNTF. Compared to DNTF crystals, CL-20 is a kind of brittle material with higher modulus, hardness, and sensitivity than that of DNTF, making the ignition response more likely to happen.

In order to study the reaction growth process of insensitive Ju En Ao Lv (JEOL) explosive after ignition under cook-off, a series of cook-off tests were carried out on JEOL explosive using a self-designed small cook-off bomb system. A thermocouple was used to measure the internal temperature of the explosive, and a camera recorded macro images of the cook-off process by Wang, X. [13]. It was found that the ignition time decreased as the heating rate increased, while the ignition temperature was not sensitive to the heating rate. When the heating rate was faster, the internal temperature gradient of the explosive was larger, and the ignition point appeared at the highest temperature position. As the heating rate decreased, the internal temperature gradient of the explosive decreased, the ignition point appeared random, and multiple ignition points appeared at the same time. The growth process of the ignition point could be divided into severe thermal decomposition, slow combustion, and violent combustion stages.

Author Contributions: Conceptualization, W.P. and R.L.; writing—original draft preparation, W.P.; writing—review and editing, R.L. and Y.W. All authors have read and agreed to the published version of the manuscript.

Conflicts of Interest: The authors declare no conflict of interest.

References

1. Qin, Y.; Yuan, J.; Sun, H.; Liu, Y.; Zhou, H.; Wu, R.; Chen, J.; Li, X. Experiment and Molecular Dynamic Simulation on Performance of 3,4-Bis(3-nitrofurazan-4-yl)furoxan (DNTF) Crystals Coated with Energetic Binder GAP. *Crystals* **2023**, *13*, 327. [CrossRef]
2. Yuan, J.; Qin, Y.; Peng, H.; Xia, T.; Liu, J.; Zhao, W.; Sun, H.; Liu, Y. Experiment and Numerical Simulation on Friction Ignition Response of HMX-Based Cast PBX Explosive. *Crystals* **2023**, *13*, 671. [CrossRef]
3. Liu, Z.; Nie, J.; Fan, W.; Tao, J.; Jiang, F.; Guo, T.; Gao, K. Simulation Analysis of the Safety of High-Energy Hydroxyl-Terminated Polybutadiene (HTPB) Engine under the Impact of Fragments. *Crystals* **2023**, *13*, 394. [CrossRef]
4. Pang, W.; Abrukov, V.; Anufrieva, D.; Chen, D. Burning Rate Prediction of Solid Rocket Propellant (SRP) with High-Energy Materials Genome (HEMG). *Crystals* **2023**, *13*, 237. [CrossRef]
5. Zheleznyi, M.; Kolchugina, N.; Kurichenko, V.; Dormidontov, N.; Prokofev, P.; Milov, Y.; Andreenko, A.; Sipin, I.; Dormidontov, A.; Bakulina, A. Micromagnetic Simulation of Increased Coercivity of (Sm, Zr) (Co, Fe, Cu)$_z$ Permanent Magnets. *Crystals* **2023**, *13*, 177. [CrossRef]
6. She, W.; Xu, Z.; Zhai, L.; Zhang, J.; Huang, J.; Pang, W.; Wang, B. Synthetic Methods towards Energetic Heterocyclic N-Oxides via Several Cyclization Reactions. *Crystals* **2022**, *12*, 1354. [CrossRef]
7. Wang, D.; Xu, G.; Tan, T.; Liu, S.; Dong, W.; Li, F.; Liu, J. The Oxidation Process and Methods for Improving Reactivity of Al. *Crystals* **2022**, *12*, 1187. [CrossRef]
8. Yu, J.; Jiang, H.; Xu, S.; Li, H.; Wang, Y.; Yao, E.; Pei, Q.; Li, M.; Zhang, Y.; Zhao, F. Preparation and Properties of RDX@FOX-7 Composites by Microfluidic Technology. *Crystals* **2023**, *13*, 167. [CrossRef]
9. Wang, K.; Deng, P.; Liu, R.; Ge, C.; Wang, H.; Chen, P. A Novel Understanding of the Thermal Reaction Behavior and Mechanism of Ni/Al Energetic Structural Materials. *Crystals* **2022**, *12*, 1632. [CrossRef]
10. Zhai, H.; Cui, X.; Gan, Y. Effect of 5-Amino-1H-Tetrazole on Combustion Pyrolysis Characteristics and Kinetics of a Combustion Tear Gas Mixture. *Crystals* **2022**, *12*, 948. [CrossRef]
11. Liang, J.; Nie, J.; Liu, R.; Han, M.; Jiao, G.; Sun, X.; Wang, X.; Huang, B. Study and Design of the Mitigation Structure of a Shell PBX Charge under Thermal Stimulation. *Crystals* **2023**, *13*, 914. [CrossRef]
12. Nan, H.; Zhu, Y.; Niu, G.; Wang, X.; Sun, P.; Jiang, F.; Bu, Y. Characterization and Analysis of Micromechanical Properties on DNTF and CL-20 Explosive Crystals. *Crystals* **2023**, *13*, 35. [CrossRef]
13. Wang, X.; Jiang, C.; Wang, Z.; Lei, W.; Fang, Y. Ignition Growth Characteristics of JEOL Explosive during Cook-Off Tests. *Crystals* **2022**, *12*, 1375. [CrossRef]

Article

Effect of 5-Amino-1H-Tetrazole on Combustion Pyrolysis Characteristics and Kinetics of a Combustion Tear Gas Mixture

Haolong Zhai, Xiaoping Cui * and Yuping Gan

Equipment Management and Guarantee Institute, Engineering University of Armed Police Force of China, Xi'an 710086, China; gcdxzhl@yeah.net (H.Z.); gcdxgyp@yeah.net (Y.G.)
* Correspondence: wjgcdxcxp@yeah.net

Abstract: Taking the combustion tear gas mixture as the research object, the system formula was optimized by adding a different mass fraction of 5-amino-1H-tetrazole(5AT). TG-DSC, a thermocouple, and a laser smoke test system were used to characterize the characteristic combustion parameters such as combustion temperature and velocity, as well as the end-point effects such as smoke concentration and particle size. Starink's method, the Flynn–Wall–Ozawa method, and the Coats–Redfern method were used to evaluate the pyrolysis kinetic parameters of the samples. The results show that when the mass fraction of 5-amino-1H-tetrazole in the system is 10%, the maximum combustion temperature of the sample decreases by nearly 70 °C and the smoke concentration increases by 12.81%. The kinetic study also found that with a different mass fraction of 5-amino-1H-tetrazole in the system, the main reaction model of the mixed agent in the first, third, and fourth stages of pyrolysis changed significantly, but for the second stage of sample pyrolysis, the main reaction model (the A4 model) showed a high degree of consistency, which can be considered as the thermal diffusion stage of the tear agent capsicum oleoresin (OC) (the temperature range is 220~350 °C), which is highly consistent with the results of the TG-DSC analysis. It was also confirmed that OC's thermal diffusion is mainly concentrated in this stage. The results of this study show that adding an appropriate amount of the combustible agent 5-amino-1H-tetrazole to the combustion tear gas mixture can improve its combustion performance and smoking performance, which provides an important, new idea for the development of a new generation of safe, efficient, and environmentally friendly tear gas mixtures.

Keywords: 5-amino-1H-tetrazole; tear gas mixture; combustible agent; combustion pyrolysis characteristics; dynamics research

Citation: Zhai, H.; Cui, X.; Gan, Y. Effect of 5-Amino-1H-Tetrazole on Combustion Pyrolysis Characteristics and Kinetics of a Combustion Tear Gas Mixture. *Crystals* **2022**, *12*, 948. https://doi.org/10.3390/cryst12070948

Academic Editors: Rui Liu, Yushi Wen, Weiqiang Pang and Qing Peng

Received: 10 June 2022
Accepted: 3 July 2022
Published: 6 July 2022

Publisher's Note: MDPI stays neutral with regard to jurisdictional claims in published maps and institutional affiliations.

1. Introduction

As the main charge of combustion tear gas, the combustion tear gas mixture plays an important role in dealing with sudden mass incidents and preventing and dealing with terrorist activities [1,2]. At present, potassium chlorate ($KClO_3$) is used as an oxidant, sucrose ($C_{12}H_{22}O_{11}$) is used as the combustible agent, and capsicum oleoresin (OC) is used as the tear agent in the formulation of this kind of mixed agent. The redox reaction of the oxidant and the combustible agent provides energy for the sublimation of the tear agent. However, due to the relatively poor thermal stability of OC, in order to maximize the functional efficiency of the tear agent in the mixed agent and improve its effective utilization rate, the energy released during the combustion of the mixed agent must be controlled. If the energy is too great, the combustion temperature will be too high, which will lead to the thermal decomposition of the tear agent in the process of heat release. On the other hand, if the energy is too small, it will delay the heat release efficiency of the tear agents in the system and even cause the release velocity to be too slow, making it difficult to reach the combat concentration in a short time, which will greatly reduce the technical and combat effectiveness of this kind of ammunition [3–7]. At the same time, the products formed by incomplete thermal diffusion will also aggravate the burden on the environment. Therefore, it is very important to study the combustion characteristics of its formula in order to improve the effective utilization of the lacrimal agent OC in the system and improve the action efficiency of this kind of mixed agent [8].

Through a literature review [9–11], it was found that 5-amino-1h-tetrazole (5AT), as an environmentally friendly combustible agent, has been greatly developed as a solid propellant and in other fields. Its greatest advantage is that, compared with other nitrogen-containing compounds, its nitrogen mass fraction is as high as 82.3%, and the combustion product is harmless N_2, with high gas production, which is more conducive to the diffusion of the functional elements in the mixture. 5AT is considered to be an ideal fuel in gas generators with a low combustion temperature. The latest research results have shown that adding an appropriate amount of 5AT instead of a sugar compound as the combustible agent in the formula of colored smoke pyrotechnic agents can significantly improve the smoke's performance in action efficiency and durability [12–15].

However, no attempt has been made to improve the formula of combustion-type tear gas mixtures. Based on the similar principle of action between combustion-type colored smoke agents and combustion-type tear gas mixtures [16], on the basis of an unchanged oxygen mass fraction coefficient (OB), this study attempted to introduce 5AT in different proportions such as 0%, 5%, 10%, 15%, and 20% into the system as the second combustible agent to obtain five groups of different formulas. TG-DSC, a thermocouple, and a laser smoke test system were used to characterize the characteristic parameters of combustion, such as the combustion temperature and velocity, as well as the end-point effects such as smoke concentration and particle size. The apparent activation energy, pre-exponential factor, and other thermodynamic parameters in the pyrolysis process were obtained by Starink's method and the Flynn–Wall–Ozawa method. At the same time, in order to further explore the pyrolysis mechanism of combustion, the possible reaction models in the pyrolysis process of the different formulations were deduced by the Coats–Redfern model-fitting method. The study provides valuable guidance for improving the performance and combustion mechanism of this kind of mixture.

2. Experiment and Method

2.1. Materials and Main Experimental Equipment

The main raw materials were chemically pure capsicum oleoresin, OC for short ($C_{18}H_{27}NO_3$) from Aldrich, St. Louis, MO, USA, and potassium chlorate ($KClO_3$), lactose ($C_{12}H_{22}O_{11}$), 5-amino-1h-tetrazole (CH_3N_5), phenolic resin ($(C_8H_6O_2)_n$) and basic magnesium carbonate ($(MgCO_3)_4 \cdot Mg(OH)_2 \cdot 5H_2O$), all of which were analytically pure and purchased from Aladdin Biochemical Technology Co., Ltd., Shanghai, China.

The main test equipment was an HS-STA-002 synchronous thermal analyzer (sensitivity: 0.01 mg) produced by Hesheng Instrument Technology Co., Ltd., Shanghai, China, with a resolution of 0.06 mV, a test temperature range of room temperature to ~1000 °C, a temperature test accuracy of ± 0.05 °C, and a calorimetric sensitivity of $\pm 0.5\%$.

Other equipment included an analytical balance (BSA224S-CW) produced by Saidoris Instrument System Co., Ltd., Gottingen, Germany; a K-type thermocouple (FLUKE53-2B) produced by Fluke company, Everett, America; an intelligent digital display vacuum drying oven (DHG-9140) produced by Donglu Instrument and Equipment Company, Shanghai, China; a high-speed camera (X8PRO) produced by Mingce Electronic Technology Company, Shanghai, China; and a smoke concentration test system (JCY-80e) produced by Chuangyi Environmental Testing Equipment Co., Ltd., Qingdao, China.

The samples were prepared according to different formulations designed by a uniform design method, as shown in Table 1. Figure 1 shows the sample preparation flowchart, and Figure 2 shows five different prepared samples.

Table 1. Formula of the mixed agents at OB = −0.18 based on a uniform design method.

Component Formula	Oxidant	Combustible Agent		Coolant	Adhesive	Tear Agent
	$KClO_3$ (wt%)	$C_{12}H_{22}O_{11}$ (wt%)	CH_3N_5 (wt%)	$(MgCO_3)_4 \cdot Mg(OH)_2$ $\cdot 5H_2O$ (wt%)	$(C_8H_6O_2)n$ (wt%)	OC (wt%)
P1	32.0	27.2	0	7.1	6.3	27.4
P2	30.4	23.8	5.0	7.1	6.3	27.4
P3	28.9	20.3	10.0	7.1	6.3	27.4
P4	27.3	16.9	15.0	7.1	6.3	27.4
P5	25.8	13.4	20.0	7.1	6.3	27.4

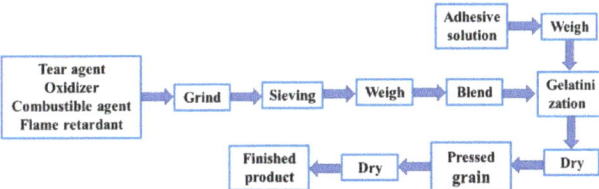

Figure 1. Sample preparation process.

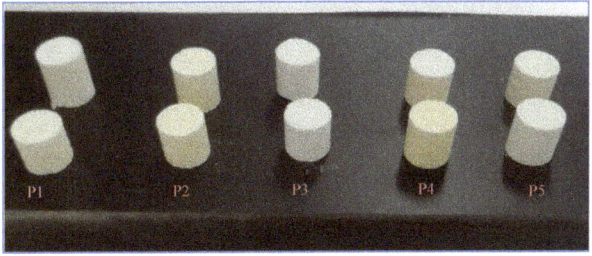

Figure 2. Tested samples with different formulations.

2.2. Test of Combustion Characteristics

2.2.1. Combustion Temperature Test

In order to reduce the influence of the oxygen concentration in the external environment on the combustion environment of the sample, the test ignited each sample in an N_2 environment, measured the temperature with a K-type thermocouple, and recorded the whole process with a high-speed camera. Three groups of tests were conducted for each group of samples, and the average value of the three groups of data was taken as the measurement result [17].

2.2.2. Burning Rate Test

The burning rate is also one of the important indexes used to measure the combat effectiveness of combustion tear gas mixtures. The burning rate can have a direct impact on the smoke effect of the agent. Linear velocity or mass velocity is usually used for pyrotechnic agents. In general, the burning rate generally refers to the linear burning rate, which refers to the displacement of the combustion wave in front of the mixed grain along its normal direction in units of time [18], and it is expressed as:

$$v = \frac{dl}{dt}(\text{mm/s}) \tag{1}$$

where v is the linear burning rate of the mixed grain, and dl is the displacement of the combustion wave of mixed grain along its normal direction in time dt (unit: mm/s). During the test, an electric igniter was used to ignite the grain, and the test was carried out in the smoke box in an N_2 environment.

2.3. Test of Combustion Smoke Characteristics

In order to characterize the combustion smoke concentration and particle size distribution of the sample, a laser smoke concentration tester was used for testing, and the data were collected and analyzed with software. Its principle is shown in Figure 3. The samples' specifications are cylinders with a diameter of 15 mm and a height of 20 mm. The specifications of the smoke collection box are 50 cm × 50 cm × 50 cm.

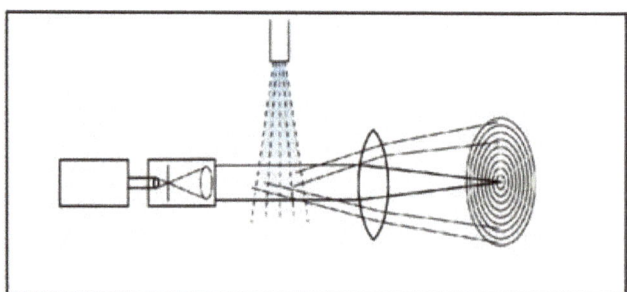

Figure 3. Schematic diagram of the laser smoke concentration tester.

2.4. Thermal Behavior Analysis

In order to study the thermal behavior of the sample, an synchronous thermal analyzer was used. Before the test, we first calibrated the differential thermal analysis baseline and temperature of the synchronous thermal analyzer and then placed about 8–10 mg of the different samples into the ceramic crucible and heated the samples from 30 °C to 600 °C at a heating rate of 5, 10, 15, and 20 °C·min^{-1}. In order to avoid environmental impact, the whole test process was carried out in an N$_2$ atmosphere, and the ventilation rate was 40 mL·min^{-1}.

2.5. Pyrolysis Kinetics

In order to further explore the reaction mechanism of each stage in the combustion process of the combustion tear gas mixture, Starink's method and the Flynn–Wall–Ozawa method with high accuracy were selected to calculate the corresponding thermal decomposition kinetic parameters [19–22] and the Coats–Redfern equation was used to predict the pyrolysis reaction model of each stage of the sample and to thus determine the reaction type of each stage so as to provide a certain theoretical basis for an in-depth study and improvement of its combustion environment.

The equation of Starink's method [23] is:

$$\ln \frac{\beta}{T^{1.8}} = -1.008 \cdot \frac{E_a}{RT} + C \tag{2}$$

The equation of the Flynn–Wall–Ozawa method [24,25] is:

$$\ln \beta + 0.4567 \frac{E_a}{RT} = C \tag{3}$$

The Coats–Redfern method equation [26] is:

$$\ln \frac{g(\alpha)}{T^2} = \ln \left[\frac{AR}{\beta E_a} \left(1 - \frac{2RT}{E_a} \right) \right] - \frac{E_a}{RT} \tag{4}$$

where β is the heating rate, T is the characteristic temperature, $E\alpha$ is the activation energy of the reaction, R is the molar gas constant, and A is the pre-exponential factor of the reaction.

By combining 17 common $g(\alpha)$ substitutes into Equation (4), we can solve the linear correlation coefficient between $\ln[g(\alpha)/T^2]$ and $1/T$. When the calculated linear correlation coefficient reaches the maximum, the corresponding reaction model of the selected $g(\alpha)$ is the reaction model of the sample at this stage. The 17 commonly used reaction models are shown in Table 2 [27].

Table 2. Thermal decomposition reaction models of 17 common solid substances.

Reaction Model	$g(\alpha)$	$f(\alpha)$	Abbreviation
Power law	α	1	P1
Power law	$\alpha^{1/2}$	$2\alpha^{1/2}$	P2
Power law	$\alpha^{1/3}$	$3\alpha^{2/3}$	P3
Power law	$\alpha^{1/4}$	$4\alpha^{3/4}$	P4
Power law	$\alpha^{3/2}$	$2/3\alpha^{1/2}$	P2/3
Avrami–Erofeev	$[-\ln(1-\alpha)]^{1/2}$	$2(1-\alpha)(-\ln(1-\alpha))^{1/2}$	A2
Avrami–Erofeev	$[-\ln(1-\alpha)]^{1/3}$	$3(1-\alpha)(-\ln(1-\alpha))^{2/3}$	A3
Avrami–Erofeev	$[-\ln(1-\alpha)]^{1/4}$	$4(1-\alpha)(-\ln(1-\alpha))^{3/4}$	A4
One-dimensional	α^2	$1/2\alpha$	D1
Two-dimensional	$(1-\alpha)\ln(1-\alpha)+\alpha$	$[-\ln(1-\alpha)]^{-1}$	D2
Three-dimensional	$[1-(1-\alpha)^{1/3}]^2$	$3/2(1-\alpha)^{2/3}[1-(1-\alpha)^{1/3}]$	D3
Ginstling-Brounshtein	$1-(2\alpha/3)-(1-\alpha)^{2/3}$	$3/2[(1-\alpha)^{-1/3}-1]^{-1}$	G-B
First-order	$-\ln(1-\alpha)$	$1-\alpha$	F1
Second-order	$(1-\alpha)^{-1}-1$	$(1-\alpha)^2$	F2
Third-order	$[(1-\alpha)^{-2}-1]/2$	$(1-\alpha)^3$	F3
Contracting area	$1-(1-\alpha)^{1/2}$	$2(1-\alpha)^{1/2}$	C2
Contracting volume	$1-(1-\alpha)^{1/3}$	$3(1-\alpha)^{2/3}$	C3

3. Results and Discussion

3.1. Analysis of the Combustion Characteristics

3.1.1. Combustion Temperature Analysis

Figure 4 shows the combustion temperature-time distribution of Samples P1–P5 measured by the thermocouple method. It can be seen that the order of the maximum combustion temperature (T_{max}) of the samples is T_{max} (P5) > T_{max} (P1) > T_{max} (P4) > T_{max} (P2) > T_{max} (P3). At the same time, it is not difficult to see that when the mass fraction of 5AT is 10%, the combustion temperature of the sample is the lowest (588 °C), but when the mass fraction of 5AT is 20%, the combustion temperature of the sample is the highest (676 °C); the difference between them is nearly 90 °C. This shows that 5AT has a great influence on the combustion temperature of the system.

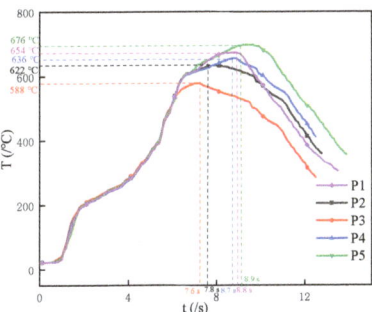

Figure 4. Combustion temperature-time distribution of Samples P1–P5 measured by a thermocouple.

Figure 5 shows the variation trend of the maximum combustion temperature of samples with different 5AT mass fractions in the system. It was found that with an increase in the 5AT mass fraction, the maximum combustion temperature first decreases and then increases. When the mass fraction of 5AT is less than 10%, the combustion temperature of the system decreases with an increase in the 5AT mass fraction, but when the mass fraction of 5AT is more than 10%, the combustion temperature of the system increases with an increase in the 5AT mass fraction. This is mainly related to the redox reaction of 5AT with the oxidant $KClO_3$ and its own pyrolysis reaction. Among these, the former is an exothermic reaction and the latter is an endothermic reaction. When the mass fraction of 5AT in the system is less than 10%, the heat released by 5AT participating in the redox reaction in the system is less than the heat absorption required for its own pyrolysis, so the overall combustion temperature of the system decreases. When the mass fraction of 5AT in the system is higher than 10%, the heat release of 5AT participating in the reaction is greater than the heat absorption required for its own pyrolysis, so the overall combustion temperature of the system will rise.

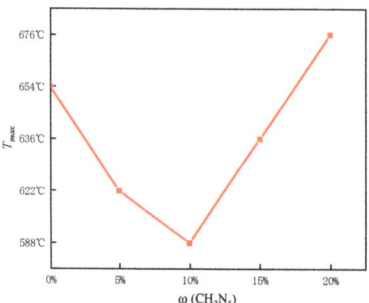

Figure 5. Variation trend of the maximum combustion temperature of samples with different 5AT mass fractions in the system.

3.1.2. Analysis of Burning Rate

Table 3 shows the burning rate of Samples P1–P5 in the same nitrogen atmosphere, with an air pressure of 0.1 MPa, room temperature T = 20 °C, and relative humidity RH = 30%. The results show that under the same atmospheric conditions, the burning rates of samples with different formulas show little difference. The maximum is 1.08 mm·s^{-1}, the minimum is 1.03 mm·s^{-1}, and the difference is only 0.05 mm·s^{-1}, which is basically the same level of burning rate. This shows that when all the other conditions are the same, the addition of 5AT to the system does not affect the overall combustion rate of this kind of mixture.

Table 3. Test results of the burning rate of samples.

Sample	Burning Rate I (mm·s^{-1})	Burning Rate II (mm·s^{-1})	Burning Rate III (mm·s^{-1})	Average Burning Rate (mm·s^{-1})	Standard Deviation (σ)
P1	1.05	1.08	1.08	1.07	0.0173
P2	1.07	1.09	1.08	1.08	0.01
P3	1.03	1.06	1.09	1.06	0.03
P4	1.03	1.05	1.07	1.05	0.02
P5	1.01	1.03	1.05	1.03	0.02

3.2. Smoke Characteristic Analysis

In order to evaluate the effect of 5AT on the thermal diffusion effect of the tear agent in combustion tear gas mixtures, the smoke concentration and particle size distribution of the samples were characterized; the results are shown in Figures 6 and 7. Figure 6 shows the particle size distribution of the combustion smoke of Samples P1–P5. It can be seen from the figure that the average particle size of the combustion smoke of Samples P1–P5 is mainly distributed between 833.4–839.8 μm. The relationship between the smoke concentration and average particle size of different samples and the mass fraction of 5AT in the system is shown in Figure 7. It is not difficult to see that when the mass fraction of 5AT in the system is less than 10%, the smoke concentration (C) and average particle size (AP) show an increasing trend. When the amount of 5AT in the system is 10%, the C and AP values of smoke reach the maximum, which are 68.59% and 839.8 μm, respectively. When the mass fraction of 5AT in the system is greater than 10%, the C and AP values of the sample smoke show a decreasing trend. When the mass fraction of 5AT in the system is 20%, the C and AP values of the smoke are the smallest: 53.58% and 833.4 μm, respectively.

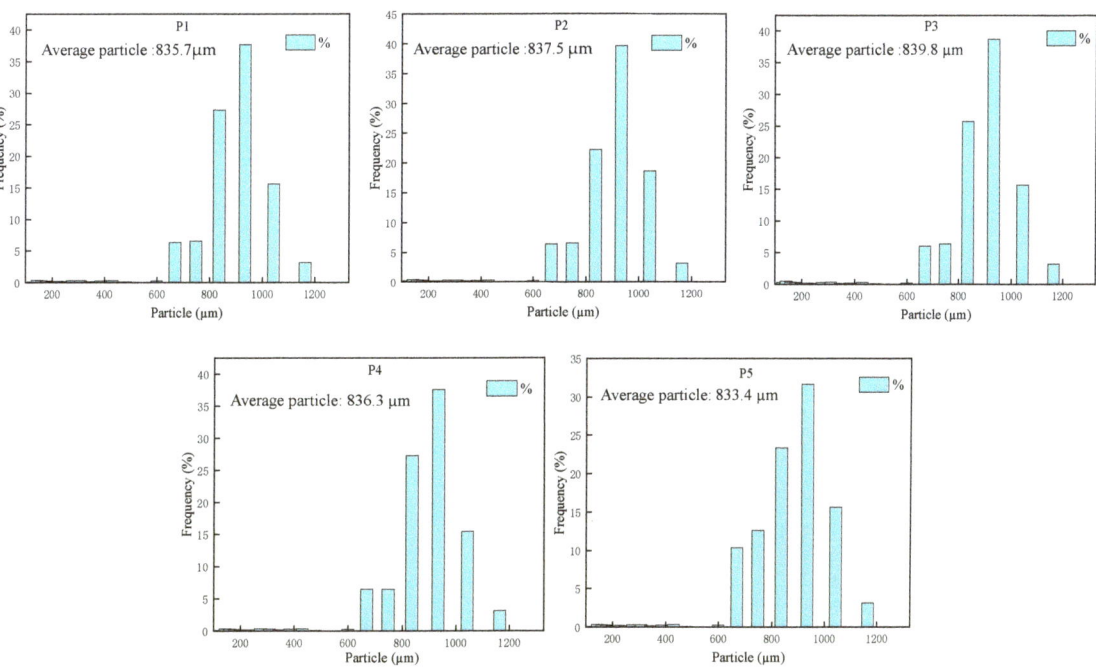

Figure 6. Particle size distribution of the combustion smoke of Samples P1–P5.

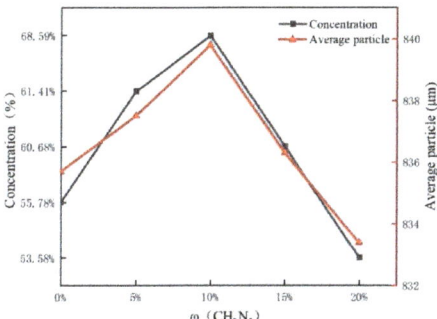

Figure 7. Variation trend of the smoke concentration and average particle size of samples with different 5AT mass fractions in the system.

Compared with the ranking of the maximum combustion temperature (T_{max}) of the different formulations measured above, the rankings for smoke concentration and the average particle size of different samples were just the opposite; that is, the higher the T_{max}, the smaller the corresponding C and AP values. On the contrary, the lower T_{max}, the greater the corresponding C and AP values. This may be related to the thermal decomposition of the tear agent OC during the combustion process of the system; that is, when the combustion temperature is higher, the amount of OC will increase, and the corresponding C value will decrease. With the thermal decomposition of the tear agent OC, the corresponding smoke AP value will decrease.

3.3. Pyrolysis Behavior Analysis and Related Kinetic Analysis

3.3.1. Thermal Behavior Analysis of Individual Components

Figure 8 shows the distribution of each individual component in the mixed reagent system as the TG-DSC-DTG curve at $\beta = 10\,^\circ C \cdot min^{-1}$. According to the TG curve, compared with other components in the system, the temperature at which the oxidant $KClO_3$ begins thermal decomposition is higher. Near 400 °C, the decomposition process is mainly one stage, and the weight loss ratio is about 30%.

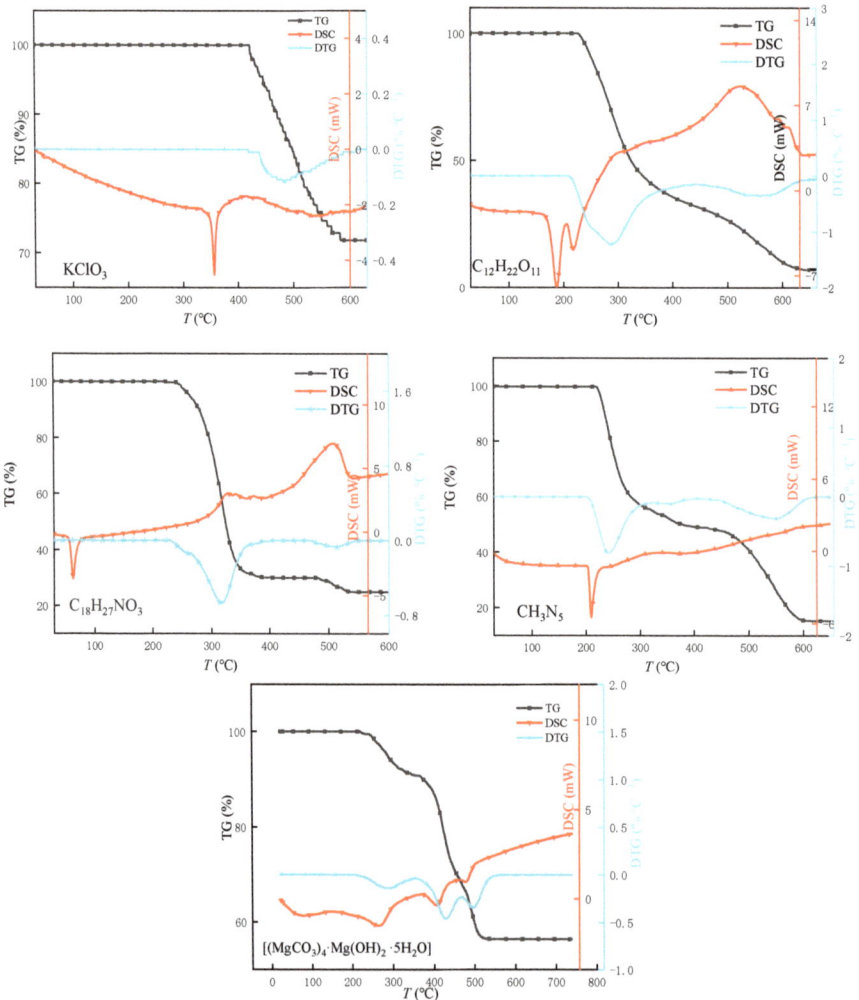

Figure 8. TG-DSC-DTG curve of each individual component in mixed reagent sample at $\beta = 10\,^\circ C \cdot min^{-1}$.

The temperature of the thermal decomposition of 5AT is the lowest, which starts near 200 °C. There is an obvious endothermic peak in the weight loss process of thermal decomposition, indicating that its thermal decomposition is mainly an endothermic process [28].

According to the weight loss trend of the TG curve, the weight loss process is mainly divided into three stages, for which the weight loss ratio is about 40%, 10%, and 30%.

Compared with sucrose, which is also a combustible agent, the temperature when sucrose starts thermal decomposition is slightly higher than that of 5AT; the weight loss begins near 210 °C, and there is an exothermic peak in the thermal decomposition process [29], indicating that the thermal decomposition of sucrose is mainly an exothermic process. According to the weight loss trend of the TG curve, the weight loss process is mainly divided into two stages: the weight loss ratio of the first stage is about 70%, and the weight loss ratio of the second stage is about 30%.

According to the TG-DSC-DTG curve of the lacrimal agent OC and previous research [8,30], the endothermic peak near 58 °C corresponds to its melting point. The weight loss phenomenon begins at around 230 °C, and the weak exothermic phenomenon does not appear until near 340 °C. In this temperature range, the DTG curve corresponds to an obvious pyrolysis weight loss peak, which is mainly considered to be the thermal diffusion process of OC. The second exothermic peak near 500 °C corresponds to the thermal decomposition of OC, and the weight loss ratio in this stage is about 10%.

According to the TG-DSC-DTG curve of the basic coolant magnesium carbonate and previous studies [31], the pyrolysis process is mainly divided into two stages. The temperature range of the first stage is 220–360 °C, and the weight loss ratio is about 16%. This is considered to mainly be the loss process of crystal water. The temperature range of the second stage is 360–500 °C, and the weight loss ratio is about 55%, which is basically consistent with the theory of complete pyrolysis to produce carbon dioxide, magnesium oxide, and water.

3.3.2. Thermal Behavior Analysis of the Samples

DSC Analysis of the Samples

Figure 9 shows the TG-DSC curve of Samples P1–P5 at a heating rate of $\beta = 10\ °C·min^{-1}$. According to the DSC curve, in the temperature range of 30–600 °C, the formulae of Samples P1–P4 mainly correspond to four thermal behaviors, which are the primary endothermic phenomenon and the tertiary exothermic phenomenon successively (the corresponding peak temperatures are T_1, T_2, T_3, and T_4).

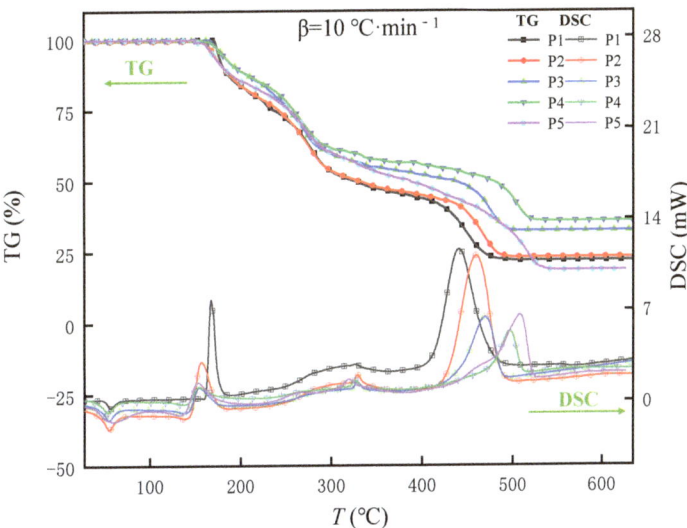

Figure 9. The TG-DSC curve of Samples P1–P5 at $\beta = 10\ °C·min^{-1}$.

However, P5 corresponds to five thermal behaviors, namely the primary endothermic phenomenon and four exothermic phenomena (the corresponding peak temperatures are T_{5-1}, T_{5-2}, T_{5-3}, T_{5-4}, and T_{5-5}), as shown in Table 4.

Table 4. Characteristic peak temperatures of samples P1–P5 corresponding to the DSC curve at $\beta = 10\,°C\cdot min^{-1}$.

Peak Temperature Formula	T_1 (°C)	T_2 (°C)	T_3 (°C)	T_4 (°C)	T_5 (°C)
P1	60.61	171.72	330.81	443.92	–
P2	59.32	160.81	332.03	462.94	–
P3	58.61	158.33	332.21	473.61	–
P4	58.22	157.63	330.04	506.73	–
P5	62.02	162.22	276.05	327.34	516.82

It can be seen from the characteristic peak temperatures in Table 4 that the thermal behavior of the formulae of Samples P1–P4 is basically the same. From P1 to P4, with the increase in 5AT mass fraction in the formula, the first exothermic peak T_2 gradually decreases and the third exothermic peak T_4 gradually increases, while the endothermic peak T_1 and the second exothermic peak T_3 have no obvious change. In combination with the TG-DSC-DTG curve of individual components in the previous section, it can be seen that the endothermic peak T_1 and the second exothermic peak T_3 of P1–P4 correspond to the melting point of OC in the system and the temperature at which pyrolysis begins. The first exothermic peak, T_2, is mainly caused by the exothermic oxidation–reduction reaction of the oxidant $KClO_3$, combustible $C_{12}H_{22}O_{11}$, and 5AT. With an increase in 5AT, the initial temperature of the reaction at this stage moves to the left, and the peak's shape gradually becomes gentle, which indicates that the addition of 5AT can slow down the intensity of the reaction, which may be related to the need to absorb some heat for the decomposition of 5AT [32,33]. Compared with P1–P4, P5 also has a weak exothermic peak in the temperature range of 200–300 °C. In combination with the changes in the components in the formula and the pyrolysis curve of each individual component, it is considered that the exothermic phenomenon is related to the fact that the heat released by 5AT participating in the reaction in the system begins to be greater than the heat absorbed by its own pyrolysis.

The exothermic peak of Samples P1–P5 near 330 °C mainly corresponds to the initial thermal decomposition of OC in the system. The exothermic enthalpy corresponding to each exothermic peak is shown in Table 5. It can be seen that with an increase in the 5AT mass fraction in the system, the exothermic enthalpy at the corresponding position first decreases and then increases. The corresponding exothermic enthalpy of P3 is the smallest, which indicates that the amount of thermal decomposition of OC in P3 is the smallest, which is the same as the smoke concentration of P3 measured above. The results are basically consistent with those of the largest average particle size.

Table 5. Exothermic enthalpy corresponding to the exothermic peak of Samples P1–P5 near 340 °C.

Formula	P1	P2	P3	P4	P5
Exothermic enthalpy (ΔH)	4.61 J/g	4.48 J/g	3.54 J/g	4.51 J/g	4.79 J/g

TG-DTG Analysis of the Samples

Figure 10 shows the TG-DTG curve of Samples P1–P5 at $\beta = 10\,°C\cdot min^{-1}$. In the DTG curve of the samples, the thermogravimetric process of the samples can be divided into several different stages according to the peak value corresponding to the mass loss rate of the samples.

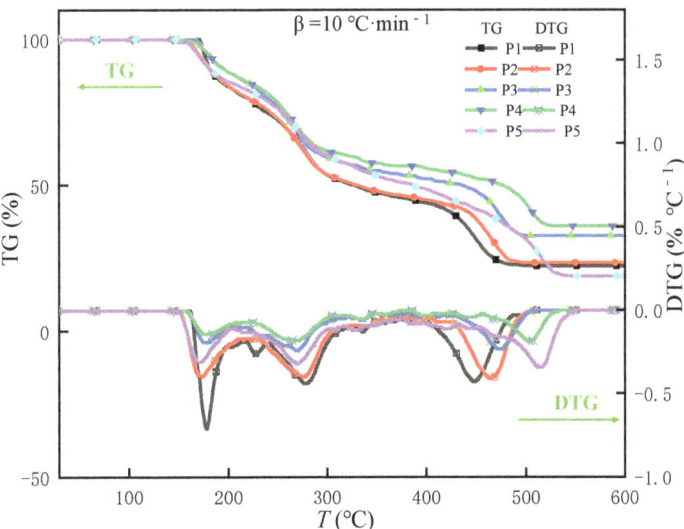

Figure 10. The TG-DTG curve of Samples P1–P5 at $\beta = 10\ °C·min^{-1}$.

Table 6 shows the characteristic values such as the initial temperature, the temperature corresponding to the DTG peak value, and the thermal weight loss ratio at each stage. From the TG-DTG curve, it can be seen that the pyrolysis weight loss process of P1–P5 is mainly divided into four stages in the temperature range of 30–600 °C. Combined with the curve in Figure 10, the first weight-loss stage is 150–220 °C, and an obvious DTG peak can be observed at this stage, in which the peak shape of P1 is the sharpest, indicating that the reaction is violent, and the peak value of the corresponding curve is 0.71% $°C^{-1}$. With an increase in the 5AT mass fraction in the system, the exothermic peak tends to be gentle. The peak values of the corresponding curves of P2–P5 are 0.4% $°C^{-1}$, 0.14% $°C^{-1}$, 0.19% $°C^{-1}$, and 0.31% $°C^{-1}$, respectively. Compared with P1, the DTG peak in the corresponding stages decreases significantly. According to Table 6, the thermal weight loss ratio M_{L1} corresponding to P1 at this stage is the largest. In combination with the previous research results of this kind of mixed agent [34], it can be considered that the redox reaction between the combustible agent and the oxidant has occurred in this stage. The peak value of P1's curve is the largest, and the thermal weight loss ratio is the largest, which is caused by the violent reaction between the combustible $C_{12}H_{22}O_{11}$ and the oxidant $KClO_3$ at this stage.

The temperature range of the second stage is 190–320 °C. At this stage, except for P5, which corresponds to a weak exothermic peak, the other formulae have no obvious heat absorption and exothermic phenomena. The DTG curve of this stage corresponds to an obvious peak, indicating that the thermal weight loss at this stage is obvious. In combination with the properties of each component in the mixed agent and relevant research results, it can be determined that this mainly corresponds to the thermal diffusion process of OC in the mixed agent; that is, the greater the weight loss ratio at this stage, the greater the amount of OC for effective thermal diffusion. When the heating rate of Samples P1–P5 is 10 °C min^{-1}, the thermal weight loss ratio of this stage is 26.7%, 27.5%, 28.9%, 27.1%, and 24.8%, respectively. It can be seen that the weight loss ratio of P3 is the largest, while the weight loss ratio of P5 is the smallest. The corresponding order is consistent with the concentration of each sample measured above. Therefore, appropriately increasing the mass fraction of 5AT in the system can improve the effective utilization rate of the lacrimal agent OC in the system.

Table 6. Starting and ending temperatures and corresponding characteristic values of Samples P1–P5 at each stage, based on the DTG curve.

(a) Eigenvalues corresponding to the first stage.

Formula	P1				P2				P3				P4				P5			
β	T_o	T_p	T_f	M_L	T_o	T_p	T_f	M_L	T_o	T_p	T_f	M_L	T_o	T_p	T_f	M_L	T_o	T_p	T_f	M_L
5	160.81	170.04	191.02	19.2%	136.51	164.51	191.93	18.5%	139.21	159.42	179.22	15.2%	134.22	162.41	178.81	15%	138.52	154.82	190.43	16.5%
10	152.92	176.33	194.71	15.7%	148.13	168.23	198.91	15.6%	144.03	172.83	194.13	10.1%	142.63	171.62	190.02	10%	147.03	165.81	203.42	14%
15	161.83	184.42	200.62	13.2%	150.42	181.54	201.14	11.7%	142.04	173.31	199.24	14%	147.02	174.11	205.03	14.3%	140.54	172.43	214.62	18.9%
20	158.93	189.33	209.83	17.4%	151.02	183.52	208.02	16.4%	144.91	178.94	214.03	17%	150.04	189.02	211.03	9.7%	148.04	182.22	219.13	16.1%

(b) Eigenvalues corresponding to the second stage.

Formula	P1				P2				P3				P4				P5			
β	T_o	T_p	T_f	M_L	T_o	T_p	T_f	M_L	T_o	T_p	T_f	M_L	T_o	T_p	T_f	M_L	T_o	T_p	T_f	M_L
5	191.02	258.82	270.02	28.4%	191.92	251.62	274.81	27.8%	179.22	250.52	273.12	35.7%	178.81	245.63	271.21	27.5%	190.42	253.62	278.02	24.2%
10	194.71	275.63	291.41	26.7%	198.93	273.04	294.22	27.5%	194.13	268.13	285.63	28.9%	190.02	264.42	284.42	27.1%	203.43	267.72	296.42	24.8%
15	200.62	279.22	303.32	27.5%	201.12	276.33	309.73	31.9%	199.22	276.54	307.42	32.6%	205.04	276.31	302.73	30.6%	214.61	283.53	307.23	25.4%
20	209.81	286.33	312.04	26.4%	208.03	283.14	311.02	29.9%	214.03	289.72	317.44	33.6%	211.05	280.82	314.01	28.71%	219.12	288.81	313.82	23%

(c) Eigenvalues corresponding to the third stage.

Formula	P1				P2				P3				P4				P5			
β	T_o	T_p	T_f	M_L	T_o	T_p	T_f	M_L	T_o	T_p	T_f	M_L	T_o	T_p	T_f	M_L	T_o	T_p	T_f	M_L
5	270.02	334.32	374.03	14.5%	274.82	329.72	391.04	11.8%	273.13	331.91	382.13	15.8%	271.21	329.62	389.23	11.2%	278.03	328.52	392.63	11.5%
10	291.43	336.43	397.31	10.9%	294.22	339.23	410.05	10.4%	285.62	334.82	390.62	9.1%	284.42	332.93	398.32	7%	296.42	329.83	410.82	13.2%
15	303.32	349.02	423.52	10.9%	309.71	340.11	414.61	9.2%	307.44	340.04	407.03	8.4%	302.71	334.72	413.23	7.6%	307.23	330.14	413.13	10.5%
20	312.01	350.92	426.04	11.5%	311.03	343.04	419.04	11.2%	317.45	341.03	414.82	8.3%	314.02	336.02	440.72	6.9%	313.81	332.51	422.05	10.6%

(d) Eigenvalues corresponding to the fourth stage.

Formula	P1				P2				P3				P4				P5			
β	T_o	T_p	T_f	M_L	T_o	T_p	T_f	M_L	T_o	T_p	T_f	M_L	T_o	T_p	T_f	M_L	T_o	T_p	T_f	M_L
5	374.03	421.62	465.05	23.5%	391.03	438.53	482.03	20.7%	382.12	456.32	486.62	26.2%	389.23	486.04	499.92	26.3%	392.61	512.32	532.21	28.5%
10	397.32	444.93	499.03	21.6%	410.01	462.52	501.72	20.6%	390.63	472.43	502.13	19.5%	398.32	496.92	512.73	19.1%	410.82	513.21	540.32	28.3%
15	423.52	453.71	512.03	17.3%	414.62	469.51	513.71	21.6%	407.04	499.04	531.03	21.3%	413.23	514.83	544.64	22.7%	413.11	529.43	573.02	28%
20	426.04	465.74	523.02	20.1%	419.03	480.24	530.04	22.2%	414.82	500.82	554.02	22.1%	440.73	526.91	562.03	19%	422.03	536.44	581.04	25.7%

Note: T_o is the initial temperature; T_P is the peak temperature; T_f is the cut-off temperature; M_L refers to the mass ratio of thermal weight loss in this stage; T_o, T_P, T_f unit: °C; M_L unit: %.

The temperature range of the third stage is 270–440 °C. A weak exothermic peak can be observed at this stage, and the temperature of this exothermic peak is consistent with the corresponding exothermic peak in the TG-DTG curve of the individual component OC. Therefore, it can be determined that this exothermic peak is related to the exothermic pyrolysis of OC.

3.3.3. Analysis of the Pyrolysis Kinetics of the Samples

In order to further explore the thermal decomposition mechanism of Samples P1–P5 and to calculate the kinetic parameters of each stage of the reaction, Starink's method and the Flynn–Wall–Ozawa method were used [35–37]. The TG-DTG curve of Samples P1–P5 at different heating rates is shown in Figure 11. The corresponding characteristic peak temperature of each stage at different heating rates for each sample is shown in Table 6.

Starink's Method

Based on the measured TG-DTG curves of P1–P5 at different heating rates, combined with the characteristic peak temperatures of the four stages in Table 6, the value of $ln(\beta/T_P^{1.8})$ and $1/T_P$ can be obtained for the sample across the four pyrolysis stages [23]. Taking $1/T_P$ as the independent variable and $ln(\beta/T_P^{1.8})$ as the dependent variable, we then carried out linear fitting to obtain the slope of the fitting line and substituted it into Equation (2) to obtain the activation energy Ea at this stage.

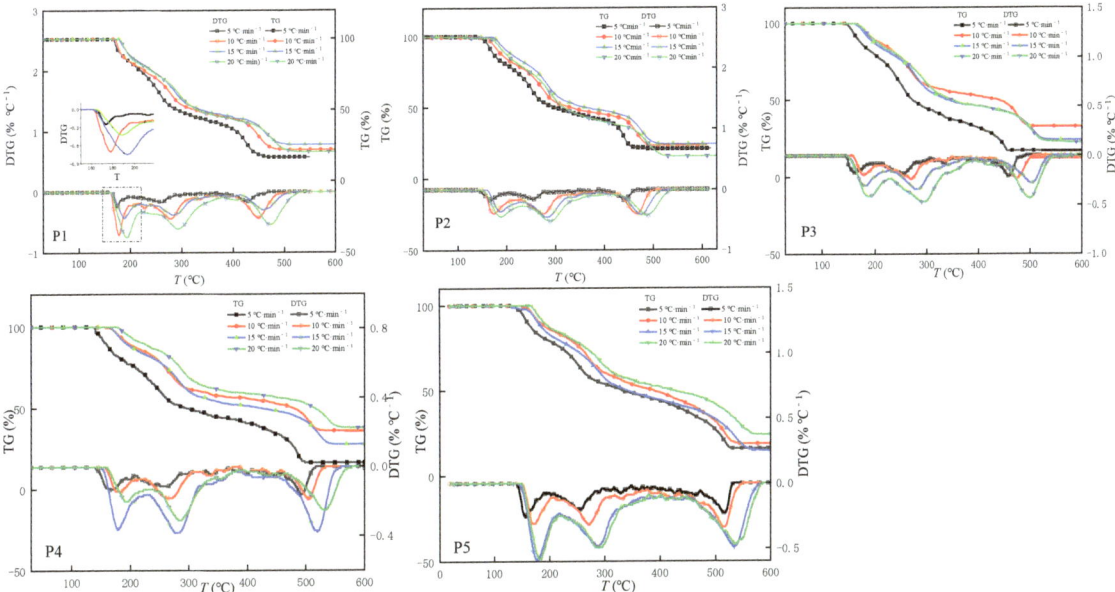

Figure 11. TG-DTG curve of Samples P1–P5 at different heating rates.

The linear fitting results of $ln(\beta/T_P^{1.8})$ and $1/T_P$ are shown in Figure 12, and the results of calculating the activation energy Ea for each stage are shown in Table 7. The correlation coefficient R^2 represents the accuracy of the fitting results, and the closer it is to 1, the higher the reliability.

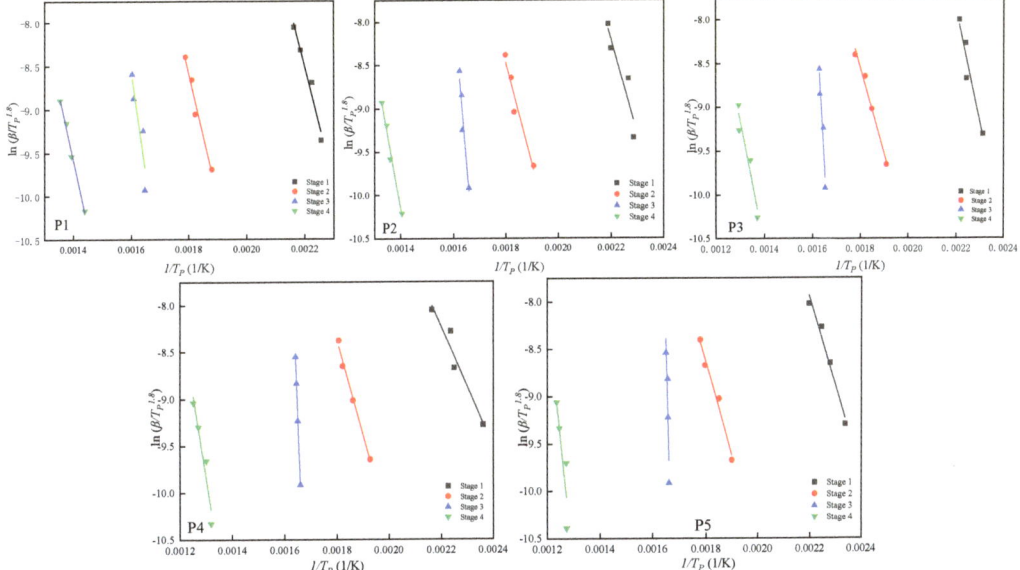

Figure 12. Activation energy curve of Samples P1–P5 at each stage of pyrolysis calculated via Starink's method.

Table 7. Reaction kinetic parameters of Samples P1–P5 at each stage of pyrolysis calculated via Starink's method.

Stage	Stage 1		Stage 2		Stage 3		Stage 4	
Formula	E_a (kJ/mol)	R^2	E_a (kJ/mol)	R^2	E_a (kJ/mol)	R^2	E_a (kJ/mol)	R^2
P1	110 ± 7	0.98223	116 ± 7	0.98417	193 ± 9	0.91128	124 ± 6	0.99518
P2	91 ± 8	0.93143	95 ± 7	0.97126	299 ± 9	0.96533	137 ± 7	0.99147
P3	106 ± 8	0.96432	80 ± 5	0.99133	406 ± 8	0.97029	115 ± 8	0.96539
P4	78 ± 5	0.96847	82 ± 7	0.99545	646 ± 7	0.99846	148 ± 6	0.96738
P5	76 ± 6	0.98638	83 ± 5	0.98949	659 ± 8	0.91439	164 ± 8	0.90127

Flynn–Wall–Ozawa method.

It can be seen from the results that with the same OB, with the addition of 5AT to the mass fraction, the activation energy of the first and second stages of P2–P5 shows a decreasing trend compared with P1, which confirms that the starting temperature of the component reaction after the addition of 5AT mentioned in the pyrolysis behavior analysis is significantly lower, which plays a certain role in promoting the reaction at this stage. In addition, if we compare the activation energies of the first and second stages of each formula, it can also be seen that only the activation energy of P3 in the second stage is significantly lower than that of the first stage, indicating that P3 can spontaneously carry out the second stage reaction after the first-stage reaction, which again confirms the reason why the weight loss ratio of P3 in the second stage is significantly higher than that of other samples. The activation energy of the third stage of P2–P5 increases significantly compared with that of P1, which may be related to a large amount of heat absorbed by the pyrolysis of 5AT. With a continual increase in the mass fraction of 5AT, the activation energy corresponding to this stage also increases accordingly.

Based on the measured TG-DTG curve data (Table 6), the corresponding temperature value $1/T$ at the same value of conversion α is an independent variable and $ln\beta$ is a dependent variable. The obtained data points were linearly fitted, then the slope of the straight line was obtained. By substitution in Formula (3), the activation energy $E\alpha$ corresponding to the reaction conversion α can be obtained at this stage [24,25]. Figure 13 shows the activation energy curve of Samples P1–P5 at each stage obtained via the Flynn–Wall–Ozawa method.

The activation energy of each stage for P1–P5 obtained by the Flynn–Wall–Ozawa method (Figure 13) is basically consistent with the activation energy of each stage obtained by Starink's method (Table 7), which further verifies the reliability of the kinetic parameters obtained by this method.

In addition, according to the activation energy curve of each stage of the mixed reagent obtained by the Flynn–Wall–Ozawa method, the reaction activation energy at the first, third, and fourth stages of thermal decomposition of Samples P1–P5 varies with α. This shows that the three-stage reaction process is a multi-step reaction, which is basically consistent with the results of the pyrolysis analysis. In the second stage of thermal decomposition, when $\alpha > 0.3$, there is an independent linear relationship between the corresponding reaction activation energy and the conversion α. This shows that the thermal decomposition process of the sample at this stage is mainly a one-step reaction, which confirms that this stage is mainly the thermal diffusion process of the tear agent OC in the analysis of the pyrolytic behavior.

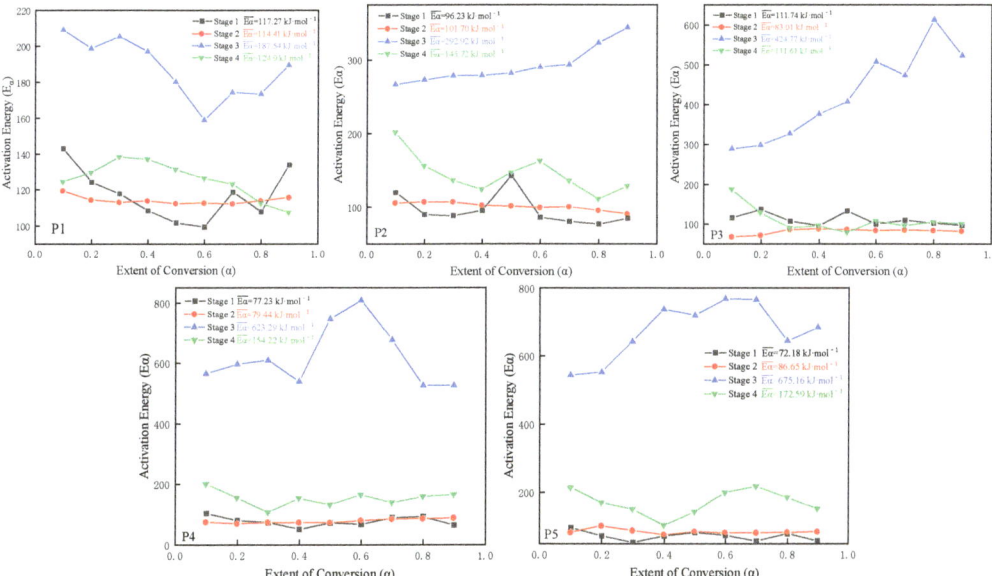

Figure 13. Activation energy curve of Samples P1–P5 at each stage obtained via the Flynn–Wall–Ozawa method.

Prediction of the Reaction Model of Samples P1–P5

In order to further explore the reaction mechanism of the thermal decomposition process of the main charge mixture, the Coats–Redfern method [26,38] was used to predict the most likely reaction model at each stage. According to 17 common reaction mechanism functions (Table 3), the linear fitting results of $ln[g(α)/T^2]$ and $1/T$ corresponding to the pyrolysis reaction at different stages are shown in Figure 14 (the maximum correlation coefficient has been marked in red in the figure). Table 8 shows the reaction models of four stages in the pyrolysis process of Samples P1–P5. In Figure 14 and Table 8, we can see the correlation coefficient obtained by fitting the data based on each reaction mechanism function and the most likely model of each stage of Samples P1–P5.

Table 8. Reaction model of four stages during the pyrolysis of Samples P1–P5.

Stage / Formula	Stage 1	Stage 2	Stage 3	Stage 4
P1	A3	A4	F3	F1
P2	F2	A4	F3	F1
P3	D3	A4	F3	D2
P4	D3	A4	F2	P2/3
P5	F3	A4	F2	D2

From these results, it can be seen that the reaction models of the first, third, and fourth stages of the sample change significantly when different amounts of 5AT are added to the sample, which shows that the reaction models of each stage can be effectively changed by adding 5AT. At the same time, it also further explains the relevant mechanism of Samples P1–P5 corresponding to their different combustion characteristics. In addition, for the second stage, which is most suitable for OC's thermal diffusion temperature range, the reaction model maintains a high degree of consistency, which further verifies the correctness of the physical thermal diffusion weight loss theory of OC in the second stage of pyrolysis weight loss. The discovery of this theory is of great significance for studying and improving the smoke characteristics of combustion tear gas mixtures.

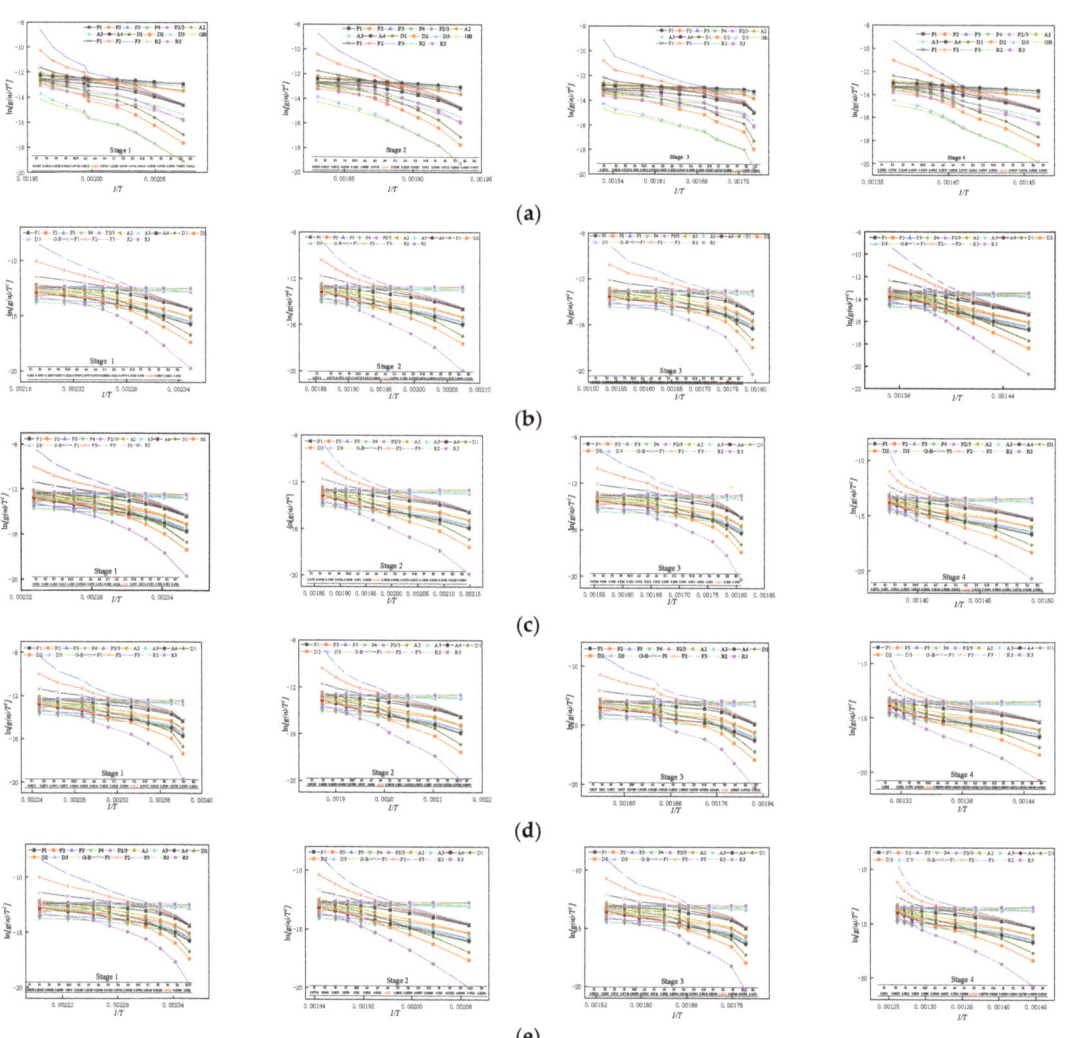

Figure 14. Fitting curve of the most likely reaction mechanism model of Samples P1–P5 formula obtained via the Coats–Redfern method. (**a**) Fitting curve of P1's reaction mechanism model. (**b**) Fitting curve of P2's reaction mechanism model. (**c**) Fitting curve of P3's reaction mechanism model. (**d**) Fitting curve of P4's reaction mechanism model. (**e**) Fitting curve of P5's reaction mechanism model.

4. Conclusions

The conclusions regarding the combustion pyrolysis characteristics and kinetic analysis of a combustion-type tear gas mixture based on 5AT are as follows:

1. Through a comparison of the maximum combustion temperature and the linear combustion rate of Samples P1–P5 with different amounts of 5AT, it was found that when the amount of 5AT is 10%, the maximum combustion temperature of the sample can be reduced by nearly 70 °C under the condition that the linear combustion rate is basically unchanged, thus improving the combustion environment of the mixture.

2. If we compare the T_{max} of Samples P1–P5, and the C and AP of smoke, it can be seen that the C and AP of smoke are inversely proportional to the T_{max} of the sample. The higher the T_{max} of the mixture, the smaller the C and AP values of the corresponding tear gas. As the combustion temperature of the mixed agent is higher, the amount of the pyrolytic tear agent OC in the agent will be greater, and the concentration and particle size of the smoke will be reduced. Combined with the exothermic enthalpy near 340 °C of the DSC curve and the weight loss ratio at the second stage of pyrolysis of Samples P1–P5, this observation is again confirmed. Therefore, adding an appropriate amount of 5AT is of great significance for improving the smoke characteristics of combustion tear gas mixtures.

3. The results of calculating the activation energy of Samples P1–P5 at each stage via Starink's method and the Flynn–Wall–Ozawa method are basically the same, which further verifies that the kinetic parameters obtained by these methods have high reliability. When the OB value in the formula's design is fixed, with the addition of the 5AT mass fraction, the activation energy of the first and second stages of Samples P2–P5 shows a decreasing trend compared with P1, which confirms that the reaction's starting temperature of the components after the addition of 5AT, as mentioned in the pyrolysis behavior analysis, is significantly lower, which plays a role in promoting the reaction at this stage.

4. According to the Coats–Redfern method, the most likely reaction models of the different formulations at each stage were predicted. It can be concluded that with different amounts of 5AT in the formulation, the reaction models of mixed agents in the first, third, and fourth stages changed significantly, indicating that the addition of 5AT can affect the reaction mechanism at some stages. In addition, for Samples P1–P5, the reaction model at the second stage of pyrolysis is the nucleation model A4, which maintains a high degree of consistency and further verifies the correctness of the physical thermal diffusion weight loss theory of the sample at the second stage of pyrolysis weight loss, which mainly corresponds to OC.

Author Contributions: H.Z. and X.C. contributed to the conception of the study; H.Z. and Y.G. performed the experiments; H.Z. and X.C. contributed significantly to analysis and manuscript preparation; H.Z. performed the data analysis and wrote the manuscript; X.C. helped perform the analysis with constructive discussions. All authors have read and agreed to the published version of the manuscript.

Funding: This work was financially supported by the equipment comprehensive research project (WJ2021 1A030013), (WJ20182A020036); basic research of university technology (WJY202146); basic frontier innovation research (WJY202236) and research on military theory (JLY2022089).

Institutional Review Board Statement: Not applicable.

Informed Consent Statement: Not applicable.

Data Availability Statement: Data are available on request from authors.

Acknowledgments: The authors gratefully acknowledge the research infrastructure provided by the equipment management and support of the College of Engineering University of the Chinese People's Armed Police Force.

Conflicts of Interest: The authors declare no conflict of interest.

References

1. Wang, H.Y.; Mu, Y.L.; Su, S.C.; Hao, H.M.; Li, G.B.; He, J. Study on the properties of a stimulating smoking agent. *J. Initiat. Pyrotech.* **2013**, *1*, 36–38. (In Chinese)
2. Shaw, A.P.; Chen, G.D. Advanced boron carbide-based visual obscurants for military smoke grenades. In Proceedings of the 40th International Pyrotechnics Seminar, Colorado Springs, CO, USA, 13–18 July 2014; pp. 170–191.
3. Haar, R.J.; Iacopino, V.; Ranadive, N.; Weiser, S.D.; Dandu, M. Health impacts of chemical irritants used for crowd control: A systematic review of the injuries and deaths caused by tear gas and pepper spray. *BMC Public Health* **2017**, *17*, 831. [CrossRef] [PubMed]

4. Hoz, S.S.; Aljuboori, Z.S.; Dolachee, A.A.; Al-Sharshahi, Z.F.; Alrawi, M.A.; Al-Smaysim, A.M. Fatal Penetrating head injuries caused by projectile tear gas canisters. *World Neurosurg.* **2020**, *138*, e119–e123. [CrossRef] [PubMed]
5. Stopyra, J.P.; Winslow, J.E., III; Johnson, J.C., III; Hill, K.D.; Bozeman, W.P. Baby Shampoo to relieve the discomfort of tear gas and pepper spray exposure: A randomized controlled trial. *West. J. Emerg. Med.* **2018**, *19*, 294. [CrossRef]
6. Rothenberg, C.; Achanta, S.; Svendsen, E.R.; Jordt, S. Tear gas: An epidemiological and mechanistic reassessment. *Ann. N. Y. Acad. Sci.* **2016**, *1378*, 96–107. [CrossRef] [PubMed]
7. Corbacıoğlu, Ş.K.; Güler, S.; Er, E.; Seviner, M.; Aslan, Ş.; Aksel, G. Rare and severe maxillofacial injury due to tear gas capsules: Report of three cases. *J. Forensic Sci.* **2016**, *61*, 551–554.
8. Zhai, H.L.; Cui, X.P. Thermal behavior test and characterization of a capsicum oleoresin (OC) composite material used for burning tear bomb. *J. Phys. Conf. Series. IOP Publ.* **2021**, *1965*, 012058.
9. Talawar, M.B.; Sivabalan, R.; Mukundan, T.; Muthurajan, H.; Sikder, A.K.; Gandhe, B.R.; Rao, A.S. Environmentally compatible next generation green energetic materials (GEMs). *J. Hazard. Mater.* **2009**, *161*, 589–607. [CrossRef]
10. Zhang, D.; Jiang, L.; Lu, S.; Cao, C.-Y.; Zhan, H.-P. Particle size effects on thermal kinetics and pyrolysis mechanisms of energetic 5-amino-1H-tetrazole. *Fuel* **2018**, *217*, 553–560. [CrossRef]
11. Zhang, D.; Lu, S.; Cao, C.Y.; Liu, C.C.; Gong, L.L.; Zhang, H.P. Impacts on combustion behavior of adding nanosized metal oxide to CH_3N_5-$Sr(NO_3)_2$ propellant. *Fuel* **2017**, *191*, 371–382. [CrossRef]
12. Küblböck, T. *New Trends in Sustainable Light-and Smoke-Generating Pyrotechnics*; Ludwig Maximilian University of Munich: München, Germany, 2020.
13. Glück, J.; Klapötke, T.M.; Küblböck, T. 5-Amino-1H-tetrazole-based multi-coloured smoke signals applying the concept of fuel mixes. *New J. Chem.* **2018**, *42*, 10670–10675. [CrossRef]
14. Glück, J.; Klapötke, T.M.; Rusan, M.; Shaw, A.P. Improved Efficiency by Adding 5-Aminotetrazole to Anthraquinone-Free New Blue and Green Colored Pyrotechnical Smoke Formulations. *Propellants Explos. Pyrotech.* **2017**, *42*, 131–141. [CrossRef]
15. Zeman, O.; Pelikan, V.; Pachman, J. *Diketopyrrolopyrrole—A Greener Alternative for Pyrotechnic Smoke Compositions*; ACS Sustainable Chemistry & Engineering: Washington, DC, USA, 2022.
16. Hemmilä, M.; Hihkiö, M.; Linnainmaa, K. Evaluation of the acute toxicity and genotoxicity of orange, red, violet and yellow pyrotechnic smokes in vitro. *Propellants Explos. Pyrotech. Int. J. Deal. Sci. Technol. Asp. Energetic Mater.* **2007**, *32*, 415–422. [CrossRef]
17. Zarko, V.; Kiskin, A.; Cheremisin, A. Contemporary methods to measure regression rate of energetic materials: A review. *Prog. Energy Combust. Sci.* **2022**, *91*, 100980. [CrossRef]
18. Ambekar, A.; Kim, M.; Lee, W.-H.; Yoh, J.J. Characterization of display pyrotechnic propellants: Burning rate. *Appl. Therm. Eng.* **2017**, *121*, 761–767. [CrossRef]
19. Yi, J.H.; Zhao, F.Q.; Wang, B.Z.; Liu, Q.; Zhou, C.; Hu, R.Z.; Ren, Y.H.; Xu, S.Y.; Xu, K.Z.; Ren, X.N. Thermal behaviors, nonisothermal decomposition reaction kinetics, thermal safety and burning rates of BTATz-CMDB propellant. *J. Hazard. Mater.* **2010**, *181*, 432–439. [CrossRef]
20. Aboulkas, A.; El, B.A. Thermal degradation behaviors of polyethylene and polypropylene. Part I: Pyrolysis kinetics and mechanisms. *Energy Convers Manag.* **2010**, *51*, 1363–1369. [CrossRef]
21. Vyazovkin, S. *Isoconversional Kinetics of Thermally Stimulated Processes*; Springer: Berlin/Heidelberg, Germany, 2015.
22. Muravyev, N.V.; Monogarov, K.A.; Asachenko, A.F.; Nechaev, M.S.; Ananyev, I.V.; Fomenkov, I.V.; Kiselev, V.G.; Pivkina, A.N. Pursuing reliable thermal analysis techniques for energetic materials: Decomposition kinetics and thermal stability of dihydroxylammonium 5,5'-bistetrazole-1,1'-diolate (TKX-50). *Phys. Chem. Chem. Phys.* **2017**, *19*, 436–449. [CrossRef]
23. Starink, M.J. The determination of activation energy from linear heating rate experiments: A comparison of the accuracy of isoconversion methods. *Thermochim. Acta* **2003**, *404*, 163–176. [CrossRef]
24. Ozawa, T. A new method of analyzing thermogravimetric data. *Bull. Chem. Soc. Jpn.* **1965**, *38*, 1881–1886. [CrossRef]
25. Flynn, J.H.; Wall, L.A. General treatment of the thermogravimetry of polymers. *J. Res. Natl. Bur. Standards. Sect. A Phys. Chem.* **1966**, *70*, 487. [CrossRef] [PubMed]
26. Coats, A.W.; Redfern, J.P. Kinetic parameters from thermogravimetric data. *Nature* **1964**, *201*, 68–69. [CrossRef]
27. Vyazovkin, S.; Burnham, A.K.; Criado, J.M.; Pérez-Maqueda, L.A.; Popescu, C.; Sbirrazzuoli, N. ICTAC Kinetics Committee recommendations for performing kinetic computations on thermal analysis data. *Thermochim. Acta* **2011**, *520*, 1–19. [CrossRef]
28. Zhang, D.; Cao, C.-Y.; Lu, S.; Cheng, Y.; Zhang, H.-P. Experimental insight into catalytic mechanism of transition metal oxide nanoparticles on combustion of 5-Amino-1H-Tetrazole energetic propellant by multi kinetics methods and TG-FTIR-MS analysis. *Fuel* **2019**, *245*, 78–88. [CrossRef]
29. Eslami, A.; Hosseini, S.G. Improving safety performance of lactose-fueled binary pyrotechnic systems of smoke dyes. *J. Therm. Anal. Calorim.* **2011**, *104*, 671–678. [CrossRef]
30. Wang, H.; Cao, L.; Chen, Y.K.; Hao, H.M.; Zhang, Z.L. Study on thermal cracking of vanillin nonanoate by pyrolysis gas chromatography-mass spectrometry. *J. Anal. Test.* **2018**, *37*, 858–861. (In Chinese) [CrossRef]
31. Wang, C.Y.; Du, Z.M.; Cong, X.M.; Tian, X.L. Study on thermal decomposition of basic magnesium carbonate. *Appl. Chem. Ind.* **2008**, *37*, 657–660. (In Chinese) [CrossRef]

32. Cao, C.Y.; Zhang, D.; Liu, C.C.; Song, L. Experimental investigation on combustion behaviors and reaction mechanisms for 5-aminotetrazole solid propellant with nanosized metal oxide additives under elevated pressure conditions. *Appl. Therm. Eng.* **2019**, *162*, 114207. [CrossRef]

33. Miyata, Y.; Date, S.; Hasue, K. Combustion mechanism of consolidated mixtures of 5-amino-1H-tetrazole with potassium nitrate or sodium nitrate. *Propellants Explos. Pyrotech. Int. J. Deal. Sci. Technol. Asp. Energetic Mater.* **2004**, *29*, 247–252. [CrossRef]

34. Tabacof, A.; Calado, V.M.A. Thermogravimetric analysis and differential scanning calorimetry for investigating the stability of yellow smoke powders. *J. Therm. Anal. Calorim.* **2017**, *128*, 387–398. [CrossRef]

35. Muravyev, N.V.; Pivkina, A.N.; Koga, N. Critical appraisal of kinetic calculation methods applied to overlapping multistep reactions. *Molecules* **2019**, *24*, 2298. [CrossRef] [PubMed]

36. Luciano, G.; Svoboda, R. Activation energy determination in case of independent complex kinetic processes. *Processes* **2019**, *7*, 738. [CrossRef]

37. Ambekar, A.; Yoh, J.J. Chemical kinetics of multi-component pyrotechnics and mechanistic deconvolution of variable activation energy. *Proc. Combust. Inst.* **2019**, *37*, 3193–3201. [CrossRef]

38. Ebrahimi-Kahrizsangi, R.; Abbasi, M.H. Evaluation of reliability of Coats-Redfern method for kinetic analysis of non-isothermal TGA. *Trans. Nonferrous Met. Soc. China* **2008**, *18*, 217–221. [CrossRef]

Review

The Oxidation Process and Methods for Improving Reactivity of Al

Deqi Wang, Guozhen Xu, Tianyu Tan, Shishuo Liu, Wei Dong, Fengsheng Li and Jie Liu *

National Special Superfine Powder Engineering Research Center of China, School of Chemistry and Chemical Engineering, Nanjing University of Science and Technology, Nanjing 210094, China
* Correspondence: jie_liu_njust@126.com; Tel.: +86-025-8431-5042

Abstract: Aluminum (Al) has been widely used in micro-electromechanical systems (MEMS), polymer bonded explosives (PBXs) and solid propellants. Its typical core-shell structure (the inside active Al core and the external alumina (Al_2O_3) shell) determines its oxidation process, which is mainly influenced by oxidant diffusion, Al_2O_3 crystal transformation and melt-dispersion of the inside active Al. Consequently, the properties of Al can be controlled by changing these factors. Metastable intermixed composites (MICs), flake Al and nano Al can improve the properties of Al by increasing the diffusion efficiency of the oxidant. Fluorine, Titanium carbide (TiC), and alloy can crack the Al_2O_3 shell to improve the properties of Al. Furthermore, those materials with good thermal conductivity can increase the heat transferred to the internal active Al, which can also improve the reactivity of Al. Now, the integration of different modification methods is employed to further improve the properties of Al. With the ever-increasing demands on the performance of MEMS, PBXs and solid propellants, Al-based composite materials with high stability during storage and transportation, and high reactivity for usage will become a new research focus in the future.

Keywords: aluminum; oxidation process; reaction mechanism; modification method; development trend

Citation: Wang, D.; Xu, G.; Tan, T.; Liu, S.; Dong, W.; Li, F.; Liu, J. The Oxidation Process and Methods for Improving Reactivity of Al. *Crystals* 2022, 12, 1187. https://doi.org/10.3390/cryst12091187

Academic Editors: Rui Liu, Yushi Wen, Weiqiang Pang and Thomas M. Klapötke

Received: 4 July 2022
Accepted: 10 August 2022
Published: 24 August 2022

Publisher's Note: MDPI stays neutral with regard to jurisdictional claims in published maps and institutional affiliations.

1. Introduction

Aluminum (Al) is an important solid metal fuel, which has been widely used in connection between crystalline silicon [1], micro-electromechanical systems (MEMS) [2–5], polymer bonded explosives (PBXs) [6–12], and solid propellants [13–18] to provide energy. As a combustion agent, Al displays obvious advantages, such as low oxidant consumption, high combustion calorific value (31,070 J/g), and high measured specific impulse [19–21]. The dense and high melting point (about 2000 °C) alumina (Al_2O_3) shell will be formed when Al is exposed to air, which can improve the safety of aluminum powder in the process of production, storage, transportation and usage [22–24]. However, the oxidation process of Al is also closely related to the Al_2O_3 shell, which causes the inside active Al to become hindered (it cannot come into contact with oxidation components) [25–28]. As a result, the ignition and combustion reaction activity of Al is limited by the Al_2O_3 shell [21,29–31]. In addition, the combustion product Al_2O_3 will wrap onto the surface of the active Al again, which further hinders the combustion chemical reaction, resulting in an incomplete combustion of Al, and reduced energy release efficiency [29].

To solve the problem of the Al_2O_3 shell limiting oxidation activity and combustion efficiency, the oxidation process of Al has been studied extensively. To describe the ignition and combustion process of Al, some mechanisms, such as the oxidant diffusion mechanism [32–34], the Al_2O_3 crystal transformation mechanism [35–41], and the melt-dispersion mechanism [42–46], have been proposed. Meanwhile, to improve the activity of Al, some new composite materials, such as Al-based metastable intermixed composites (MICs), flake Al-based composite materials, alloying and so on, have been found and used gradually [12,47–57].

A large number of reviews on Al have been published. Xiang Zhou [58] summarized the synthesis, ignition and combustion modeling, and applications of Al-based nanocomposites. Wei He [59] introduced the preparation and characterization of Al-based MICs. In addition, Xiaoxia Ma [60] focused on the preparation and fundamental properties of the Al-based core–shell structured nanoenergetic. However, they discussed the preparation and characterization methods, and hardly any reviews systematically summarized the relationship between the modification method and the different ignition mechanism of Al during the past decades. This review examines systematically the key influencing factors and the mechanisms during the Al oxidation process, and classified modification methods by the influencing factors of its transformation. Furthermore, it provides ideas for selecting modification methods of Al under different application conditions. What is more, the new trend of Al-based composite materials has been indicated.

2. The Key Influencing Factors and the Reaction Mechanism during the Al Oxidation Process

The Al particle is a typical core-shell structure with active Al wrapped in dense Al_2O_3. Therefore, the oxidation process of the Al particle is the evolution process of the oxidant, the alumina shell, and the inside active Al. Furthermore, the diffusion of the oxidant, the growth and rupture of the Al_2O_3 shell, and the melting and gasification of the inside active Al are the key influencing factors of the Al oxidation process.

2.1. Effect of Diffusion of the Oxidant on the Al Oxidation Process

The diffusion of the oxidant is a necessary process for Al oxidation. Zachariah [32–34] proposed a three-step oxidation process of Al (oxidant diffusion mechanism) on the basis of a well-known idealized "shrinking core" mechanism for spherical particles, which involves a reaction front moving radially inward separating an unreacted core with a completely reacted ash outer layer. As shown in Figure 1, the diffusion of the oxidant is influenced by three factors: the Al_2O_3 shell (generated on the surface in air), the Al_2O_3 ash (the product of oxidation reaction) and the chemical reaction of the oxidant with the inside active Al.

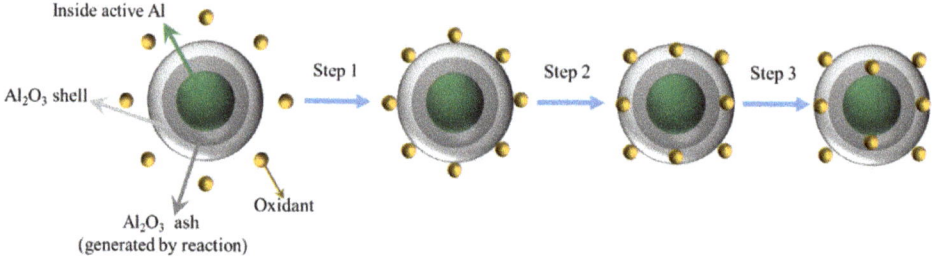

Figure 1. The three-step oxidation process of Al proposed by Zachariah.

Step 1: Diffusion of the oxidant through the Al_2O_3 shell to ash (Al_2O_3 generated by reaction, Al_2O_3 shell resistance).

Step 2: Diffusion of the oxidant through the ash (Al_2O_3 generated by reaction) to the surface of the inside active Al (ash resistance).

Step 3: Chemical reaction of the oxidant with the inside active Al at the unreacted core surface (chemical reaction resistance).

Since ash (Al_2O_3 generated by reaction) forms very rapidly, step 1 (Al_2O_3 shell resistance) can be ignored. This still leaves steps 2 and 3 from which to deduce the rate-limiting step. Then it is proved by experiments that the reaction of Al is controlled by diffusion. Therefore, the diffusion of the oxidant through the Al_2O_3 ash to the inside active Al is the rate-limiting step.

2.2. Effect of the Al_2O_3 Shell on the Al Oxidation Process

The Al_2O_3 shell on the surface of Al has a crucial effect on the oxidation process of Al. Trunov [35] analyzed the processes of simultaneous growth and phase transformations of Al_2O_3 during oxidation of the Al particle (Al_2O_3 crystal transformation mechanism). The process of Al oxidation can be divided into four stages, which can be seen in Figure 2. Firstly, amorphous Al_2O_3 on the Al surface gradually grows thicker and the reaction rate is controlled by the outward diffusion of Al cations [36]. Secondly, amorphous Al_2O_3 on the Al surface transforms to γ- Al_2O_3 when the critical thickness is approached, or when the temperature becomes sufficiently high [37–40]. Since the density of γ-Al_2O_3 exceeds that of the amorphous Al_2O_3, the Al_2O_3 shell on the Al surface ruptures. In addition, the inside active Al can come into contact with oxide, which greatly increases the reaction rate. With the growth and healing of the γ-Al_2O_3 shell, the reaction rate decreases significantly. Eventually, a regular polycrystalline layer of γ-Al_2O_3 forms by the end of the second stage. In stage three, the growth of γ-Al_2O_3 continues. In the meantime, γ-Al_2O_3 transforms the crystal phase to δ-Al_2O_3 or θ-Al_2O_3. Due to the density of δ-Al_2O_3, and θ-Al_2O_3 is similar to that of γ-Al_2O_3, the shell of Al_2O_3 will not rupture and the reaction will not change significantly. Additionally, the reaction rate is limited by the inward grain boundary diffusion of oxidant anions in stage three [36,41]. When the stable and denser α-Al_2O_3 forms by increased temperature, stage three ends. Stage four starts when Al_2O_3 is completely transformed to α-Al_2O_3. In stage three, the thickness of the γ-Al_2O_3 layer decreases, and the oxidation rate increases momentarily. Once most of the oxide layer is transformed to coarse and dense α-Al_2O_3, the contact between the internal active aluminum and the oxidized components is completely blocked, and the reaction rate decreases rapidly.

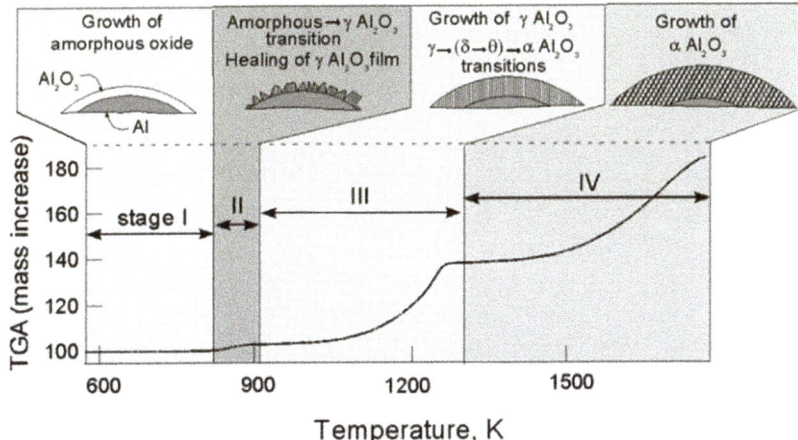

Figure 2. The change of the Al_2O_3 shell during the oxidation process of the Al particle, reprinted/adapted with permission from Ref. [35]. 2006, Dreizin, E. L.

2.3. Effect of the Inside Active Al on the Al Oxidation Process

Levitas [42–46] proposed a melt-dispersion mechanism when Al is heated rapidly. As shown in Figure 3, the volume change (6% volume expansion) due to the melting of the inside active Al can make the pressure of the Al_2O_3 shell reach about 11 GPa. That can cause spallation of the Al_2O_3 shell and the inside active Al can come into contact with the oxidant. Furthermore, the pressure inside the molten active Al will disperse the active Al into small droplets. Then the oxidation process of the active Al would occur rapidly.

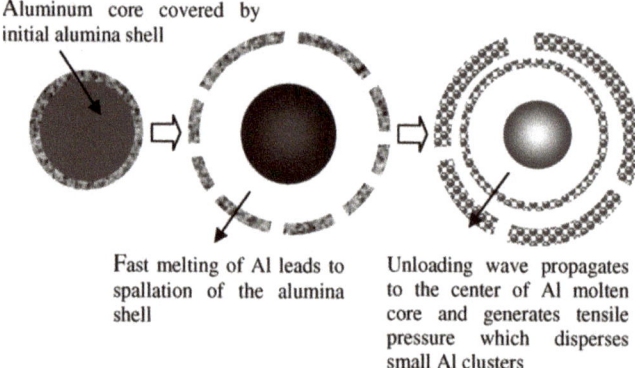

Aluminum core covered by initial alumina shell

Fast melting of Al leads to spallation of the alumina shell

Unloading wave propagates to the center of Al molten core and generates tensile pressure which disperses small Al clusters

Figure 3. The process of melt and dispersion of the active Al during the oxidation process of the Al particle, reprinted/adapted with permission from Ref. [46]. 2008, Levitas Valery I.

From the above discussion, it can be seen that the diffusion of the oxidant, the crystallization and the growth of the Al_2O_3 shell, and the melting and dispersion of the internal active Al, all have a crucial impact on the oxidation process of Al. The leading factors will change with the particle size and the heating rate of Al. Therefore, the oxidation process of Al under different conditions is not the same. At a slow heating rate, the diffusion of the oxidant plays a leading role, and the oxidation process of nano Al, which has a large specific surface area, conforms to the three-step oxidation process (proposed by Zachariah [32,33]). At the slow heating rate, the transformation and growth of the Al_2O_3 shell have the greatest influence on the oxidation process of micron Al, so the oxidation process of micron Al conforms to the Al_2O_3 shell transformation and growth mechanism (proposed by Trunov [35]). At a fast heating rate, the melting and dispersion of the internal active Al have the greatest influence on the oxidation process of nano Al, so the oxidation process of nano Al conforms to the melting diffusion mechanism (proposed by Levitas [42–46]).

The three crucial factors that determine the oxidation process of Al are: the diffusion of the oxidant, the transformation and growth of the Al_2O_3 shell, and the melting and dispersion of the internal active Al. As the key factors affecting the oxidation process of Al have been investigated, the oxidation process of Al under different conditions can be accurately described. Then the method of changing the Al oxidation process can be found, and the properties of Al can be improved.

2.4. Effect of the Gas Phase Reaction on the Al Oxidation Process

The gasification temperature of the internal active Al is lower than that of the Al_2O_3 shell, which leads to the gas phase reaction involving gas phase Al during Al oxidation [61]. As the models of the combustion process of micron Al, which were introduced both by Law [62] and Glassman [61], and later expanded upon by Brooks and Beckstead [63], the gas phase flame can drive the surface reactions of the Al particle. However, it is difficult to accurately characterize the process and products of gas phase reactions. Lynch [64,65] found that volatile suboxides existed during Al combustion, and Tappan [66] found the reaction between the gas phase Al and the Al_2O_3 shell. Although the gas phase reaction has been found, the reaction process under the different reaction conditions has not been accurately confirmed, and its effect on the oxidation of Al continues to be studied.

3. Methods for Improving Al Properties

To improve the properties of Al, the following methods have been investigated: metastable intermixed composites (MICs, reducing the diffusion of the oxidant), nano Al and flake Al (increasing the specific surface area of Al) for improving diffusion efficiency, fluorine modification [67], Titanium carbide (TiC) modification and alloying for cracking the Al_2O_3 shell and the modification of good thermal conductivity materials (Ag, graphene) for improving heat transfer efficiency.

3.1. Improving Diffusion Efficiency

3.1.1. MICs

Metastable intermixed composites (MICs) are usually composed of a metal fuel and an oxidizer. The Los Alamos National Laboratory in the US was the first to study the combustion performance of MICs. Aumann [68] prepared the Al/MoO_3 nano Al, which bulk energy density can reach 16 kJ/cm^3, and the combustion rate is more than 1000 times that of the traditional thermit. The oxidizer of the Al-based MICs is usually metal oxide (bismuth oxide (Bi_2O_3), copper oxide (CuO), cuprous oxide (Cu_2O), molybdenum oxide (MoO_3), ferric oxide (Fe_2O_3), nickel oxide (NiO), tungsten trioxide (WO_3)), halate (potassium perchlorate ($KClO_4$), potassium periodate (KIO_4), iron iodate ($Fe(IO_3)_3$), copper iodate ($Cu(IO_3)_2$), bismuth iodate ($Bi(IO_3)_3$)), and persulfide (potassium persulfate($K_2S_2O_8$)) etc. Al-based MICs effectively reduce the diffusion distance of oxidation, which has significant advantages in volume energy density, ignition, and burning rate [59]. Although, MICs have not been able to achieve efficient, safe and low cost batch preparation, and the reaction process of MICs is difficult to accurately control. Therefore, MICs are difficult to apply in solid propellants and PBXs these days [69].

Ludovic Glavier [70] prepared Al-based MICs by Al, Bi_2O_3, CuO and MoO_3. Among them, as shown in Table 1, Al/Bi_2O_3 have the shorter the ignition delay time (5 µs), the highest burning rate (420 m/s), and the fastest pressurization rate (5762 kPa/µs). Egor A. Lebedev [71] prepared the MICs layer composed of Al, CuO and Cu_2O by electrophoretic deposition, and the maximum heat release of MICs layer is 1954 J/g. Aifeng Jiang [72] prepared Fe_2O_3/nano Al by ball milling. The initial combustion temperature of Fe_2O_3/nano is about 600 °C. Ning Wang [73] prepared a Al@NiO core-shell structure composite microunit; the ignition temperature can be advanced to 531.5 °C, the heat release is 1410.2 J/g. Chunpei Yu [74] prepared 3D ordered macroporous Al/NiO MICs. Its heat releases up to 2462.27 J/g. Wei He [75] prepared Al/Energetic metal organic frameworks (EMOF) MICs, which can activate Al by eliminating Al_2O_3 shell and produce metal oxide by decomposing of EMOF. The heat release of Al/EMOF is 3464 J/g; the burning rate is 2.8 m/s.

Table 1. The thermal and combustion properties of Al-based MICs, reprinted/adapted with permission from Ref. [70]. 2015, Rossi Carole.

Sample	Heat Release J/g	Delay Time µs	Burning Rate m/s	Pressurization Rate kPa/µs
Al/Bi_2O_3	1541	5	420	5762
Al/CuO	1057	15	340	172
Al/MoO_3	1883	110	100	35

W. Lee Perry [76] prepared $WO_3 \cdot H_2O/Al$ MICs, which had an energy release of approximately 1.8 MJ/kg at a rate of approximately 215 GW/m^2. They found that the enhanced behavior of the hydrated MICs formulation resulted from the reaction of Al with the interstitially bound H_2O, which had additional energy release and generated hydrogen gas.

Ahmed Fahd [77] compared and analyzed the thermal behavior of different nanothermite tertiary compositions based on nano Al, graphene oxide (GO), various salt and metallic oxidizers (as shown in Figure 4). The addition of GO enhances the reactivity of nanothermites with both salt and metallic oxidizers by reducing the reaction onset temperature, activation energy and increasing the heat release. For nanothermites with oxidizing salts, the heterogeneous solid–gas reaction mechanism plays a more important role than the condensed phase reactions. In general, nanothermites based on oxidizing salts are more reactive than those with metallic ones, as indicated in both theoretical and experimental data. Among them, the $GO/Al/KClO_4$ nanothermite exhibits the highest heat release (9614 J/g), while the GO/Al/K$_2$S$_2$O$_8$ nanothermite shows the lowest onset temperature and activation energy (380 °C and 105 kJ/mol^{-1}).

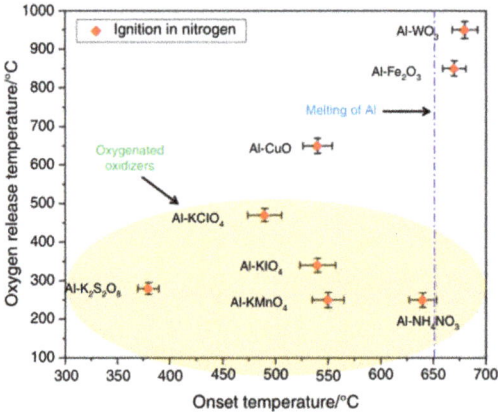

Figure 4. Relation between the oxidant release temperature of different oxidizers and the onset temperature of their nanothermites, reprinted/adapted with permission from Ref. [77]. 2022, Dubois, Charles.

3.1.2. Flake Al and Nano Al

Flake Al and nano Al have larger specific surface areas than that of sphere Al, which can improve the diffusion efficiency of oxidation [12,49–51,78–82]. However, the large-scale preparation process of nano Al and the effective Al content are still the key problems restricting its application.

Al has good ductility when it is subjected to external force, which would deform firstly. So flake Al is usually prepared by ball milling. As shown in Figure 5, in the process of ball milling, the sphere Al is extruded and sheared by ball milling beads. Under the action of force, sphere Al is changed to cake-like, firstly. In the second stage, the caked Al continues to be subjected to force and becomes flake Al; in the third stage, the flake Al would be broken and becomes smaller flake Al. So different thicknesses of flake Al can be prepared by controlling the parameters of ball milling.

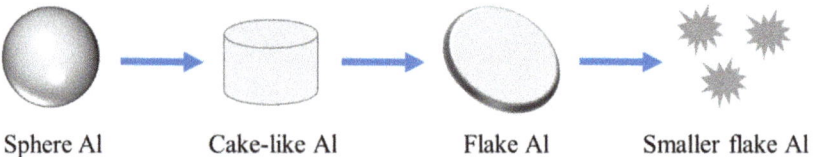

| Sphere Al | Cake-like Al | Flake Al | Smaller flake Al |

Figure 5. The deformation process of Al during ball milling.

Lei Xiao [51] prepared flake Al powder by ball milling, which could increase the detonation heat of Al-containing explosives by 6.48%. William Wilson [49] found that the flake Al became small particles more easily by shock waves during combustion. Qingming Liu [78] found that the flake Al dust–air mixture could be ignited, and self-sustained detonation by an electric spark of 40 J. DeQi Wang [81] applied the flake Al powder to a solid propellant, which increased the burning rate of the propellant by 5.5%. A.L. Kuhl [50,82] found the flake-Al could increase the impulse of the TNT composite charges by the fast combustion of the flake-Al.

The main methods of preparing nano Al include the electric explosion of wires method [83] and the plasma-arc recondensation [84]. In order to protect the nano Al from oxidation during storage, it is usually necessary to form a coating film on the surface of the nano Al. Deluca L T [21,85] prepared a nano-coated Al with stearic acid (L-ALEX), palmitic acid (P-ALEX), and trihydroperfluoro-undecyl alcohol (F-ALEX), et al. (as shown in Figure 6).

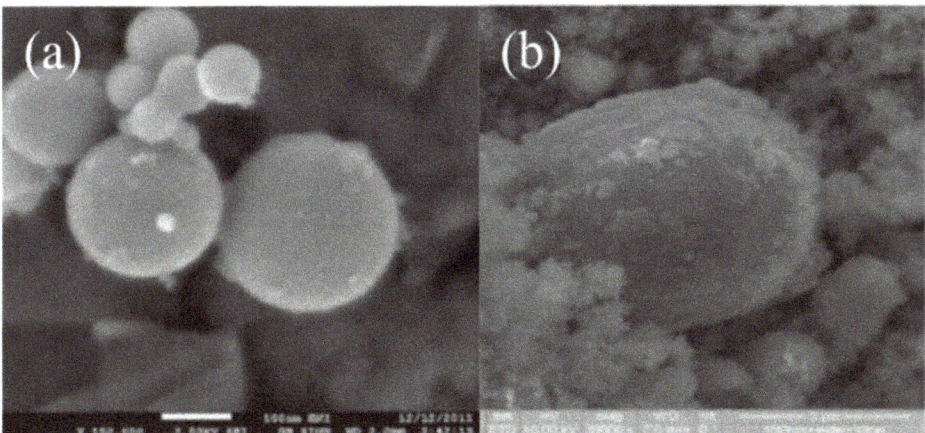

Figure 6. The SEM image of P-ALEX (**a**) and F-ALEX (**b**), reprinted/adapted with permission from Ref. [21]. 2017, Deluca, L.T.

Nano Al is widely used in MICs and solid propellants, etc [86]. Deyun Tang [87] used tannic acid (TA) to coat on nano Al as an interfacial layer to bind with ($Fe(IO_3)_3$), copper iodate ($Cu(IO_3)_2$), bismuth iodate ($Bi(IO_3)_3$), respectively. For the energy release, $Fe(IO_3)_3$-based MICs can be increased to 24.1 kJ/cm^3 (14.5% higher), whereas the $Cu(IO_3)_2$-based MICs to 22.8 kJ/cm^3 (19.4%), and $Bi(IO_3)_3$-based MICs to 20.2 kJ/cm^3 (3.1%). Deluca L [85] found that 100–200 nm Al can clearly increase the burning rate of the propellant.

Nano Al has high reactivity, which also produces some problems such as a high safety risk, a low content of active aluminum and sintering during combustion. Uncoated nano Al is classified by the International Air Transport Association as a highly flammable solid [88]. Weismiller [52] researched the effect of the particle size of Al and the oxidant on the properties of Al/CuO and Al/MoO_3. As shown in Table 2, the properties of Al/CuO and Al/MoO_3 are greatly improved due to the rapid decomposition of nano oxidants. However, due to the high Al_2O_3 content of nano Al, the nano Al does not show theoretical property advantages.

Table 2. Properties for Al/CuO and Al/MoO_3, reprinted/adapted with permission from Ref. [52]. 2011, Weismiller, M. R.

Sample	Linear Burning Rate m/s	Mass Burning Rate kg/s	Pressurization Rate MPa/μs
Nano Al/nano CuO	980	3.8	0.67
Micron Al/nano CuO	660	4.8	1.82
Nano Al/micron CuO	200	1.3	0.28
Micron Al/micron CuO	180	2.0	0.11
Nano Al/nano MoO_3	680	2.0	0.68
Micron Al/nano MoO_3	360	1.5	0.44
Nano Al/micron MoO_3	150	0.45	0.20
Micron Al/micron MoO_3	47	0.52	0.17

Zachariah [89] found that the combustion product of nanothermites has two distinct populations of particles, as shown in Figure 7. The large particles include aluminum, oxidant, and reduced metal while the nano-sized particle is composed of reduced metal/metal oxide. As such large particles cannot be formed from the vapor phase condensation during the available transit time to the substrate, they must be formed in the condensed state as molten material.

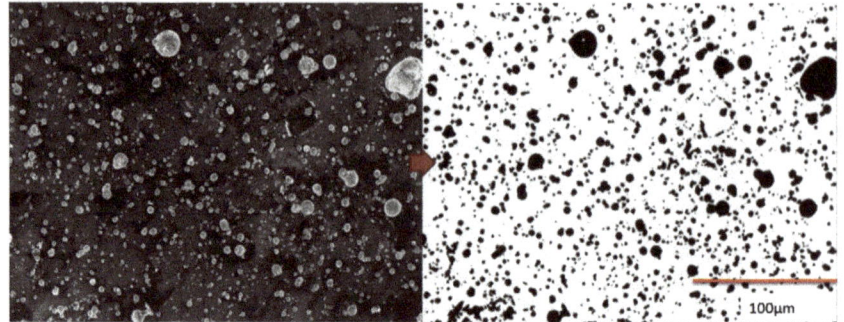

Figure 7. The combustion product of nanothermites, reprinted/adapted with permission from Ref. [89]. 2015, Zachariah, Michael R.

3.2. Cracking the Al$_2$O$_3$ Shell

3.2.1. Fluorine Modification

Fluoropolymers and fluorides can crack the Al$_2$O$_3$ shell of Al by the reaction between fluorine and Al$_2$O$_3$ shell. The curve of DSC will appear as a small exothermic peak before the main exothermic peak, caused by oxidation of Al. This phenomenon is known as preignition reaction (PIR), which is found by Osborne [90], and is verified by Zachariah by a quadrupole mass spectrometer and TG-DSC-MS coupling techniques [91]. The properties of the fluorine-modified Al are greatly improved due to the crack of the Al$_2$O$_3$ shell by PIR. Fluorine-modified Al can be used in MEMS and PBXs, but compatibility of fluoride with a propellant system is a key problem to realize its application in the propellant system.

Siva K. Valluri [92] prepared composite micro-units containing Al and NiF$_2$ by reaction inhibition ball milling, which shortened the ignition delay time and improved the combustion efficiency of Al powder. Sergey Matveev [48] prepared Al/BiF$_3$, which could produce 3200 K temperatures. Aifeng Jiang [72] prepared FeF$_3$/Al@vinyltrimethoxysilane by ball milling, which had a large specific surface area (26.33 m^2/g) and could be well-preserved from the air atmosphere and water. Additionally, the maximum heat release of FeF$_3$/nano Al@vinyltrimethoxysilane can go up to 12,852 J/g.

Jena McCollum [93] composited Al with perfluoropolyethylene to advance the ignition temperature of Al to about 330 °C. Dong Won Kim [94] coated polyvinylidene fluoride (PVDF) onto the surface of Al (Al@PVDF), and the heat release of Al@PVDF at 900 °C was 11,040 J/g. Jun Wang from the Institute of Chemical Materials [95] prepared Carbon nanotubes/polytetrafluoroethylene (PTFE)/Al nanocomposites, which had a lower initial reaction temperature (reduced by 80 °C) and a shorter ignition delay time (reduced by 0.21 ms). Xiang Ke [96] coated PVDF onto the surface of Al to prepare reactive film materials, which appear PIR at 430 °C.

3.2.2. TiC Modification

The reaction (as shown in Equation (1)) of Al$_2$O$_3$ and Titanium carbide (TiC) has been found during the preparation process of ceramics, which also can generate gas [97–99]. If this reaction can take place before the oxidation reaction of Al, the Al$_2$O$_3$ shell can be removed effectively, and the generated gas can also break the Al into small particles under high pressure. DeQi Wang [56] prepared the thick flake Al/TiC, which can crack the Al$_2$O$_3$ shell in-situ before 633 °C. As shown in Figure 8, the reaction of TiC with the Al$_2$O$_3$ shell on the Al surface to produce gas has been experimentally confirmed. The heat release of it can reach 21,419 J/g, and this powdered material has good application prospects in solid propellants and polymer bonded explosives (PBXs). This provides a new research idea for solving the limitation of the Al_2O_3 shell on the Al oxidation process. However, the application of it in PBXs and the propellant system needs further exploration.

$$Al_2O_3(s) + TiC(s) \rightarrow Al_2O(g) + TiO(g) + CO(g) \tag{1}$$

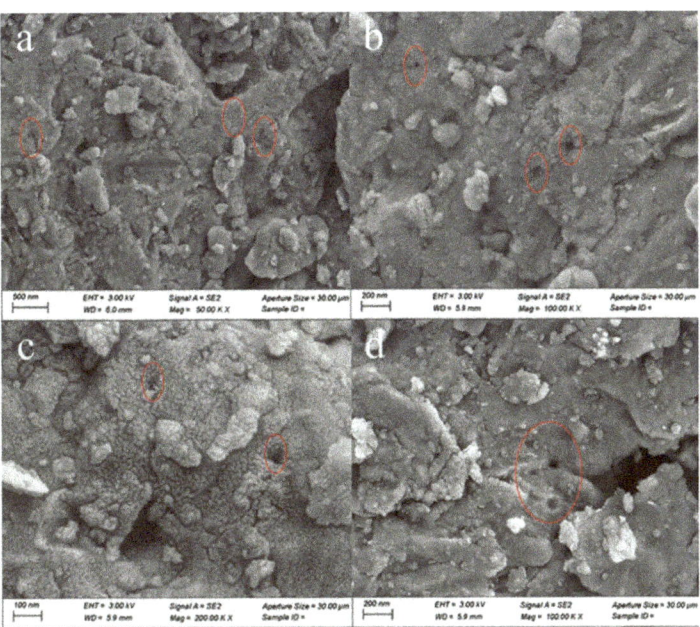

Figure 8. The pores morphology of TF-Al/5%TiC surface at different temperature: 633 °C (**a**), 650 °C (**b**), 670 °C (**c**) and 690 °C (**d**), reprinted/adapted with permission from Ref. [56]. 2021, Jie Liu.

3.2.3. Alloying

The oxide film on the surface of the Al-alloy particle may not be pure Al_2O_3 shell, which may have higher transmittance to oxidation. Therefore, the Al-alloy can be oxidized at lower temperatures. The oxidation product is not a dense structure like the Al_2O_3 shell. Accordingly, alloying is one of the effective ways to destroy the structure of the Al_2O_3 shell. Some metals (such as Li, Mg and Zr et al.) can be oxidized at a lower temperature than Al and can provide activation energy for the oxidation of Al, so that oxidation activity of Al-alloy is higher [53–55]. The preparation method of the high-density alloy is the key factor restricting the application of this modification method.

Hao Fu [55] prepared the Al- europium (Eu) alloy, which could be oxidized at 1065 °C and the heat release of the Al-Eu alloy is almost 5 times that of pure Al. Aobo Hu [53,54] prepared the Al-Zr alloy and the Al-W alloy. New alloy phases, $ZrAl_3$, formed in the Al-Zr alloy, which changed the oxidation process (as shown in Figure 9a). Therefore, the Al-Zr alloy can complete combustion under high pressure. Furthermore, the Al-W alloy is almost completely oxidized in air at 1500 °C. In addition, as shown in Figure 9b,c, the gas product, WO_3, increases the contact area between the active Al and the oxidant. As a result, the properties of the Al-W alloy have been improved greatly. Fahad Noor [100] prepared the Al-Cu alloy, which had the lower ignition temperature (565 °C).

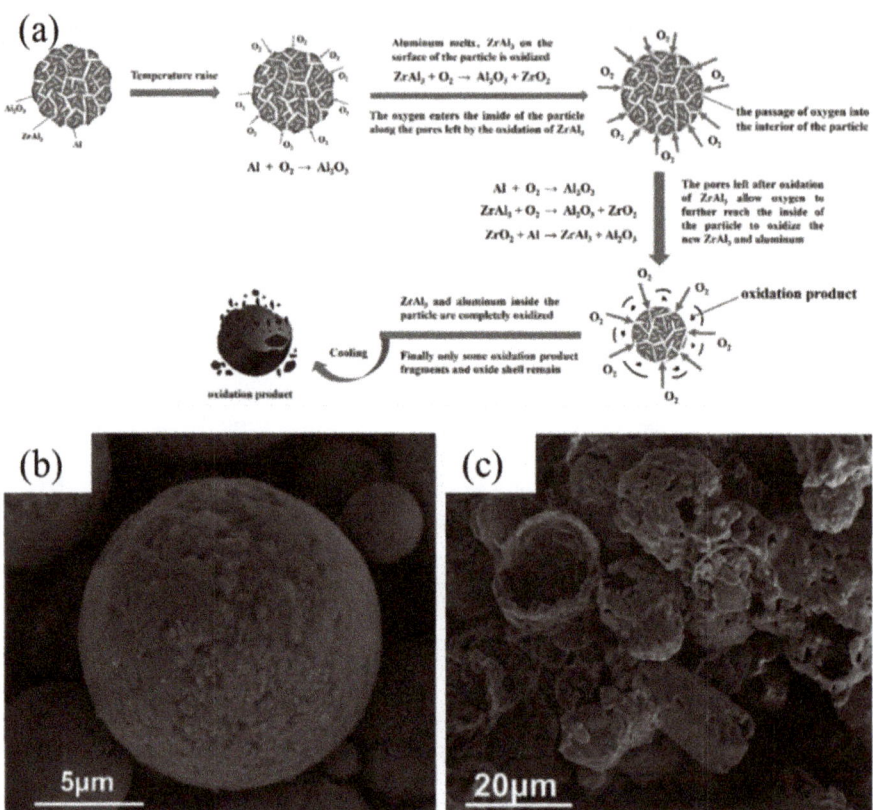

Figure 9. The oxidation process of Al-Zr alloy (**a**), Al-W alloy (**b**) and Al-W alloy after combustion (**c**), reprinted/adapted with permission from Refs. [53,54]. 2019 and 2021, Ai-min Pang and Shuizhou Cai.

3.3. Accelerating the Melting of the inside Active Al

The composite made of high thermal conductivity material and Al can increase the rate of Al absorbing heat, which can accelerate the melting of active Al inside and break through the limitation of Al_2O_3 shell. Consequently, the inside active Al can make contact with the oxidant and the reactivity of Al will be improved. However, as with MICs, how to achieve an efficient, safe and low cost batch preparation is the key problem when it is used in PBXs and propellants.

Jinpeng Shen [101] investigated the effects of nano-Ag on the combustion wave behavior of Al/CuO. The experimental observations confirm that the presence of nano-Ag particles improves the heat transfer efficiency, and the first exothermic peak temperature decreases from 607.8 °C to 567.6 °C.

Ahmed Fahd [77] compared and analyzed the thermal behavior of different nanothermite tertiary compositions based on Al, graphene oxide (GO), and various salt and metallic oxidizers. The addition of GO enhances the reactivity of nanothermites with both salt and metallic oxidizers by reducing the reaction onset temperature, activation energy and increasing the heat release.

4. New Trend of Improving Reactivity of Aluminum Powder

With the deepening of research on the Al oxidation process, various factors restricting Al oxidation have been found, gradually. The methods of improving the reactivity of Al have been integrated, which can simultaneously change multiple conditions in the oxidation process of aluminum powder to improve the properties of Al from multiple perspectives.

Jena McCollum [102] investigated the reactivity of Al/MoO$_3$@perfluoropolyethers (PFPE) and Al/CuO@PFPE. As shown in Table 3, fluorine—Al-based surface reaction can improve the reactivity of Al/MoO$_3$. However, the reactivity of CuO reduces when the PFPE concentration is increased. Lei Xiao [47] successfully assembled Al/CuO/PVDF/RDX, and the combustion properties of microspheres is mainly affected by the content of RDX. Aifeng Jiang [72] prepared Fe$_2$O$_3$/nano Al and FeF$_3$/nano Al@vinyltrimethoxysilane by ball milling. The initial combustion temperature of Fe$_2$O$_3$/nano is about 600 °C. FeF$_3$/nano Al@vinyltrimethoxysilane had a large specific surface area (26.33 m^2/g) and could be well-preserved from air atmosphere and water. Furthermore, the maximum heat release of FeF$_3$/nano Al@vinyltrimethoxysilane can go up to 12,852 J/g.

Table 3. The data DSC of Al/MoO$_3$@PFPE and Al/CuO@PFPE, reprinted/adapted with permission from Ref. [102]. 2015, Michelle L. Pantoya.

Sample	PIR Onset Temperature °C	PIR Heat Release J/g	Thermite Reaction Onset Temperature °C	Thermite Heat Release J/g
Al@PFPE	315	19.80	561	133
Al/MoO$_3$	-	-	508	2078
Al/MoO$_3$@5%PFPE	298	21.79	534	1370
Al/MoO$_3$@10%PFPE	301	35.17	541	1672
Al/MoO$_3$@20%PFPE	305	103.1	566	1889
Al/CuO	-	-	517	763
Al/CuO@PFPE	298	19.56	569	1658
Al/CuO@10%PFPE	299	29.70	581	1305
Al/CuO@20%PFPE	303	51.37	583	843

In the meantime, the safety of Al during storage, transport and usage also should be addressed when improving the properties of Al. The design and preparation of Al composites with high stability during storage and transportation, and high reactivity during usage, will become a research focus in the future.

DeQi Wang [56] designed flake Al/TiC, with TiC embedded on the surface of flake Al, which retains the Al$_2$O$_3$ shell to keep Al stable during storage and transportation, and cracks the Al$_2$O$_3$ shell to improve the reaction activity and combustion efficiency by the reaction of Al$_2$O$_3$ and TiC before 633 °C. In addition, the heat release of flake Al/TiC is 21,419 J/g at 985.6 °C.

5. Summary and Prospect

With increasing research, the oxidation process of Al has been gradually revealed, which is closely related to the reaction conditions. The oxidation process of nano Al (large specific surface area) is mainly restricted by oxidant diffusion under a slow heating rate, and conforms to a three-step oxidation process, proposed by Zachariah. The Al$_2$O$_3$ shell has the greatest influence on the micron Al oxidation process and Al$_2$O$_3$ shell transformation, and the growth mechanism can well explain this process, which was proposed by Trunov under the slow heating rate. The melting and dispersion of the internal active Al has a great impact on the oxidation process of nano Al under a fast heating rate, which can be described by the melting diffusion mechanism proposed by Levitas.

Therefore, to improve the properties of Al, the key influencing factors of the Al oxidation process need to be changed. MICs, flake Al and nano Al can improve the diffusion efficiency of the oxidant during Al oxidation, which can influence the properties of Al under a slow heating rate. Since the Al$_2$O$_3$ shell can be cracked by fluorine, TiC, and alloy, the properties of Al under a fast heating rate can be changed by them. Those materials with good thermal conductivity can increase the heat transferred to the internal active Al, so the properties of Al under a fast heating rate can be improved by good thermal conductivity materials modification.

In order to improve the properties of Al more comprehensively, an integration of different modification methods has been employed, such as fluoride coated nano Al-based MICs, flake Al/TiC, various material-covered Al and so on. Furthermore, the safety of Al during storage, transport and usage also needs to be addressed when improving the properties of Al. The design and preparation of Al composites with high stability during storage and transportation, and high reactivity during usage, will become a research focus in the future.

Author Contributions: Conceptualization, D.W.; formal analysis, D.W.; investigation, G.X. and W.D.; writing—original draft preparation, T.T.; writing—review and editing, S.L.; supervision, F.L.; project administration, J.L. All authors have read and agreed to the published version of the manuscript.

Funding: This research received no external funding.

Data Availability Statement: Not applicable.

Conflicts of Interest: The authors declare no conflict of interest.

References

1. Sui, H.; Huda, N.; Shen, Z.; Wen, J.Z. Al–NiO energetic composites as heat source for joining silicon wafer. *J. Mater. Processing Technol.* **2020**, *279*, 116572. [CrossRef]
2. Nagarjuna, Y.; Lin, J.C.; Wang, S.C.; Hsiao, W.T.; Hsiao, Y.J. AZO-Based ZnO Nanosheet MEMS Sensor with Different Al Concentrations for Enhanced H2S Gas Sensing. *Nanomaterials* **2021**, *11*, 3377. [CrossRef]
3. Kabra, S.; Gharde, S.; Gore, P.M.; Jain, S.; Khire, V.H.; Kandasubramanian, B. Recent trends in nanothermites: Fabrication, characteristics and applications. *Nano Express* **2020**, *1*, 032001. [CrossRef]
4. Guo, X.; Sun, Q.; Liang, T.; Giwa, A.S. Controllable Electrically Guided Nano-Al/MoO₃ Energetic-Film Formation on a Semiconductor Bridge with High Reactivity and Combustion Performance. *Nanomaterials* **2020**, *10*, 955. [CrossRef]
5. Ke, X.; Zhou, X.; Gao, H.; Hao, G.; Xiao, L.; Chen, T.; Liu, J.; Jiang, W. Surface functionalized core/shell structured CuO/Al nanothermite with long-term storage stability and steady combustion performance. *Mater. Des.* **2018**, *140*, 179–187. [CrossRef]
6. Cheng, W.; Mu, J.; Li, K.; Xie, Z.; Zhang, P.; An, C.; Ye, B.; Wang, J. Evolution of HTPB/RDX/Al/DOA mixed explosives with 90% solid loading in resonance acoustic mixing process. *J. Energetic Mater.* **2021**, 1–20. [CrossRef]
7. Warren, A.D.; Lawrence, G.W.; Jouet, R.J.; Elert, M.; Furnish, M.D.; Chau, R.; Holmes, N.; Nguyen, J. *Investigation of Formulations Containing Perfluorocoated Oxide-Free Nano-Aluminum*; American Institute of Physics: New York, NY, USA, 2007; pp. 1018–1021.
8. Hardin, D.B.; Zhou, M.; Horie, Y. Ignition Behavior of an Aluminum-Bonded Explosive (ABX). In *AIP Conference Proceedings*; AIP Publishing LLC: Melville, NY, USA, 2017; p. 040013.
9. Guo, X.L.; Cao, W.; Duan, Y.L.; Han, Y.; Ran, J.L.; Lu, X.J. Experimental study and numerical simulation of the corner turning of TATB based and CL-20 based polymer bonded explosives. *Combust. Explos. Shock. Waves* **2016**, *52*, 719–726. [CrossRef]
10. Elia, T.; Baudin, G.; Genetier, M.; Lefrançois, A.; Osmont, A.; Catoire, L. Shock to Detonation Transition of Plastic Bonded Aluminized Explosives. *Propellants Explos. Pyrotech.* **2020**, *45*, 554–567. [CrossRef]
11. Yeager, J.D.; Bowden, P.R.; Guildenbecher, D.R.; Olles, J.D. Characterization of hypervelocity metal fragments for explosive initiation. *J. Appl. Phys.* **2017**, *122*, 035901. [CrossRef]
12. Rumchik, C.G.; Jordan, J.L.; Elert, M.; Furnish, M.D.; Chau, R.; Holmes, N.; Nguyen, J. Effect of Aluminum Particle Size on the High Strain Rate Properties of Pressed Aluminized Explosives. *Shock. Compress. Condens. Matter* **2007**, *955*, 795–798.
13. Ao, W.; Wen, Z.; Liu, L.; Wang, Y.; Zhang, Y.; Liu, P.; Qin, Z.; Li, L.K.B. Combustion and agglomeration characteristics of aluminized propellants containing Al/CuO/PVDF metastable intermolecular composites: A highly adjustable functional catalyst. *Combust. Flame* **2022**, *241*, 112110. [CrossRef]
14. Ji, J.; Zhu, W. Thermal decomposition of core–shell structured HMX@Al nanoparticle simulated by reactive molecular dynamics. *Comput. Mater. Sci.* **2022**, *209*, 111405. [CrossRef]
15. Chalghoum, F.; Trache, D.; Benziane, M.; Chelouche, S. Effect of complex metal hydride on the thermal decomposition behavior of AP/HTPB-based aluminized solid rocket propellant. *J. Therm. Anal. Calorim.* **2022**, 1–28. [CrossRef]
16. Tang, W.; Yang, R.; Zeng, T.; Li, J.; Hu, J.; Zhou, X.; Jiang, E.; Zhang, Y. Positive effects of organic fluoride on reduction of slag accumulation in static testing of solid rocket motors of different diameters. *Acta Astronaut.* **2022**, *194*, 277–285. [CrossRef]
17. Wang, D.; Guo, X.; Wang, Z.; Wang, S.; Wu, C.; Zhao, W.; Fang, H. On the Promotion of TKX-50 Thermal Activity with Aluminum. *Propellants Explos. Pyrotech.* **2022**, *47*, e202200012. [CrossRef]
18. Tian, H.; Wang, Z.; Guo, Z.; Yu, R.; Cai, G.; Zhang, Y. Effect of metal and metalloid solid-fuel additives on performance and nozzle ablation in a hydroxy-terminated polybutadiene based hybrid rocket motor. *Aerosp. Sci. Technol.* **2022**, *123*, 107493. [CrossRef]
19. Maggi, F.; Gariani, G.; Galfetti, L.; DeLuca, L.T. Theoretical analysis of hydrides in solid and hybrid rocket propulsion. *Int. J. Hydrog. Energy* **2012**, *37*, 1760–1769. [CrossRef]

20. Yavor, Y.; Rosenband, V.; Gany, A. Reduced agglomeration in solid propellants containing porous aluminum. *Proc. Inst. Mech. Eng. Part G J. Aerosp. Eng.* **2014**, *228*, 1857–1862.

21. Deluca, L.T.; Shimada, T.; Sinditskii, V.P.; Calabro, M. *Chemical Rocket Propulsion—A Comprehensive Survey of Energetic Materials*; Springer International Publishing: Cham, Switzerland, 2017.

22. Joshi, N.; Mathur, N.; Mane, T.; Sundaram, D. Size effect on melting temperatures of alumina nanocrystals: Molecular dynamics simulations and thermodynamic modeling. *Comput. Mater. Sci.* **2018**, *145*, 140–153. [CrossRef]

23. Nazarenko, O.B.; Amelkovich, Y.A.; Sechin, A.I. Characterization of aluminum nanopowders after long-term storage. *Appl. Surf. Sci.* **2014**, *321*, 475–480. [CrossRef]

24. Jin, X.; Li, S.; Yang, Y.; Yang, Y.; Huang, X. Effect of Heating Rate on Ignition Characteristics of Newly Prepared and Aged Aluminum Nanoparticles. *Propellants Explos. Pyrotech.* **2020**, *45*, 1428–1435. [CrossRef]

25. Dreizin, E.L.; Allen, D.J.; Glumac, N.G. Depression of melting point for protective aluminum oxide films. *Chem. Phys. Lett.* **2015**, *618*, 63–65. [CrossRef]

26. Tang, Y.; Kong, C.; Zong, Y.; Li, S.; Zhuo, J.; Yao, Q. Combustion of aluminum nanoparticle agglomerates: From mild oxidation to microexplosion. *Proc. Combust. Inst.* **2017**, *36*, 2325–2332. [CrossRef]

27. Sundaram, D.S.; Puri, P.; Yang, V. A general theory of ignition and combustion of nano- and micron-sized aluminum particles. *Combust. Flame* **2016**, *169*, 94–109. [CrossRef]

28. Wu, B.; Wu, F.; Zhu, Y.; He, A.; Wang, P.; Wu, H. Fast reaction of aluminum nanoparticles promoted by oxide shell. *J. Appl. Phys.* **2019**, *126*, 144305. [CrossRef]

29. Braconnier, A.; Chauveau, C.; Halter, F.; Gallier, S. Experimental investigation of the aluminum combustion in different O_2 oxidizing mixtures: Effect of the diluent gases. *Exp. Therm. Fluid Sci.* **2020**, *117*, 110110. [CrossRef]

30. Liang, L.; Guo, X.; Liao, X.; Chang, Z. Improve the interfacial adhesion, corrosion resistance and combustion properties of aluminum powder by modification of nickel and dopamine. *Appl. Surf. Sci.* **2020**, *508*, 144790. [CrossRef]

31. Liang, D.; Xiao, R.; Liu, J.; Wang, Y. Ignition and heterogeneous combustion of aluminum boride and boron–aluminum blend. *Aerosp. Sci. Technol.* **2019**, *84*, 1081–1091. [CrossRef]

32. Rai, A.; Park, K.; Zhou, L.; Zachariah, M.R. Understanding the mechanism of aluminium nanoparticle oxidation. *Combust. Theory Model.* **2006**, *10*, 843–859. [CrossRef]

33. Park, K.; Lee, D.; Rai, A.; Mukherjee, D.; Zachariah, M.R. Size-Resolved Kinetic Measurements of Aluminum Nanoparticle Oxidation with Single Particle Mass Spectrometry. *J. Phys. Chem. B* **2005**, *109*, 7290–7299. [CrossRef]

34. Chowdhury, S.; Sullivan, K.; Piekiel, N.; Zhou, L.; Zachariah, M.R. Diffusive vs Explosive Reaction at the Nanoscale. *J. Phys. Chem. C* **2010**, *114*, 9191–9195. [CrossRef]

35. Trunov, M.A.; Schoenitz, M.; Dreizin, E.L. Effect of polymorphic phase transformations in alumina layer on ignition of aluminium particles. *Combust. Theory Model.* **2006**, *10*, 603–623. [CrossRef]

36. Jeurgens, L.P.H.; Sloof, W.G.; Tichelaar, F.D.; Mittemeijer, E.J. Growth kinetics and mechanisms of aluminum-oxide films formed by thermal oxidation of aluminum. *J. Appl. Phys.* **2002**, *92*, 1649–1656. [CrossRef]

37. Jeurgens, L.P.H.; Sloof, W.G.; Tichelaar, F.D.; Mittemeijer, E.J. Thermodynamic stability of amorphous oxide films on metals: Application to aluminum oxide films on aluminum substrates. *Phys. Rev. B* **2000**, *62*, 4707–4719. [CrossRef]

38. Jeurgens, L.P.H.; Sloof, W.G.; Tichelaar, F.D.; Mittemeijer, E.J. Structure and morphology of aluminium-oxide films formed by thermal oxidation of aluminium. *Thin Solid Film.* **2002**, *418*, 89–101. [CrossRef]

39. Levin, I.; Brandon, D. Metastable Alumina Polymorphs: Crystal Structures and Transition Sequences. *J. Am. Ceram. Soc.* **1998**, *81*, 1995–2012. [CrossRef]

40. Dwivedi, R.K.; Gowda, G. Thermal stability of aluminium oxides prepared from gel. *J. Mater. Sci. Lett.* **1985**, *4*, 331–334. [CrossRef]

41. Ruano, O.A.; Wadsworth, J.; Sherby, O.D. Deformation of fine-grained alumina by grain boundary sliding accommodated by slip. *Acta Mater.* **2003**, *51*, 3617–3634. [CrossRef]

42. Levitas, V.I.; Asay, B.W.; Son, S.F.; Pantoya, M. Melt dispersion mechanism for fast reaction of nanothermites. *Appl. Phys. Lett.* **2006**, *89*, 071909. [CrossRef]

43. Levitas, V.I.; Pantoya, M.L.; Chauhan, G.; Rivero, I. Effect of the Alumina Shell on the Melting Temperature Depression for Aluminum Nanoparticles. *J. Phys. Chem. C* **2009**, *113*, 14088–14096. [CrossRef]

44. Levitas, V.I.; Pantoya, M.L.; Dean, S. Melt dispersion mechanism for fast reaction of aluminum nano- and micron-scale particles: Flame propagation and SEM studies. *Combust. Flame* **2014**, *161*, 1668–1677. [CrossRef]

45. Levitas, V.I.; Pantoya, M.L.; Dikici, B. Melt dispersion versus diffusive oxidation mechanism for aluminum nanoparticles: Critical experiments and controlling parameters. *Appl. Phys. Lett.* **2008**, *92*, 011921. [CrossRef]

46. Levitas, V. Burn time of aluminum nanoparticles: Strong effect of the heating rate and melt-dispersion mechanism. *Combust. Flame* **2009**, *156*, 543–546. [CrossRef]

47. Xiao, L.; Zhao, L.; Ke, X.; Zhang, T.; Hao, G.; Hu, Y.; Zhang, G.; Guo, H.; Jiang, W. Energetic metastable Al/CuO/PVDF/RDX microspheres with enhanced combustion performance. *Chem. Eng. Sci.* **2021**, *231*, 116302. [CrossRef]

48. Matveev, S.; Dlott, D.D.; Valluri, S.K.; Mursalat, M.; Dreizin, E.L. Fast energy release from reactive materials under shock compression. *Appl. Phys. Lett.* **2021**, *118*, 101902. [CrossRef]

49. Yoshinaka, A.; Zhang, F.; Wilson, W.; Elert, M.; Furnish, M.D.; Chau, R.; Holmes, N.; Nguyen, J. *Effect of Shock Compression on Aluminum Particles in Condensed Media*; American Institute of Physics: New York, NY, USA, 2008; pp. 1057–1060.

50. Kuhl, A.L.; Neuwald, P.; Reichenbach, H. Effectiveness of combustion of shock-dispersed fuels in calorimeters of various volumes. *Combust. Explos. Shock. Waves* **2006**, *42*, 731–734. [CrossRef]

51. Xiao, L.; Liu, J.; Hao, G.; Ke, X.; Chen, T.; Gao, H.; Rong, Y.-b.; Jin, C.-s.; Li, J.-l.; Jiang, W. Preparation and study of ultrafine flake-aluminum with high reactivity. *Def. Technol.* **2017**, *13*, 234–238. [CrossRef]

52. Weismiller, M.R.; Malchi, J.Y.; Lee, J.G.; Yetter, R.A.; Foley, T.J. Effects of fuel and oxidizer particle dimensions on the propagation of aluminum containing thermites. *Proc. Combust. Inst.* **2011**, *33*, 1989–1996. [CrossRef]

53. Hu, A.; Zou, H.; Shi, W.; Pang, A.m.; Cai, S. Preparation, Microstructure and Thermal Property of ZrAl₃/Al Composite Fuels. *Propellants Explos. Pyrotech.* **2019**, *44*, 1454–1465. [CrossRef]

54. Hu, A.; Cai, S. Research on the novel Al–W alloy powder with high volumetric combustion enthalpy. *J. Mater. Res. Technol.* **2021**, *13*, 311–320. [CrossRef]

55. Fu, H.; Zou, H.; Cai, S.-z. The role of microstructure refinement in improving the thermal behavior of gas atomized Al-Eu alloy powder. *Adv. Powder Technol.* **2016**, *27*, 1898–1904. [CrossRef]

56. Wang, D.; Cao, X.; Liu, J.; Zhang, Z.; Jin, X.; Gao, J.; Yu, H.; Sun, S.; Li, F. TF-Al/TiC highly reactive composite particle for application potential in solid propellants. *Chem. Eng. J.* **2021**, *425*, 130674. [CrossRef]

57. Yao, E.; Zhao, N.; Qin, Z.; Ma, H.; Li, H.; Xu, S.; An, T.; Yi, J.; Zhao, F. Thermal Decomposition Behavior and Thermal Safety of Nitrocellulose with Different Shape CuO and Al/CuO Nanothermites. *Nanomaterials* **2020**, *10*, 725. [CrossRef]

58. Zhou, X.; Torabi, M.; Lu, J.; Shen, R.; Zhang, K. Nanostructured energetic composites: Synthesis, ignition/combustion modeling, and applications. *ACS Appl. Mater. Interfaces* **2014**, *6*, 3058–3074. [CrossRef]

59. He, W.; Liu, P.-J.; He, G.-Q.; Gozin, M.; Yan, Q.-L. Highly Reactive Metastable Intermixed Composites (MICs): Preparation and Characterization. *Adv. Mater.* **2018**, *30*, e1706293. [CrossRef]

60. Ma, X.; Li, Y.; Hussain, I.; Shen, R.; Yang, G.; Zhang, K. Core-Shell Structured Nanoenergetic Materials: Preparation and Fundamental Properties. *Adv. Mater.* **2020**, *32*, e2001291. [CrossRef]

61. Irvin, G.; Yetter, R.A. *Combustion*, 4th ed.; Academic Press: San Diego, CA, USA, 2008.

62. Law, C.K. A Simplified Theoretical Model for the Vapor-Phase Combustion of Metal Particles. *Combust. Sci. Technol.* **1973**, *7*, 197–212. [CrossRef]

63. Brooks, K.P.; Beckstead, M.W. Dynamics of aluminum combustion. *J. Propuls. Power* **1995**, *11*, 769–780. [CrossRef]

64. Lynch, P.; Krier, H.; Glumac, N. A correlation for burn time of aluminum particles in the transition regime. *Proc. Combust. Inst.* **2009**, *32*, 1887–1893. [CrossRef]

65. Lynch, P.; Fiore, G.; Krier, H.; Glumac, N. Gas-Phase Reaction in Nanoaluminum Combustion. *Combust. Sci. Technol.* **2010**, *182*, 842–857. [CrossRef]

66. Tappan, B.C.; Dirmyer, M.R.; Risha, G.A. Evidence of a kinetic isotope effect in nanoaluminum and water combustion. *Angew. Chem. Int. Ed. Engl.* **2014**, *53*, 9218–9221. [CrossRef]

67. Ji, Y.; Sun, Y.; Zhu, B.; Liu, J.; Wu, Y. Calcium fluoride promoting the combustion of aluminum powder. *Energy* **2022**, *250*, 123772. [CrossRef]

68. Aumann, C.E.; Skofronick, G.L.; Martin, J.A. Oxidation behavior of aluminum nanopowders. *J. Vac. Sci. Technol. B Microelectron. Nanometer Struct.* **1995**, *13*. [CrossRef]

69. Zaky, M.; Elbeih, A.; Elshenawy, T. Review of Nano-thermites; a Pathway to Enhanced Energetic Materials. *Cent. Eur. J. Energetic Mater.* **2021**, *18*, 63–85. [CrossRef]

70. Glavier, L.; Taton, G.; Ducéré, J.-M.; Baijot, V.; Pinon, S.; Calais, T.; Estève, A.; Djafari Rouhani, M.; Rossi, C. Nanoenergetics as pressure generator for nontoxic impact primers: Comparison of Al/Bi₂O₃, Al/CuO, Al/MoO₃ nanothermites and Al/PTFE. *Combust. Flame* **2015**, *162*, 1813–1820. [CrossRef]

71. Lebedev, E.A.; Sorokina, L.I.; Trifonov, A.Y.; Ryazanov, R.M.; Pereverzeva, S.Y.; Gavrilov, S.A.; Gromov, D.G. Influence of Composition on Energetic Properties of Copper Oxide—Aluminum Powder Nanothermite Materials Formed by Electrophoretic Deposition. *Propellants Explos. Pyrotech.* **2021**, *47*, e202100292. [CrossRef]

72. Jiang, A.; Xia, D.; Li, M.; Qiang, L.; Fan, R.; Lin, K.; Yang, Y. Ball Milling Produced FeF₃-Containing Nanothermites: Investigations of Its Thermal and Inflaming Properties. *ChemistrySelect* **2019**, *4*, 12662–12667. [CrossRef]

73. Wang, N.; Hu, Y.; Ke, X.; Xiao, L.; Zhou, X.; Peng, S.; Hao, G.; Jiang, W. Enhanced-absorption template method for preparation of double-shell NiO hollow nanospheres with controllable particle size for nanothermite application. *Chem. Eng. J.* **2020**, *379*, 122330. [CrossRef]

74. Yu, C.; Zhang, W.; Shen, R.; Xu, X.; Cheng, J.; Ye, J.; Qin, Z.; Chao, Y. 3D ordered macroporous NiO/Al nanothermite film with significantly improved higher heat output, lower ignition temperature and less gas production. *Mater. Des.* **2016**, *110*, 304–310. [CrossRef]

75. He, W.; Li, Z.-H.; Chen, S.; Yang, G.; Yang, Z.; Liu, P.-J.; Yan, Q.-L. Energetic metastable n-Al@PVDF/EMOF composite nanofibers with improved combustion performances. *Chem. Eng. J.* **2020**, *383*, 123146. [CrossRef]

76. Lee Perry, W.; Tappan, B.C.; Reardon, B.L.; Sanders, V.E.; Son, S.F. Energy release characteristics of the nanoscale aluminum-tungsten oxide hydrate metastable intermolecular composite. *J. Appl. Phys.* **2007**, *101*, 064313. [CrossRef]

77. Fahd, A.; Dubois, C.; Chaouki, J.; Wen, J.Z. Combustion behaviour and reaction kinetics of GO/Al/oxidizing salts ternary nanothermites. *J. Therm. Anal. Calorim.* **2022**, 1–13. [CrossRef]

78. Liu, Q.; Li, X.; Bai, C. Deflagration to detonation transition in aluminum dust–air mixture under weak ignition condition. *Combust. Flame* **2009**, *156*, 914–921. [CrossRef]
79. Antipina, S.A.; Zmanovskii, S.V.; Gromov, A.A.; Teipel, U. Air and water oxidation of aluminum flake particles. *Powder Technol.* **2017**, *307*, 184–189. [CrossRef]
80. Houim, R.W.; Boyd, E. Combustion of aluminum flakes in the post-flame zone of a hencken burner. *Int. J. Energetic Mater. Chem. Propuls.* **2008**, *7*, 55–71. [CrossRef]
81. Wang, D.Q.; Yu, H.M.; Liu, J.; Li, F.S.; Jin, X.X.; Zheng, S.J.; Zheng, T.T.; Li, Y.; Zhang, Z.J.; Li, D.; et al. Preparation and Properties of a Flake Aluminum Powder in an Ammonium-Perchlorate-Based Composite Modified Double-Base Propellant. *Combust. Explos. Shock. Waves* **2020**, *56*, 691–696. [CrossRef]
82. Kuhl, A.L.; Reichenbach, H. Combustion effects in confined explosions. *Proc. Combust. Inst.* **2009**, *32*, 2291–2298. [CrossRef]
83. Kotov, Y.A. Electric Explosion of Wires as a Method for Preparation of Nanopowders. *J. Nanoparticle Res.* **2003**, *5*, 539–550. [CrossRef]
84. Lerner, M.; Vorozhtsov, A.; Guseinov, S.; Storozhenko, P. Metal Nanopowders Production. *Met. Nanopowders* **2014**, 79–106. [CrossRef]
85. DeLuca, L.; Galfetti, L. Burning of Metalized Composite Solid Rocket Propellants: From Micrometric to Nanometric Aluminum Size. In Proceedings of the Asian Joint Conference on Propulsion and Power, Gyeongju, Korea, 6–8 March 2008; pp. 6–8.
86. Trubert, J.-F.; Hommel, J.; Lambert, D.; Fabignon, Y.; Orlandi, O. New HTPB/AP/Al propellant combustion process in the presence of aluminum nano-particles. *Int. J. Energetic Mater. Chem. Propuls.* **2008**, *7*, 99–122. [CrossRef]
87. Tang, D.-Y.; Lyu, J.; He, W.; Chen, J.; Yang, G.; Liu, P.-J.; Yan, Q.-L. Metastable intermixed Core-shell Al@M(IO_3)x nanocomposites with improved combustion efficiency by using tannic acid as a functional interfacial layer. *Chem. Eng. J.* **2020**, *384*, 123369. [CrossRef]
88. Lerner, M.; Vorozhtsov, A.; Eisenreich, N. Safety Aspects of Metal Nanopowders. *Met. Nanopowders* **2014**, 153–162. [CrossRef]
89. Jacob, R.J.; Jian, G.; Guerieri, P.M.; Zachariah, M.R. Energy release pathways in nanothermites follow through the condensed state. *Combust. Flame* **2015**, *162*, 258–264. [CrossRef]
90. Osborne, D.T.; Pantoya, M.L. Effect of Al Particle Size on the Thermal Degradation of Al/Teflon Mixtures. *Combust. Sci. Technol.* **2007**, *179*, 1467–1480. [CrossRef]
91. DeLisio, J.B.; Hu, X.; Wu, T.; Egan, G.C.; Young, G.; Zachariah, M.R. Probing the Reaction Mechanism of Aluminum/Poly(vinylidene fluoride) Composites. *J. Phys. Chem. B* **2016**, *120*, 5534–5542. [CrossRef]
92. Valluri, S.K.; Bushiri, D.; Schoenitz, M.; Dreizin, E. Fuel-rich aluminum–nickel fluoride reactive composites. *Combust. Flame* **2019**, *210*, 439–453. [CrossRef]
93. McCollum, J.; Pantoya, M.L.; Iacono, S.T. Catalyzing aluminum particle reactivity with a fluorine oligomer surface coating for energy generating applications. *J. Fluor. Chem.* **2015**, *180*, 265–271. [CrossRef]
94. Kim, D.W.; Kim, K.T.; Min, T.S.; Kim, K.J.; Kim, S.H. Improved Energetic-Behaviors of Spontaneously Surface-Mediated Al Particles. *Sci. Rep.* **2017**, *7*, 4659. [CrossRef]
95. Wang, J.; Zeng, C.; Zhan, C.; Zhang, L. Tuning the reactivity and combustion characteristics of PTFE/Al through carbon nanotubes and grapheme. *Thermochim. Acta* **2019**, *676*, 276–281. [CrossRef]
96. Ke, X.; Guo, S.; Zhang, G.; Zhou, X.; Xiao, L.; Hao, G.; Wang, N.; Jiang, W. Safe preparation, energetic performance and reaction mechanism of corrosion-resistant Al/PVDF nanocomposite films. *J. Mater. Chem. A* **2018**, *6*, 17713–17723. [CrossRef]
97. Meir, S.; Kalabukhov, S.; Hayun, S. Low temperature spark plasma sintering of Al_2O_3–TiC composites. *Ceram. Int.* **2014**, *40*, 12187–12192. [CrossRef]
98. Cai, K.F.; McLachlan, D.S.; Axen, N.; Manyatsa, R. Preparation, microstructures and properties of Al_2O_3–TiC composites. *Ceram. Int.* **2002**, *28*, 217–222. [CrossRef]
99. Cheng, Y.; Sun, S.; Hu, H. Preparation of Al_2O_3/TiC micro-composite ceramic tool materials by microwave sintering and their microstructure and properties. *Ceram. Int.* **2014**, *40*, 16761–16766. [CrossRef]
100. Noor, F.; Vorozhtsov, A.; Lerner, M.; Bandara Filho, E.P.; Wen, D. Thermal-Chemical Characteristics of Al–Cu Alloy Nanoparticles. *J. Phys. Chem. C* **2015**, *119*, 14001–14009. [CrossRef]
101. Shen, J.; Qiao, Z.; Zhang, K.; Wang, J.; Li, R.; Xu, H.; Yang, G.; Nie, F. Effects of nano-Ag on the combustion process of Al–CuO metastable intermolecular composite. *Appl. Therm. Eng.* **2014**, *62*, 732–737. [CrossRef]
102. McCollum, J.; Pantoya, M.L.; Iacono, S.T. Activating Aluminum Reactivity with Fluoropolymer Coatings for Improved Energetic Composite Combustion. *ACS Appl. Mater. Interfaces* **2015**, *7*, 18742–18749. [CrossRef]

 *crystals*

Review

Synthetic Methods towards Energetic Heterocyclic N-Oxides via Several Cyclization Reactions

Weiqing She [1,2], Zhenzhen Xu [3], Lianjie Zhai [2,*], Junlin Zhang [2], Jie Huang [1], Weiqiang Pang [2] and Bozhou Wang [2,*]

1   School of Chemical Engineering, Northwest University, Xi'an 710069, China
2   Xi'an Modern Chemistry Research Institute, Xi'an 710065, China
3   China International Engineering Consulting Corporation, Beijing 100048, China
*   Correspondence: trihever0210@126.com (L.Z.); wbz600@163.com (B.W.)

Abstract: Due to the introduction of oxygen atoms, N-oxide energetic compounds have a unique oxygen balance, excellent detonation properties, and a high energy density, attracting the extensive attention of researchers all over the world. N-oxides are classified into two categories based on the structural characteristics of their skeletons: azine N-oxides and azole N-oxides, whose N→O coordination bonds are formed during cyclization. There are six kinds of azine N-oxides, namely 1,2,3,4-tetrazine-1,3-dioxide, 1,2,3,5-tetrazine-2-oxide, 1,2,3-triazine-3-oxide, 1,2,3-triazine-2-oxide, pyridazine-1,2-dioxide, and pyrazine-1-oxide. Azole N-oxides include 1,2,5-oxadiazole-2-oxide, pyrazole-1-oxide, and triazole-1-oxide. Synthetic strategies towards these two categories of N-oxides are fully reviewed. Corresponding reaction mechanisms towards the aromatic N-oxide frameworks and examples that use the frameworks to create high-energy substances are discussed. Moreover, the energetic properties of N-oxide energetic compounds are compared and summarized.

Keywords: N-oxides; cyclization reaction; synthesis; reaction mechanism; detonation performance

Citation: She, W.; Xu, Z.; Zhai, L.; Zhang, J.; Huang, J.; Pang, W.; Wang, B. Synthetic Methods towards Energetic Heterocyclic N-Oxides via Several Cyclization Reactions. *Crystals* **2022**, *12*, 1354. https://doi.org/10.3390/cryst12101354

Academic Editor: Thomas M. Klapötke

Received: 23 August 2022
Accepted: 21 September 2022
Published: 25 September 2022

Publisher's Note: MDPI stays neutral with regard to jurisdictional claims in published maps and institutional affiliations.

1. Introduction

Research on energetic materials continuously develops energetic materials with higher detonation performance and energy density, taking it as an eternal quest. Since the advent of conventional energetic materials such as trinitrotoluene (TNT), RDX, and HMX, researchers worldwide have been extensively exploring energetic materials with chain, ring, and cage parent nucleus structures composed of C, H, and N as the main elements as well as new energetic parent frameworks such as tetrazine, triazole, and furazan [1–11]. Meanwhile, they strive to improve the performance of the compounds by introducing energetic groups, including nitroamino groups, trinitroethyl groups, fluoroacyl dinitroethyl groups, and azide groups [12–17]. As a typical energetic group for improving the detonation performance of compounds, the nitro group (-NO$_2$) could not only increase the energy density of CHON-based compounds but also enhance the oxygen balance of the energetic compounds, which is conducive to energy release. However, excessive introduction of nitro groups could also lead to problems such as increased sensitivity, complex synthetic process and low yield [18]. In addition, there is an upper limit on the energy density theory of CHON-based energetic compounds represented by nitro groups. For example, hexanitrohexaazaisowurtzitane (CL-20) exhibits high detonation performance and energy density. However, to further improve its performance, its synthetic strategy needs to be changed to break through the energy limit of CHON-based energetic materials [19,20].

The energy density of energetic materials can be effectively increased by forming NO coordination bonds through the oxidation of N atom in the heterocyclic aromatic framework. The formation of N→O bonds at a suitable position in the heterocyclic aromatic ring could not only enhance the energy density but could also improve the oxygen balance of energetic compounds. More importantly, the N→O bond with the characteristics of a coordination double bond could eliminate the electron repulsion of nitrogen in a N-heterocyclic system and promote σ–π orbital separation to stabilize molecules [21,22]. To study the effect of an N→O bond on the structures and properties of the energetic compounds, Lai Weipeng et al. employed the density functional theory to conduct theoretical calculations of pyrazines, 1,2,3,4-tetrazines, tetrazinofuranzans, and their N-oxides [23]. The results showed that the N→O bond shortens the length of the C–C bond in the ring, lengthens the C–N bond close to the N→O bond and improves the detonation performance of most energetic compounds. It was also found that the improvement is positively correlated with the number of N→O bonds in the energetic compounds (Figure 1).

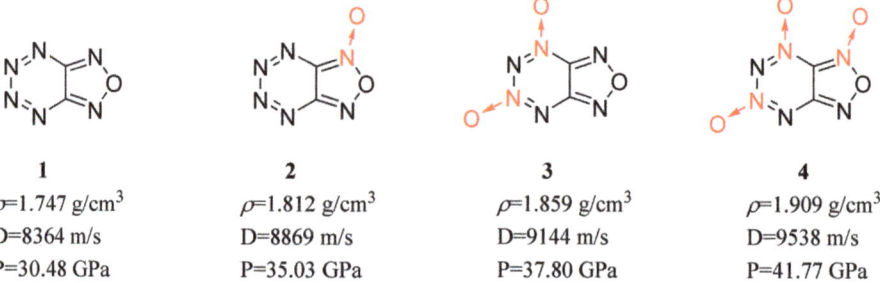

1

ρ=1.747 g/cm^3

D=8364 m/s

P=30.48 GPa

2

ρ=1.812 g/cm^3

D=8869 m/s

P=35.03 GPa

3

ρ=1.859 g/cm^3

D=9144 m/s

P=37.80 GPa

4

ρ=1.909 g/cm^3

D=9538 m/s

P=41.77 GPa

Figure 1. Calculated performance of compounds **1–4**.

Using the N-heterocyclic skeleton as the precursor, an N→O coordination bond could be directly formed by the oxidation of hypochlorous acid (HOF), H$_2$O$_2$/H$_2$SO$_4$ and other oxidation systems. Most of the N-heterocyclic aromatic frameworks could be converted to N-oxides through direct oxidation, yet N-heterocyclic skeletons require certain energy and the corresponding oxidation systems need to be adopted [24]. In the process of constructing heterocyclic skeletons, an N→O coordination bond could be directly generated via the cyclization reaction, thereby bypassing the step of oxidation, shortening the reaction process, and improving production efficiency.

The synthetic methods towards energetic heterocyclic N-oxides, which are simultaneously synthetized during cyclization, are elaborated in this study. The N-oxides could be classified into nine categories based on the structural characteristics of the N-heterocyclic aromatic skeleton. The categories are: 1,2,3,4-tetrazine-1,3-dioxide, 1,2,3,5-tetrazine-2-oxide, 1,2,3-triazine-3-oxide, 1,2,3-triazine-2-oxide, pyridazine-1,2-dioxide, pyrazine-1-oxide, 1,2,5-oxadiazole-2-oxide (furoxan), pyrazole N-oxide, and triazole N-oxide. Some typical cyclization reaction mechanisms are also discussed. The physicochemical properties and detonation performance of the N-oxides are introduced. Direct synthetic strategies towards N-oxides through cyclization reaction would provide a theoretical guidance for the design and development of new energetic compounds.

2. Energetic Heterocyclic N-Oxides via Different Cyclization Reaction

2.1. Azine N-Oxides

2.1.1. 1,2,3,4-Tetrazine-1,3-Dioxide

1,2,3,4-Tetrazine-1,3-dioxide was originally derived from 1,2,3,4-tetrazine compounds. According to Churakov et al. [25,26], converting tetrazine into its N-oxide might improve its energetic properties, while the benzene ring was introduced to increase the stability of the tetrazine N-oxide. Benzo-1,2,3,4-tetrazine dioxide (**8, BTDO**) was synthesized by diazotization or nitration from an intermediate O-tert-butyl-*NNO*-azoxyaniline (**7**) [27]. This method is also applicable for the synthesis of 1,2,3,4-tetrazine dioxide. Using o-nitrosonitrobenzene (**5**) as the raw material, and the intermediate **7** was obtained by a two-step reaction. First, nitroso was transformed into tert-butyl-NNO-azoxy group under the action of N, N-dibromo -t-butylamine. Then, the nitro group was reduced to amino group by $SnCl_2$ to produce the intermediate **7** with a yield of 77%. **BTDO** was obtained by the diazotization oxidation reaction of intermediate **7** with $NOBF_4$ and benzoic acid peroxide or direct nitrification reaction with excess N_2O_5. By direct nitration, several nitro-substituted benzo-1,2,3,4-tetrazine dioxides have been obtained via the nitration of benzene in excessive N_2O_5 (Scheme 1 **9–11**).

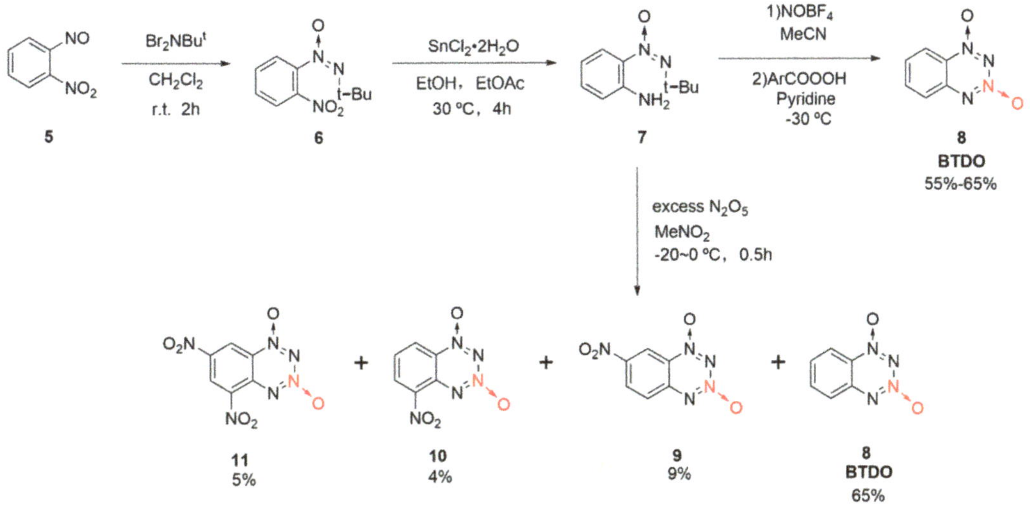

Scheme 1. Synthetic route for **BTDO**.

For the synthesis reaction of **BTDO** by diazotization oxidation and nitration, the reaction mechanism was also examined [25]. Based on the diazotization reaction of nitroso positive ions from $NOBF_4$ and amino group on the benzene ring of intermediate **7**, diazonium salt intermediate **12** was generated, and an unstable intermediate **13** formed from **12** and benzoic acid peroxide. The intermediate attacked the N atom connected with tert-butyl to the N atom connected with oxygen atom, and the product **BTDO** was finally obtained from intramolecular electron transfer after eliminating the benzoic acid anion and tert-butyl cation (Scheme 2a). An intermediate **14** of N, N-dinitroamine was formed by direct nitration of compound **7** using excess N_2O_5 as the nitrating agent. *ONN*-azoxy nitrate compound **15** was produced by intramolecular rearrangement. Similarly, the O atom connected with the tert-butyl group attacked the N atom connected with nitrate for addition elimination, and the product **BDTO** was obtained after removing the tert-butyl cation and nitrate ion (Scheme 2b).

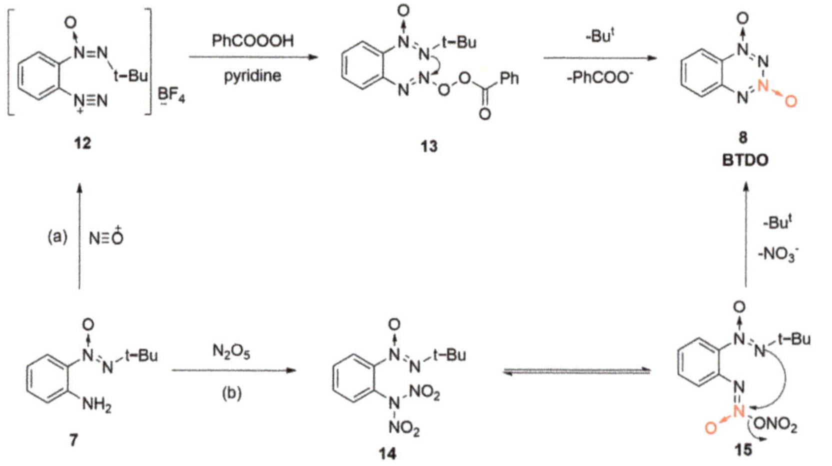

Scheme 2. Different reaction mechanism (**a**) and (**b**) of BTDO.

The by-product **11** was produced by nitrating compound **7** with N_2O_5 for further improving the detonation performance of the compound through introducing two nitro groups into the benzene ring. Using o-nitroaniline (**16**) as the raw material, compound **7** was synthetized by Klapotke et al. [28] as follows: The compound **5** was first obtained by the nitration of the amino group with potassium persulfate and sulfuric acid, and then compound **7** was formed by the condensation of dibromo-t-butylamine with nitroso group and a further reduction reaction using $SnCl_2$. Using N_2O_5/dichloromethane as the nitration system, compound **11** as yellow crystal was finally synthetized (Scheme 3). The density of compound **11** was 1.896 g/cm^3, and its sensitivity (*IS* = 5 J) was higher than that of RDX (*IS* = 7.5 J). Based on theoretical calculation, its detonation performance (D = 8411 m/s, P = 33.0 GPa) was equal to that of RDX (D = 8748 m/s, P = 34.9 GPa).

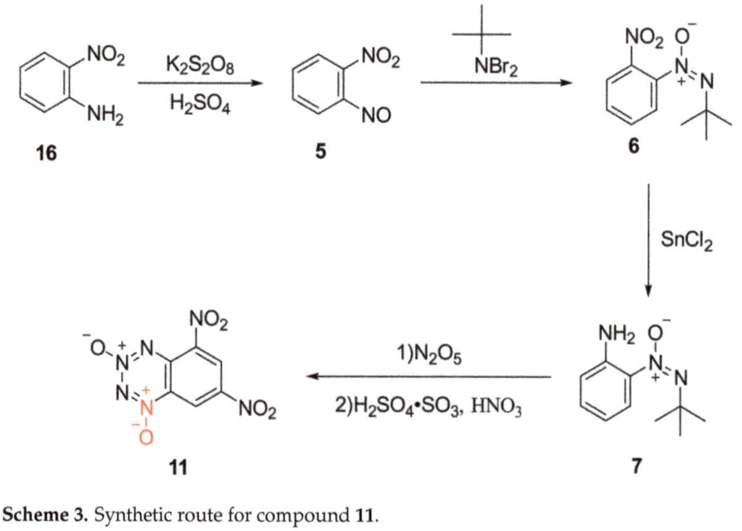

Scheme 3. Synthetic route for compound **11**.

A large number of theoretical calculations [29–32] showed that [1–4]tetrazino [5,6-e][1–4]tetrazino-1,2,3,4-tetraoxide (**25, TTTO**) is an energetic material with a high energy density (ρ = 1.899 g/cm^3) and high detonation performances (D = 9710 m/s, P= 43.2 GPa), superior to CL-20 (D = 9420 m/s, P= 42.0 GPa). Its symmetrical "butterfly" structure has also attracted extensive attention from researchers. Using 2,2-bis(tert-butyl-*NNO*-azoxy)acetonitrile (**17**) as the raw material, **TTTO** was synthetized by Klenov et al. [33] based on a 10-step synthetic strategy with a total yield of 1.1%(Scheme 4). The synthesis process was too complex and the yield was too low, and chromatographic separation was required for purifying key intermediates. Therefore, yet the synthesis of **TTTO** has great theoretical value, there is currently no prospect for industrial applications.

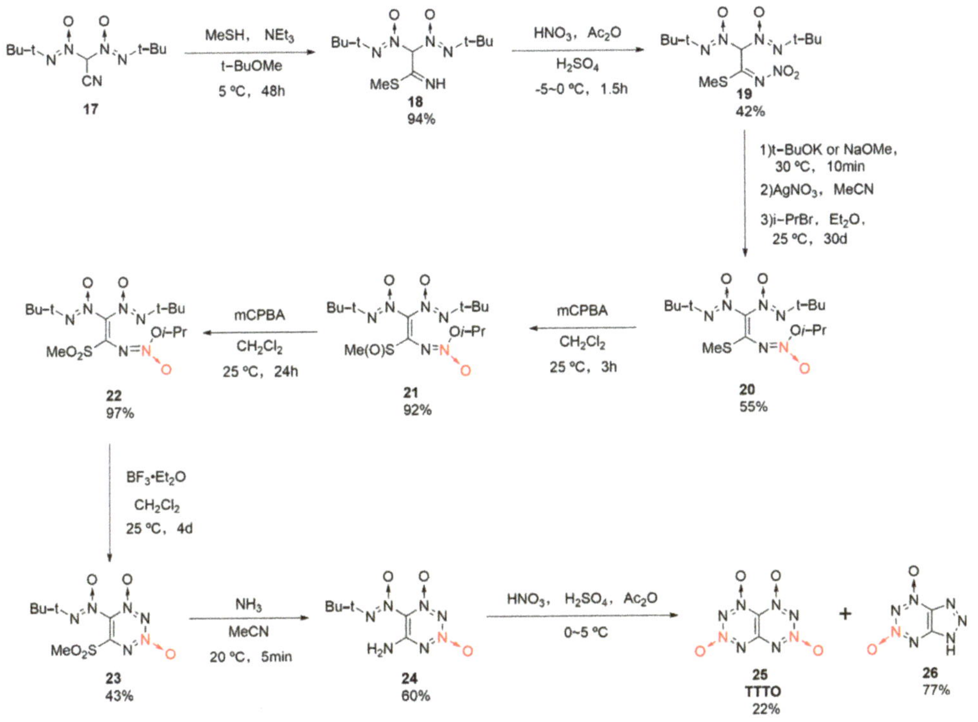

Scheme 4. Synthetic route for **TTTO**.

The reaction process of compound **24** forming **TTTO** and by-product **26** was also reasonably speculated [33]. Under the action of nitric acid, the mononitroamine compound **27** was generated from compound **24**, and compound **28** created by further acetylation of the acetic anhydride. Compound **28** was protoned in sulfuric acid after removing an acetic acid molecule to form oxidized diazonium ion intermediate **29**, which could be cyclized in two different ways. One was to conduct electron transfer between the oxidized diazo positive ion and the N atom connected with tert-butyl group, and then **TTTO** was formed by cyclization after removing the tert-butyl positive ion (Scheme 5a). For the other method, compound **31** was obtained by cyclization based on the nucleophilic attack of the O atom in the oxidizing diazo positive ions. However, compound **31** was unstable, so diazo ketone compound **33** was obtained after the ring-opening reaction of **31** and by removing the tert-butyl positive ions. Eventually, by-product **26** was produced by further cyclization and hydrolysis (Scheme 5b)).

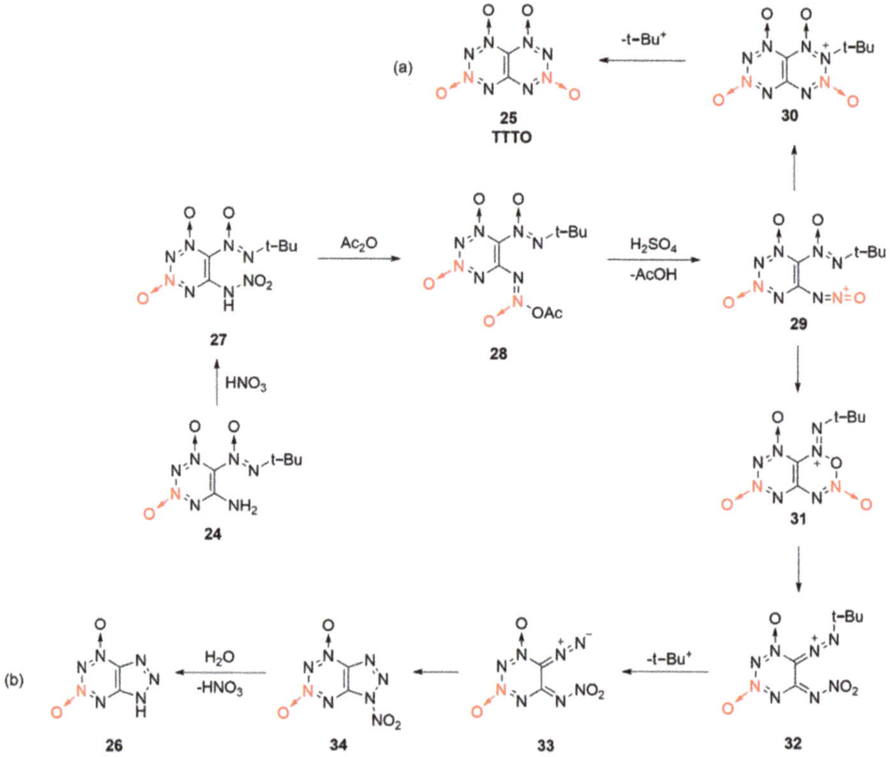

Scheme 5. Possible synthetic routes (**a**) and (**b**) for compound **26** and **TTTO**.

Based on the structuring idea of using anthracene ring and phenanthrene ring, two fused tetrazine dioxides, 1,2,3,4-tetrazino [5,6-f]benzo-1,2,3,4-tetrazino-1,3,7,9-tetraoxide (**38**) and 1,2,3,4-tetrazino [5,6-g]benzo-1,2,3,4-tetrazino-1,3,7,9-tetraoxide (**42**), were synthetized by Frumkin et al. [34,35] from different raw materials. Using compound **35** as the starting material, compound **36** was obtained in the first cyclization reaction using nitration with nitric anhydride. Then, compound **38** was obtained by amino substitution and the second cyclization reaction via nitric anhydride. The chlorine atom on compound **38** was easily replaced by some nucleophiles to form other derivatives, such as compound **39**, and the properties of the substituted compounds were improved, such as a higher thermal decomposition temperature (T_d = 210 °C) than that of compound **38** (T_d = 140 °C). Using compound **40** as the starting material, compound **43** was synthetized through nitration using nitric anhydride, amino substitution, and secondary nitration-cyclization. By-product **44** with a nitro group on the benzene ring would also be produced. Meanwhile, compound **43** could be synthesized by one-step nitration from compound **45**(Scheme 6).

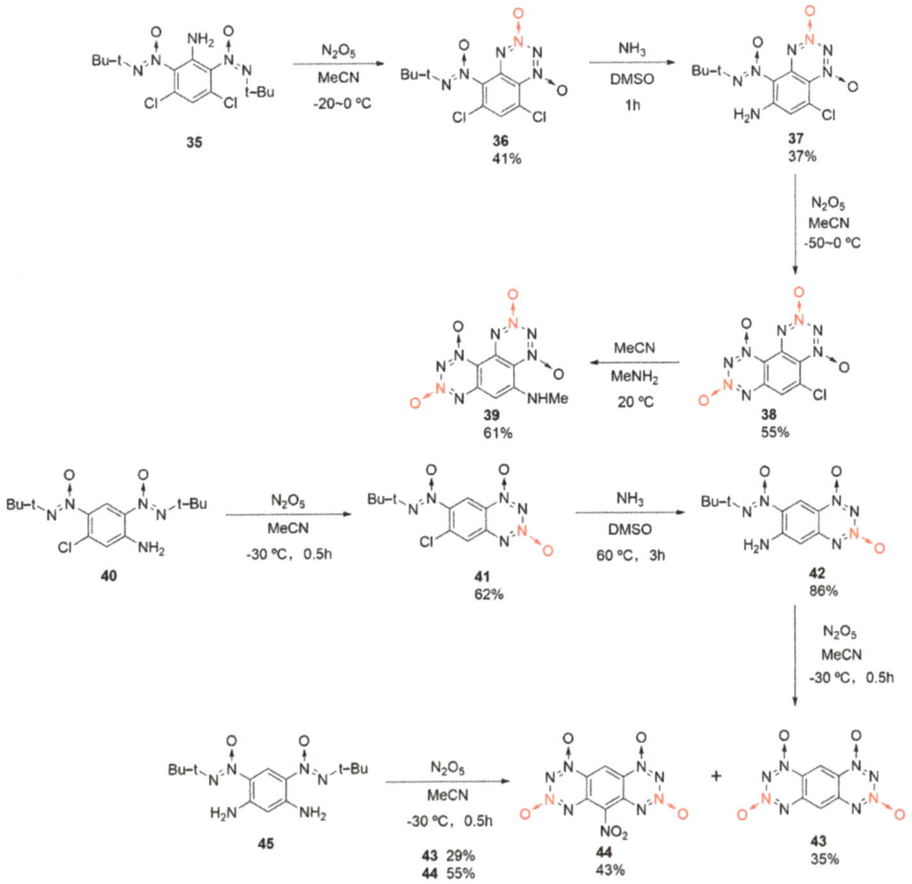

Scheme 6. Synthetic routes for compounds **38** and **43**.

Based on the fusing furazan with tetrazine dioxide, an energetic compound furazano [3,4-e]-1,2,3,4-tetrazine-4,6-dioxide(**50, FTDO**) was synthetized by Churakov et al. [36] for the first time. **FDTO** displays a broad prospect with a measured density of 1.85 g/cm^3, formation enthalpy of 4.23 MJ·kg, theoretical detonation velocity of 9802 m/s, and explosion pressure of 44.78 GPa, arousing widespread research interests [37,38]. It was synthesized through the nitration-cyclization of compound **48** by using tetrafluoroborate nitrate as the nitration reagent. Yet, the price of tetrafluoroborate nitrate was too expensive, and the nitrated product was difficult to purify. Using 3,4-diaminofurazan (**46**) as the raw material, compound **48** was obtained by Li Xiangzhi et al. [39] via oxidation and condensation, followed by the nitration of 100% HNO$_3$ and cyclization of P$_2$O$_5$, leading to the high-purity product **FDTO** (Scheme 7). The effects of N$_2$O$_5$, NO$_2$BF$_4$, 100% HNO$_3$, etc. nitration systems on the product and purity were also compared. It was concluded that using 100% HNO$_3$ as the nitrating agent, a maximum yield of 99.54% and purity of 99.18% could be achieved.

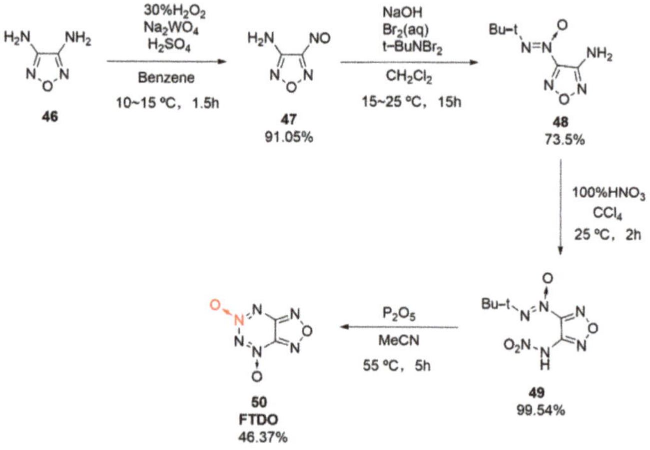

Scheme 7. Synthetic route for **FTDO**.

In this reference [39], the cyclization mechanism under the action of P$_2$O$_5$ was discussed (Scheme 8). The nitroamino group in compound **49** was isomerized to produce compound **51** containing a N-hydroxyl group. P$_2$O$_5$ was esterified with the hydroxyl group on compound **51** to produce 3-(tert-butyl-*NNO*-azoxy)-furazan-4-azoxy alcohol phosphite (**52**). The N atom connected with tert-butyl group in compound **52** attacked the nucleophilic N atom connected to phosphite. Moreover, the metaphosphate group was removed while electron transfer was performed. The tert-butyl positive ion was then removed to finally obtain **FDTO**.

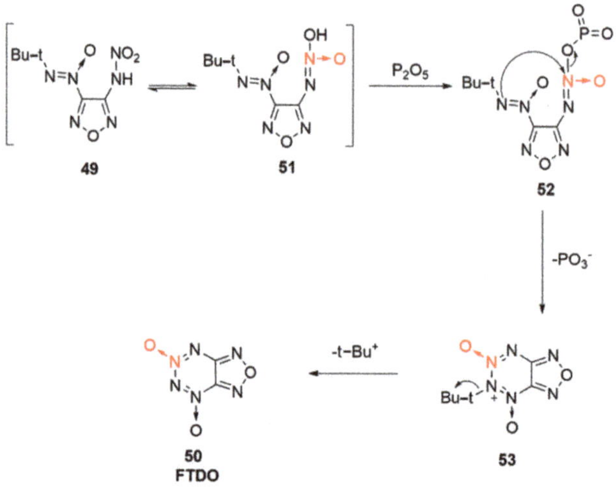

Scheme 8. Cyclization mechanism of compound **49** under the action of P$_2$O$_5$.

Although the fusion of a five-membered furazan ring with tetrazine dioxide has achieved great success, its low decomposition temperature of 112 °C for **FTDO** still limits its application as an energetic material [40]. Based on the fusing the five-membered triazole ring with tetrazine dioxide, after introducing alkyl into the triazole ring, a series of triazole tetrazine dioxides (Scheme 9 **55a–c, 57a–c**) with stable thermal properties were obtained by Voronin et al. [41]. Energetic compounds with excellent detonation properties were also obtained by introducing energetic groups (-N$_3$, -NO$_2$, others) into the triazole ring [42].

Among them, the thermal decomposition temperature of compound **55a** is 199 °C, while that of compound **57a-c** is within 208 ~ 230°C. Due to the introduction of a benzofurazan ring into the triazole ring, compound **60** as a potential energetic compound was further obtained by Shvets et al. [43]. It exhibits an energy density of 1.84 g/cm³, a thermal decomposition temperature of 190 °C, and a calculated enthalpy of formation of 1005 kJ/mol.

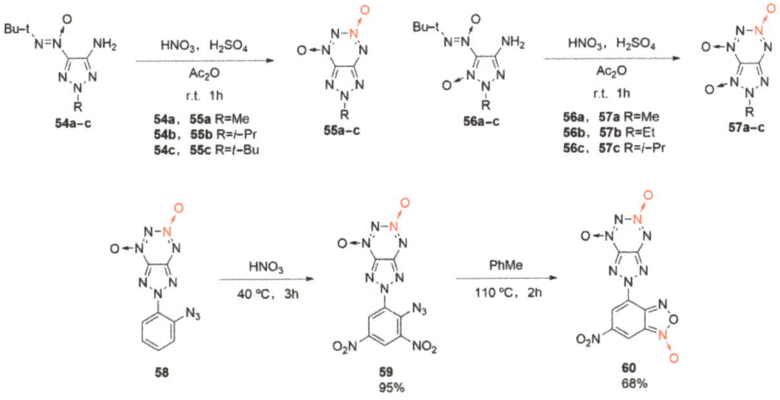

Scheme 9. Synthetic routes for compounds **55a-c**, **57a-c** and **60**.

2.1.2. 1,2,3,5-Tetrazine-2-Oxides

Bian Chengming et al. were the first to study the synthetic methods for 1,2,3,5-tetrazine-2-dioxide [44]. After fusing 1,2,3,5-tetrazine-2-dioxide with triazole, 7-nitro-4-one-4,8-dihydro-1,2,4-triazolo [5,1-d]-1,2,3,5-tetrazine-2-oxide (**63**), a series of energetic ionic salts were designed and synthetized (Scheme 10 **65–70**). Using 5-amino-3-nitro-1,2,4-triazole (**61**) as the raw material [45], a tetrazole ring was introduced under the action of cyanogen bromide and sodium azide to obtain compound **62**. Nitric acid and fuming sulfuric acid were subjected to nitration and cyclization, and the sodium salt of compound **63** was obtained after extraction and washing with brine. At last, the energetic ionic salts **65–70** were obtained by an ion exchange reaction for the sodium salt of compound **63** and silver nitrate. Compounds **65–70** display good thermal stability, high density, high detonation performance, and low sensitivity (Table 1). Among them, compound **66** with a hydroxylaminium cation shows the highest potential to be used as an energetic material.

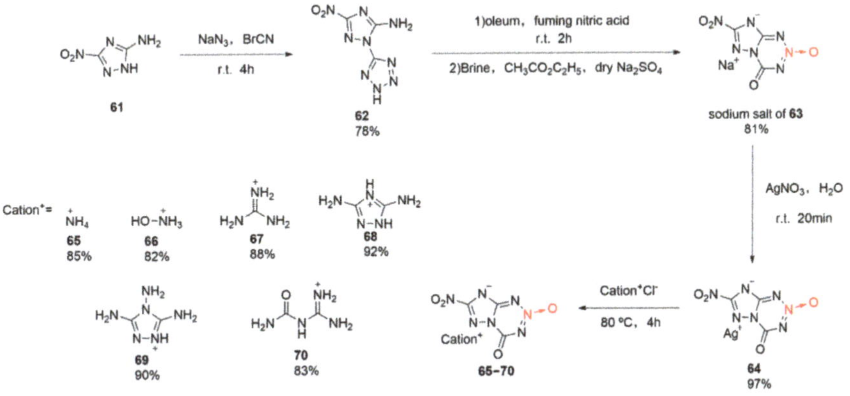

Scheme 10. Synthetic routes for compounds **65–70**.

Table 1. Physiochemical properties and detonation parameters of energetic salts **65–70**.

Compound	ρ/g·cm^{-3}	D/m·s^{-1}	P/GPa	IS/J	FS/N	T$_d$/°C
65	1.77	8252	29.0	>40	324	249
66	1.97	9069	39.5	>40	324	197
67	1.78	8113	27.1	>40	324	269
68	1.89	8463	30.7	>40	360	252
69	1.81	8373	29.2	>40	324	237
70	1.80	7856	25.2	>40	>360	241
TATB [43]	1.93	8114	31.2	50	>360	~360
RDX [43]	1.82	8748	34.9	7	120	230

During the nitration reaction of **62**, a white precipitate AgN$_3$ was generated from the reaction of AgNO$_3$ and the released gas, which was later verified to be azide acid. According to the infrared spectrum of compound **63**, a strong peak appears at 1787 cm^{-1} and no obvious absorption peak is found at 3600~2500 cm^{-1}. It was verified that the 4-position in compound **63** was acyl group rather than hydroxyl group. Based on the experimental facts, the cyclization mechanism of compound **62** was reasonably speculated [44]: Compound **62** was nitrated in the HNO$_3$/H$_2$SO$_4$ mixed acid system to produce nitramine intermediate **71**. The tetrazole in intermediate **71** carried out electron transfer. Then, a molecule of HN$_3$ was removed and a C = N bond was created to form intermediate **72**. The O atom on the nitro group initiated a nucleophilic attack on the C atom on the C = N bond, reducing the density of the electron cloud on the N atom of the nitro group to make it appear deficiently electronic. The N atom on the C = N bond attacked the N atom of the nitro group. With the electron transfer and proton removal, a tetrazine ring was basically formed for producing intermediate **73**. The intermediate **73** was aromatized to perform electron and proton transfer, and the energetic compound **63** was finally obtained (Scheme 11).

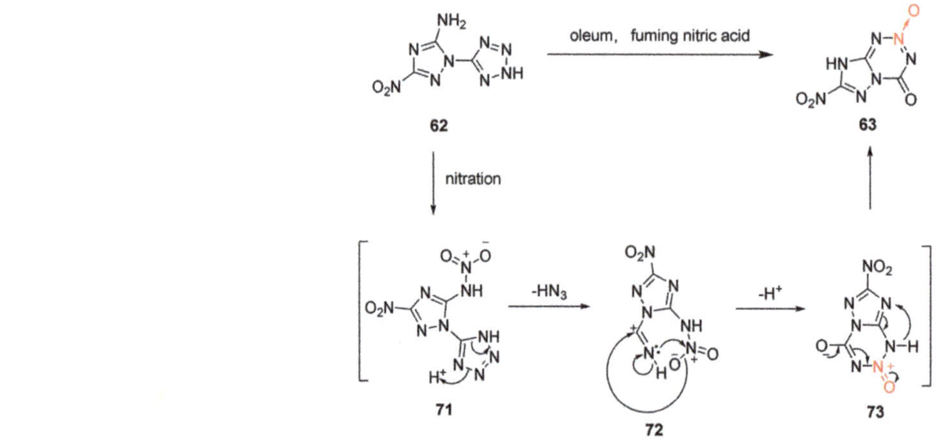

Scheme 11. Proposed synthetic route for compound **63**.

By replacing the triazole ring in compound **63** with a pyrazole ring and introducing another nitro group into the pyrazole ring, Zhao et al. [46] synthesized 7, 8-dinitro-4-keto-4,6-dihydropyrazolo [5,1-d]-1,2,3,5-tetraazine-2-oxide(**75**). Due to the acidity of **75**, a series of energetic ionic salts (Scheme 12 **77–86**) was synthesized by the same ion exchange method. After analyzing and comparing the measured density with the theoretically calculated detonation performance, it was found that the hydroxylaminium salt of compound **78** also display excellent properties in all aspects. Its density and detonation performance (ρ = 1.95 g/cm^3, D = 9228 m/s, P = 39.4 GPa) are similar to those of HMX (ρ = 1.91 g/cm^3, D = 9186 m/s, P = 39.7 GPa) and low sensitivity is also presented (IS = 19 J, FS = 360 N).

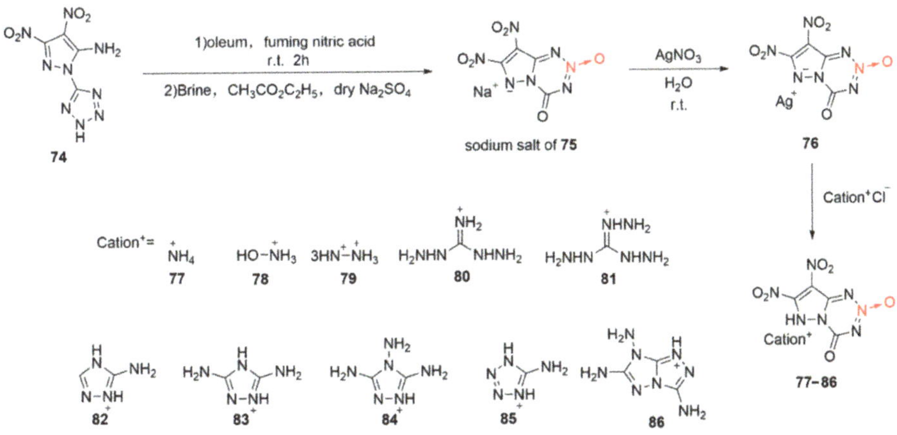

Scheme 12. Synthetic routes for compounds **77–86**.

According to the experimental fact that gas was produced when nitration reacted with AgNO$_3$ solution to obtain the white precipitate AgN$_3$[46], the nitration-cyclization mechanism of compound **74** was reasonably described as follows (Scheme 13). The amino group (-NH$_2$) on compound **74** first reacted with the NO$_2^+$ to obtain the nitroamine intermediate **87**, and the nitroamine intermediate was unstable. After the H atom on the nitroamine group was captured by HSO$_4^-$ in the system, the O atom on the nitro group attacked the C atom of the tetrazole ring, and the tetrazole ring carried out electron transfer, the intermediate **88** was obtained by removing a molecule of HN$_3$ and cyclization reaction between the remaining C = N unit of the tetrazole ring and the nitro oxygen atom. The intermediate **88** was rearranged under the action of HSO$_4^-$ to produce a fused ring intermediate **89** containing 1,2,3,5-tetrazine-2-oxide. Further aromatization was carried out to obtain the product compound **75** via intramolecular protons and electron transfer.

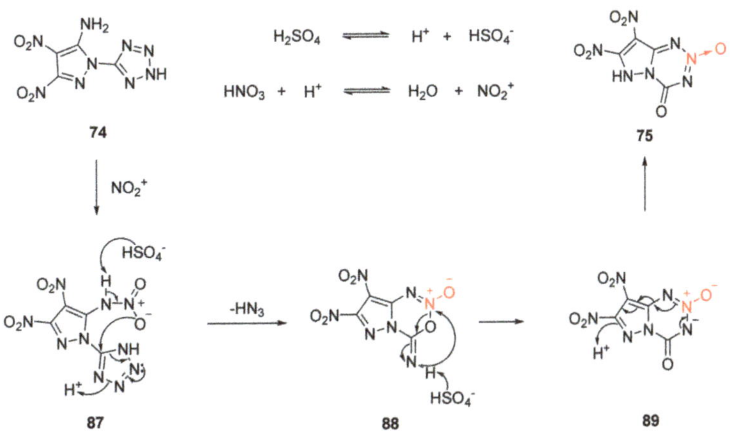

Scheme 13. Proposed synthetic route for compound **75**.

Compound **63** and compound **75** both exhibit excellent detonation properties, but the separation is difficult due to their strong moisture absorption. Moreover, the reagent required for the synthetic process is toxic and the gas produced during reaction is dangerous, limiting their applications to a great extent. An efficient and safe synthesis route for 1,2,3,5-tetrazine-2-oxide was independently developed by Lei et al. [47]. Based on the reactivity of the hydrazino group, two pyrazolo 1,2,3,5-tetrazine-2-oxides with no hygroscopicity, high density, excellent detonation performance, and low sensitivity were obtained (Scheme 14 **92, 94**). Using ethoxymethylene malononitrile (**90**) as the raw material, the two compounds **92** and **94** were synthesized via different cyclization and nitration. Based on theoretical calculation and practical analysis, compound **94** display high density (ρ = 1.874 g/cm^3), high detonation velocity (D = 8983 m/s), high detonation pressure (P = 34.5 GPa), and low sensitivity (IS = 20 J, FS > 360 N). At the same time, its high thermal decomposition temperature (T_d = 302 °C) endows it with the potential to become a heat-resistant explosive.

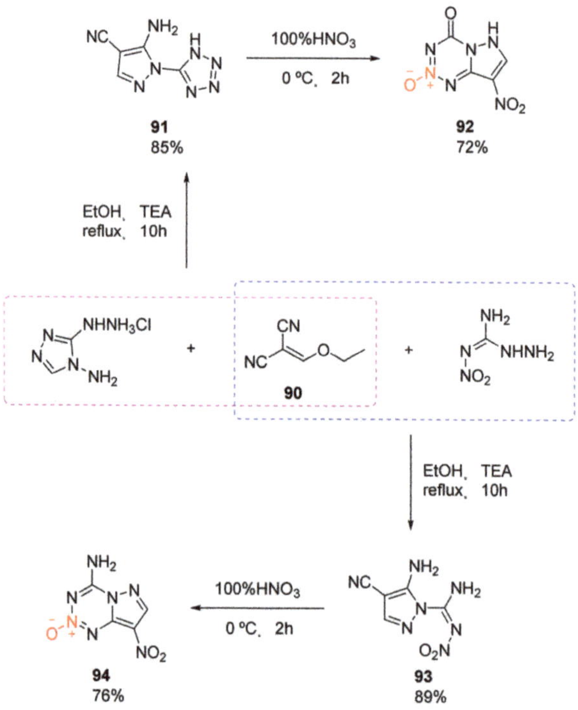

Scheme 14. Synthetic routes for compounds 92 and 94.

The possible cyclization mechanism for compound **91** and compound **93** was described as follows [47] (Scheme 15). The amino group of compound **91** was nitrated to a nitroamine group, and the o-tetrazole group exposed C = N moiety after removing a molecule of HN$_3$ by electron transfer. A cyclization was carried out after the C = N moiety and nitro group had a nucleophilic reaction. With the electron induction of ring, the cyano group was replaced by a nitro group in the nitration reaction to produce compound **92**. Compound **93** was cyclized by the dehydration condensation of nitro and amino groups in the imine structure, also under the induction effect of 1,2,3,5-tetrazine-2-oxycycle. Then, the cyano group was substituted with nitro to obtain compound **94**.

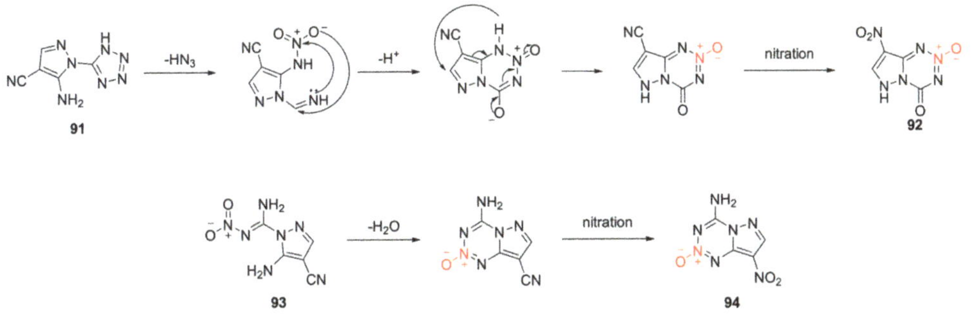

Scheme 15. Possible synthetic routes for compounds **92** and **94**.

2.1.3. 1,2,3-Triazine-3-Oxides

Using o-cyano N-heterocyclic aromatic amine as the starting material, 1,2,3-triazine-3-oxide skeleton was formed by the addition, diazotization and cyclization reaction via the reactivity of an amino group with the o-cyano group. Based on the above design ideas of 1,2,3-triazine-3-oxide skeleton, three energetic compounds (Scheme 16 **97**, **103**, **106**) in which 1,2,3-triazine-3-oxide was fused with triazole, imidazole, and pyrazole, were synthetized by Tang et al. [48] for the first time. Since triazole and pyrazole are acidic under the action of strong electron absorbing groups, five energetic ionic salts (**98–100**, **103** and **106**) were obtained by the neutralization reaction from three fused energetic compounds, namely **97**, **103, and 106.** Based on theoretical calculation and experimental study, it was found that compound **99** displayed the best detonation performance (D = 9358 m/s, P = 33.6 GPa), even exceeding RDX (D = 8795 m/s) in detonation velocity (Table 2).

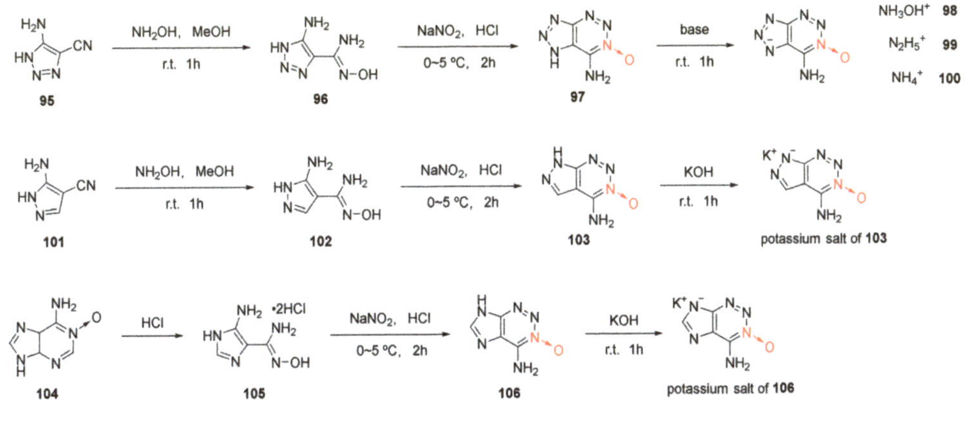

Scheme 16. Synthetic routes for ionic salts **98–100** and compounds **97**, **103**, **107**.

Table 2. Energetic properties and detonation parameters of ionic salts **98–100** and compounds **97**, **103**, **107**.

Compound	ρ/g·cm^{-3}	D/m·s^{-1}	P/GPa	IS/J	FS/N	T_d/°C
97	1.815	8792	30.3	30	360	194
98	1.772	8670	28.0	20	>360	137
99	1.800	9358	33.6	25	>360	160
100	1.740	8744	28.2	30	>360	180
103	1.772	8271	25.6	>40	>360	221
106	1.743	8008	23.3	>40	>360	227
RDX [47]	1.800	8795	34.9	7.5	120	204

In the nitrosation reaction of compound **102** with dilute hydrochloric acid, a white precipitate was firstly produced, which turned yellow after being exposed to air for several days or being washed with water. Its crystal structure was confirmed by X-ray single-crystal diffraction. It was a hydrochloride intermediate (**107**) formed after amino diazotization and cyclization with a hydroxime group. According to the experimental results of the synthetic compound **103**, the cyclization mechanism of 1,2,3-triazine-3-oxide was proposed as follows [48] (Scheme 17): The diazonium positive ions was generated by the nitrosation reaction of the amino group. The hydrochloride intermediate **107** was obtained by further cyclization with a hydroxime group. Finally, compound **103** was obtained after removing HCl and H_2O.

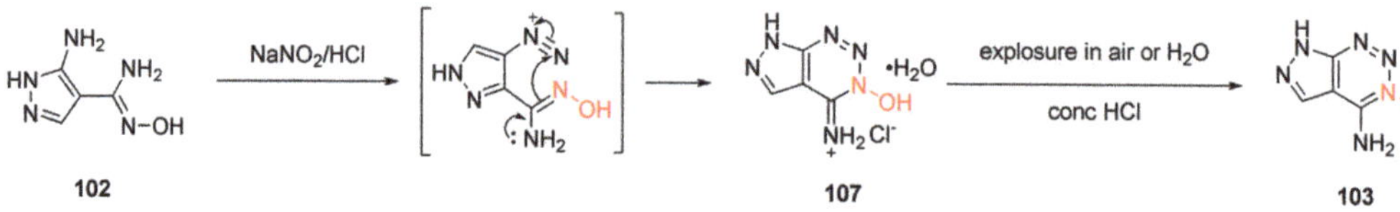

Scheme 17. Proposed synthetic route for compound **103**.

2.1.4. 1,2,3-Triazine-2-Oxides

High detonation performance, appropriate mechanical sensitivity, and high chemical and thermal stability are the basic characteristics of green detonators. A fused ring energetic compound 6-nitro-7-azido-pyrazolo [3,4-d]-1,2,3-triazine-2-oxide (**110**, **ICM-103**) was synthetized by Deng Mucong et al. [49,50] from industrial 3-amino-4-cyanopyrazole(**108**). First, under the action of the catalyst $ZnCl_2$, the cyano group on the pyrazole ring reacted with azide anion to form a tetrazole ring. Then, the nitration reaction was cyclized to form 1,2,3-triazine-2-oxide(Scheme 18). The total yield of the two-step reaction was 77.4%. Especially, by only one nitration, azido and nitro groups were formed and cyclized, laying a foundation for the industrial mass production of **ICM-103**. Additionally, a high measured density and excellent calculated detonation performance (ρ= 1.86 g/cm^3, D = 9111 m/s, P = 35.14 GPa), **ICM-103** shows an appropriate mechanical sensitivity (IS = 4 J, FS = 60 N), and its flame sensitivity (Flame S > 60 cm) even exceeds that of the widely used azide dinitrophenol (DDNP) (Flame S = 17 cm). More importantly, **ICM-103** does not contain metal components and is environmentally friendly. Thus, its synthesis is of great significance for the research of green primary explosives.

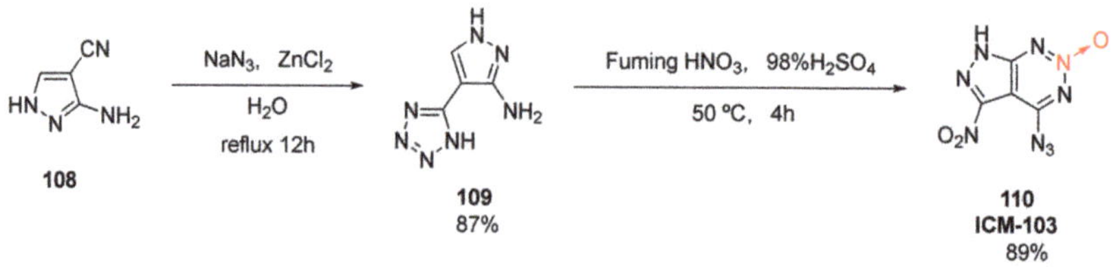

Scheme 18. Synthetic route for **ICM-103**.

Deng Mucong proposed a reasonable cyclization mechanism for compound **109** based on the theoretical calculation of the Gibbs free energy of various intermediates encountered in the reaction process [49] (Scheme 19). Compound **109** was first nitrated with mixed acid to produce an unstable nitroamine intermediate **111**, and its Gibbs free energy was defined as 0 (ΔG= 0 kcal.mol^{-1}). With the induction of the intramolecular hydrogen bond, the N-H in the tetrazole ring reacted with the N = O in the nitro group and intermediate **113** was generated via the transition state **112**. After an intramolecular elimination reaction was conducted, a fused ring intermediate **114** was obtained by removing a molecule of water. The tetrazole ring on the intermediate **114** was isomerized to form a azido group and obtain 7-azido-pyrazolo-[3,4-d]-1,2,3-triazine-2-oxide(**115**). **IMC-103** was obtained by further nitration and introduction of a nitro group at the position 6 of **115**.

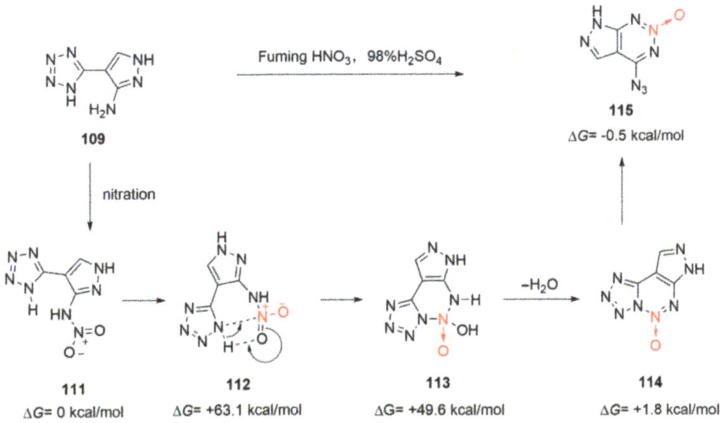

Scheme 19. Possible synthetic route for **ICM-103**.

Using ethoxymethylene malononitrile (**90**) and carbamoyl hydrazide as raw materials, compound **116** was obtained by the condensation cyclization reaction in the solution of triethylamine and ethanol, and pyrazolo-1,2,3-triazine-2-oxide(**117**)(ρ= 1.825 g/cm^3, D = 8323 m/s, P = 27.7 GPa, *IS* > 20 J, *FS* > 360 N) was synthetized by Lei et al. [47] via further nitration and cyclization using fuming nitric acid. Due to the high thermal decomposition temperature (T$_d$ = 275 °C), compound **117** could be used as a heat-resistant explosive. The cyclization mechanism of compound **116** was basically the same as that of compound **109**: By removing the amide group on the pyrazole ring using compound **116** under the action of nitric acid, while hydrolyzing the cyano group on the ring to an amide group, the amino group was nitrated to a nitroamino group. Similarly, with the induction of a hydrogen bond, the nitro group and the amino group underwent electrophilic addition and elimination reaction to remove a molecule of water. Then, further intramolecular electron transfer was performed to obtain compound **117**(Scheme 20).

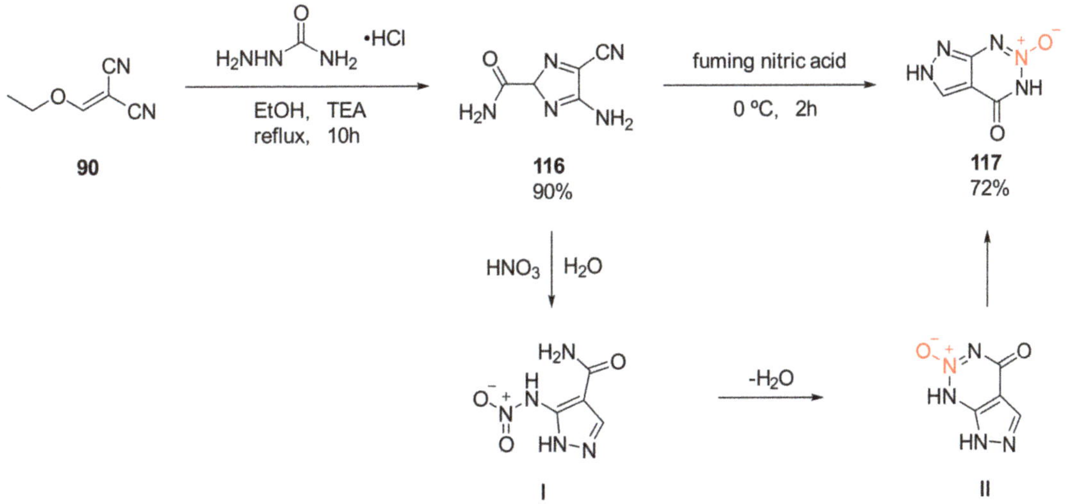

Scheme 20. Possible synthetic route for compound **117**.

2.1.5. Pyridazine-1,2-Dioxide

According to the characteristics of NO donor drugs, 4,6-dinitro-furazan [3,4-d]-1,5,6-trioxide(**120**) was obtained by Ogurstov et al. [51] by fusing furoxan with pyridazine-1,2-dioxide. Using 3,4-dihydroxyfurazan(**118**) as the raw material, a 1:1 stable complex of furazan dinitroxime acid (**119**) with 1,4-dioxane was obtained by the nitration of N_2O_4 in Et_2O solution. Then, compound **120** (m.p. = 50–52 °C) was synthetized in the nitration system of HNO_3 and CF_3COOH(Scheme 21). As compound **120** could not exist stably in organic solvents such as CH_2Cl_2 and MeCN, it could not be detected by elemental analysis and mass spectrometry. Its structure, however, could still be observed by infrared spectrum in KBr and carbon spectrum in deuterated acetone. Compound **120** has a density as high as 1.98 g/cm^3 and a calculated detonation velocity of 9.52 km/s, making it comparable to CL-20. Yet, its applications as an energetic material are largely limited by its poor thermal stability.

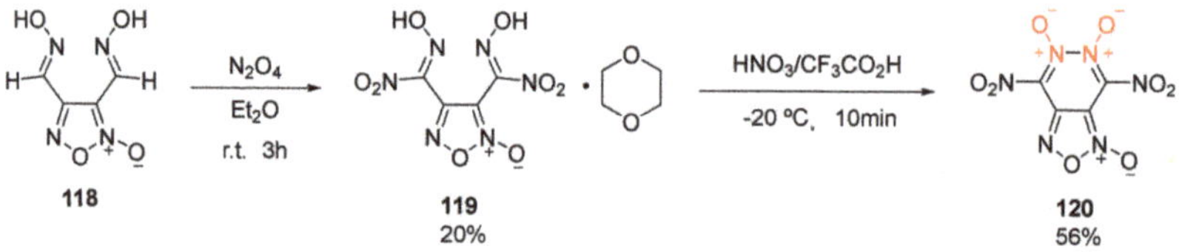

Scheme 21. Synthetic route for compound **120**.

2.1.6. Pyrazine-1-Oxide

The 2,6-diamino-3,5-dinitro-1-oxide (**125**, **LLM-105**) has a crystal density of 1.918 g/cm^3, experimental detonation velocity [52] of 8560 m/s, and DSC exothermic peak of 342 °C. It exhibits good thermal stability, an impact sensitivity of 117 cm, high sensitivity, and excellent comprehensive properties, thus being widely used in the field of energetic materials. It could not only be used as the main charge and insensitive initiator of weapons but it also has potential in civil engineering, such as oil exploration. Jing Suming et al. used iminodiacetonitrile (**121**) as the raw material to synthetize **LLM-105** by a three-step process of substitution, condensation, and nitration (Scheme 22a) [52]. **LLM-105** was also synthetized by Zhao Xiaofeng et al. [53] using a novel synthesis method of nitrosation, condensation, and nitration (Scheme 22b), which shows potential for industrial applications.

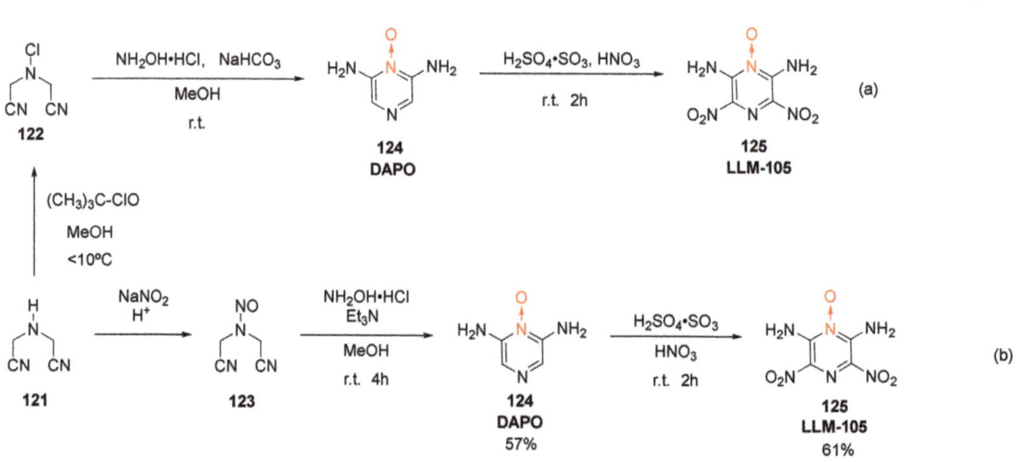

Scheme 22. Synthetic routes (**a**) and (**b**) for **LLM-105**.

Zhao Xiaofeng et al. discussed the reaction mechanism for synthesizing **DAPO** [54]. Compound **123** was stripped of a proton under the alkaline action of Et₃N, and the nitrogen oxide anion (NO⁻) was removed after electron transfer to form intermediate **126**. Intermediate **126** was then condensed and cyclized with free hydroxylamine to obtain compound **127**, which was further aromatized to obtain compound **128**. Finally, the product **DAPO** was obtained by intramolecular proton transfer (Scheme 23a). On such basis, a new possible reaction mechanism was proposed by Wang Jinmin et al. [55], who used sodium hydroxide as alkaline medium for the preparation of **DAPO**. A cyano group in compound **123** first reacted with hydroxylamine to obtain compound **129**. Under the alkaline condition of NaOH, HNO (H⁺ and NO⁻) was removed to obtain intermediate **130**, and the hydroxylamino group of the intermediate condensed with another cyano group to obtain compound **127**. After further intramolecular electron and proton transfer, the product **DAPO** was obtained (Scheme 23b). Li et al. [56] found that compound **131** and compound **134** (Scheme 23c) could be separated from the product by changing the ratio of compound **123**, hydroxylamine hydrochloride, and alkaline medium. Then, the cyclization mechanism of **DAPO** was divided into two stages. In the first stage, the basic reagent reacted with hydroxylamine hydrochloride to obtain free hydroxylamine and condensed with compound **123** to form compound **129**. In the second stage, Et₃N grabbed protons to separate nitrogen oxide anions and then condensed and cyclized to obtain the product **DAPO**.

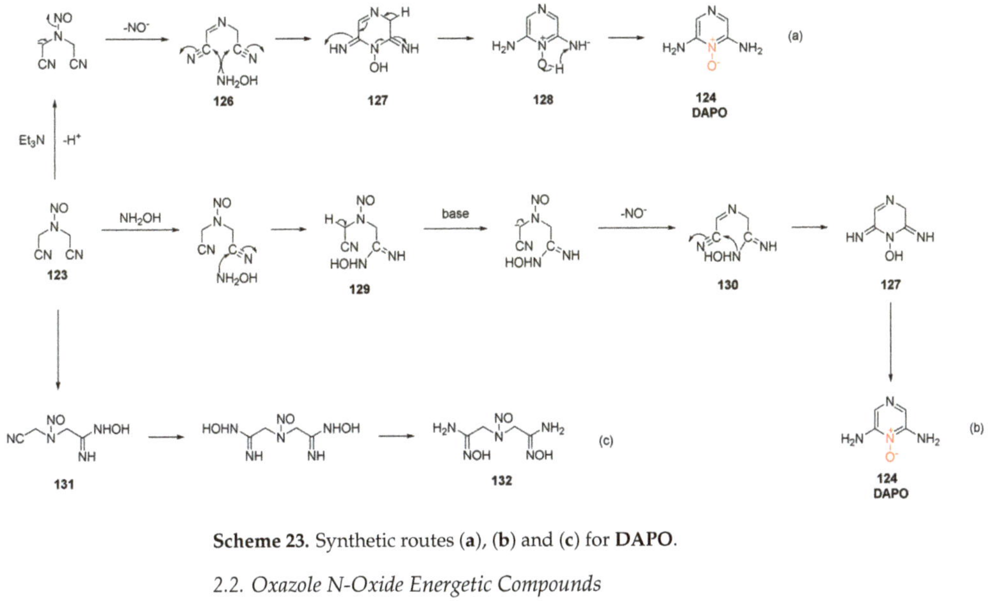

Scheme 23. Synthetic routes (**a**), (**b**) and (**c**) for **DAPO**.

2.2. Oxazole N-Oxide Energetic Compounds

2.2.1. 1,2,5–Oxadiazole-2-Oxides

1,2,5-Oxadiazole-2-oxide is also called furoxan. Due to the coordinated oxygen atom on its ring, it has a unique "potential nitro" structure, and it can be used as an energetic group to effectively improve the detonation performance and energy density of compounds, thus attracting wide research interests. Using malondinitrile (**133**) as the starting material, 3,3'-dicyano-4,4'-azofuroxan was synthetized by Luo Yifen et al. [57] via nitrosation, NH₂OH addition, and PbO₂ and KMnO₄ secondary oxidation. The four-step reaction had a yield of 30.2% (Scheme 24).

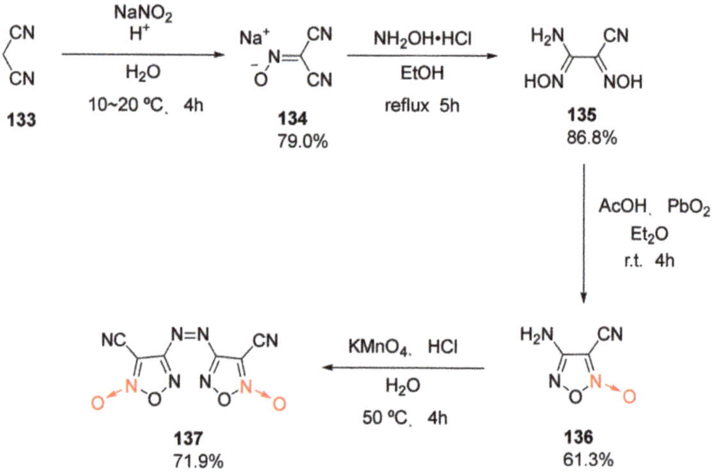

Scheme 24. Synthetic route for compound **137**.

The oxidation-cyclization mechanism under the action of lead peroxide was analyzed [57] as follows (Scheme 25). First, lead peroxide reacted with acetic acid to obtain lead tetraacetate. Then, the hydroxyl in compound **135** reacted with lead tetraacetate to obtain intermediate **138**. Under the action of the intramolecular hydrogen bond, intermediate **138** experienced intramolecular coupling and intramolecular electron transfer via a macrocyclic transition state. Compound **136** was obtained by the cyclization reaction after removing a molecule of lead acetate and a molecule of acetic acid.

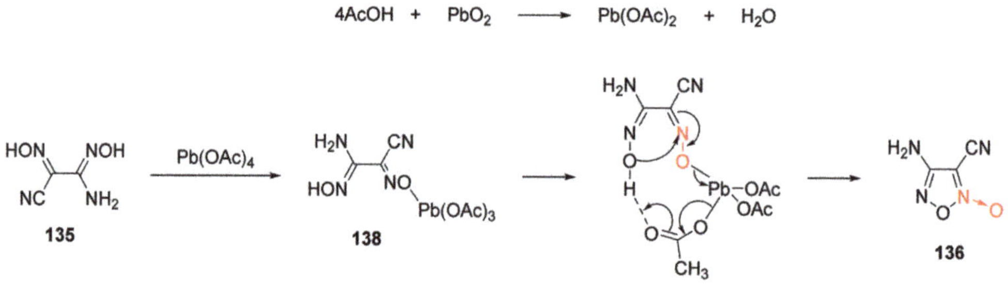

Scheme 25. Possible synthetic route for compound **136**.

Using malonyl monohydrazide monopotassium salt (**139**) as the raw material, dinitroazofurazan (**144**) was synthetized by nitrosation-nitration, first hydrolysis, oxidative coupling, second hydrolysis and oxidation [58] (Scheme 26). Dinitroazofurazan has a density as high as 2.002 g/cm^3 and a measured detonation velocity of 10 km/s. Such characteristics as high density and high detonation velocity make it an important component for energetic composite solid propellants.

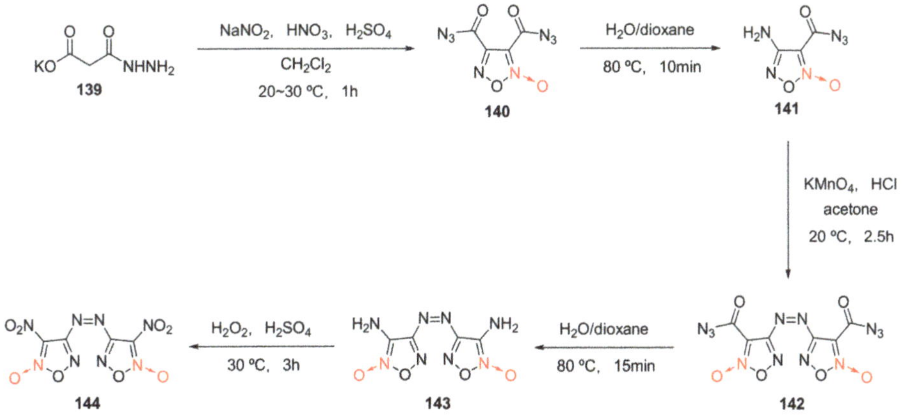

Scheme 26. Synthetic route for compound **144**.

The reaction mechanism of compound **140** was discussed as follows [58] (Scheme 27). Compound **139** was first nitrated to produce intermediate **145** with a resonance structure **146**. The intermolecular hydrogen bond of compound **146** increased the stability of the compound. After Curtius rearrangement, compound **147** was generated. The hydrazido group in compound **147** was transformed into an azido group after nitrosation to obtain compound **140**.

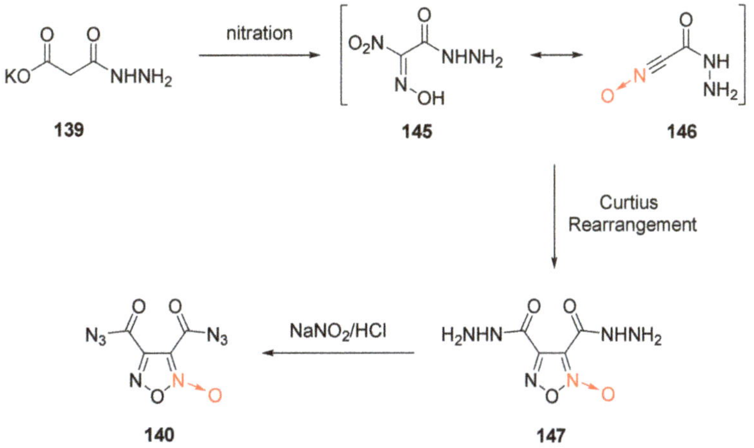

Scheme 27. Synthetic route for compound **140**.

Ma et al. synthesized a fused ring energetic compound 4-amino-5-nitro-1,2,5-oxadiazolo [3,4-e]tetrazolo [1,5-a]pyridine-3-oxide (**150**) from industrial 4-amino-2,6-dichloropyridine (**148**) via two steps, nitration and substitution-cyclization [59] (Scheme 28). The existence of compound **151** could be detected during the second step of the reaction, indicating that the azide group was not introduced into the pyridine ring at one time but was divided into two steps. The first introduced azide group was cyclized with a nitro group and a molecule of N_2 was removed to generate compound **151**. After the second azide group was introduced, the tautomerization of azide-tetrazole occurred due to the electron deficiency of the pyridine ring, resulting in compound **150**. With a density of 1.921 g/cm^3, a calculated detonation velocity of 8838 m/s, and a detonation pressure of 36.01 GPa, compound **150** could be an energetic material with an excellent detonation performance.

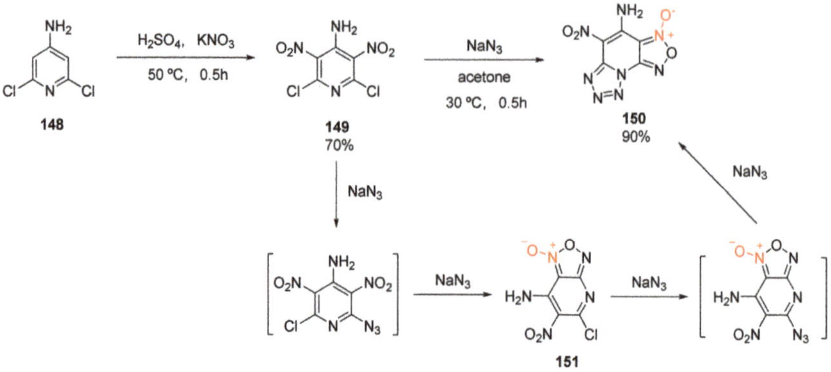

Scheme 28. Synthetic route for compound **150**.

Using 4-amino-3-amidoximo furoxan(**152**) as a starting material, 3,4-bis (4'-aminofuroxan-3'-yl)furoxan(**156**) (ρ = 1.787 g/cm^3, D = 8480 m/s, P = 31.0 GPa) was synthetized by He et al. [60] via amino protection, diazotization-chlorination, intermolecular condensation cyclization, and hydrolysis. Then, compounds **157** (ρ = 1.895 g/cm^3, D = 9417 m/s, P = 39.6 GPa), and **158** (ρ = 1.914 g/cm^3, D = 9503 m/s, P = 40.8 GPa) were obtained through oxidative coupling or oxidation (Scheme 29). The three compounds show good detonation performance and energy density. Compound **158** exhibit a positive oxygen balance (OB = +18.6%) with good detonation performance, and its sensitivity (IS = 3 J, FS = 40 N) is close to that of CL-20. Due to its outstanding properties, it could be used as an energetic component in solid rocket propellants.

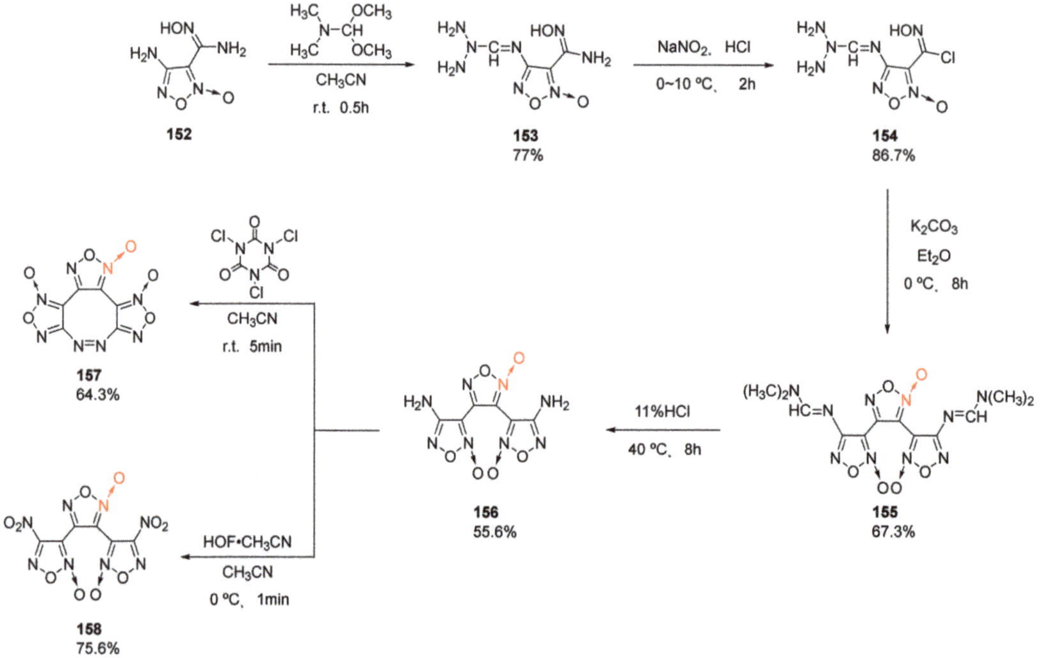

Scheme 29. Synthetic routes for compounds **156–158**.

Using on-line infrared technology to monitor the synthesis of 3,4-bis(4'- aminofuroxan-3'-yl)furoxan, Sun Kunlun et al. observed the infrared spectrum of the pure substance of the intermediate component produced in the reaction to deduce the reasonable reaction mechanism of furoxan construction from chloro-oxime via dimerization and cyclization [61] (Scheme 30). Under alkaline conditions, the O–H bond and C–Cl bond of the chloro-oxime group broke and removed a molecule of HCl. Then, intramolecular electron transfer was carried out to obtain intermediate **159**. As the reaction progressed, intermediate **159** generated the resonance structure of intermediate **160**, which was dimerized with intermediate **159** to obtain product **156**.

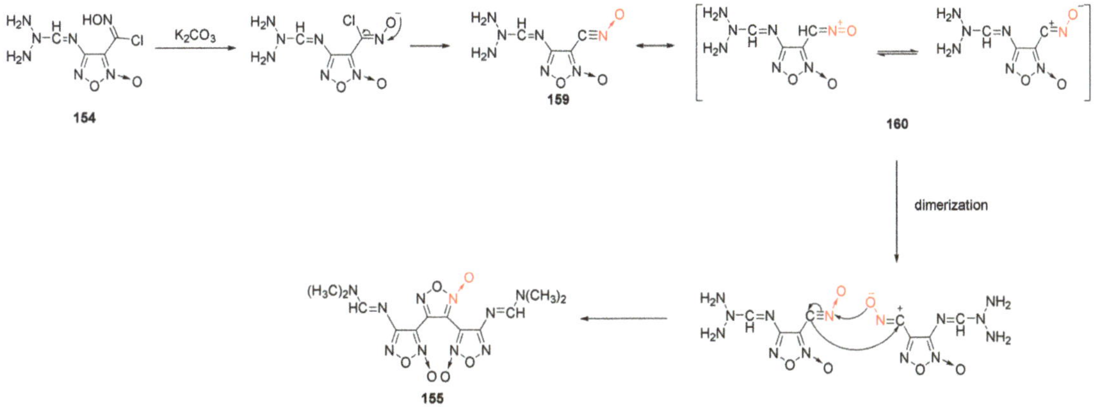

Scheme 30. Synthetic route for compound **155**.

2.2.2. Pyrazole-1-Oxide

4-Amino-3-chlorooximofurazan (**162**) can be oxidized to be the dipotassium salt of 3-dinitromethyl-4-nitroaminofurazan. Yet, when Li et al. used this synthetic approach, the potassium salt of 6-nitro-pyrazolo [3,4-c]furazan-5-oxide (**164**) was accidentally isolated [62]. This unique reaction process was thoroughly studied by Tang et al., [63]. Using malonic nitrile as the starting material, 4-amino-razan-3-chlorooxime (**162**) was obtained by three steps, namely nitration, addition, and cyclization. Then, the 100% HNO$_3$/TFAA system was used for nitration. Compound **162** was first nitrated to be intermediate **173** during nitration, and intermediate **173** experienced an intramolecular electron transfer to remove HNO$_3$ for generating compound **163**. Compound **163** was dissolved in methanol and was ion-exchanged with KI to obtain the potassium salt of compound **164**. Based on the reaction characteristics of **164**, a series of energetic ionic salts (**165–172**) were designed and synthesized by the neutralization or ion exchange method (Scheme 31). Through theoretical calculation and experimental measurement, the potassium salt of compound **164** was found as high as 2.036 g/cm^3, the sensitivity of compound **172** was satisfying (*IS* = 20 J, *FS* = 360 N), and the calculated detonation velocity and detonation pressure of hydroxylaminium salt (**166**) were the highest among the products (9174 m/s and 39.1 GPa). The hydroxylaminium salt **166** with high sensitivity (*IS* = 2 J, *FS* = 40 N) also has potential for becoming a green primary explosive.

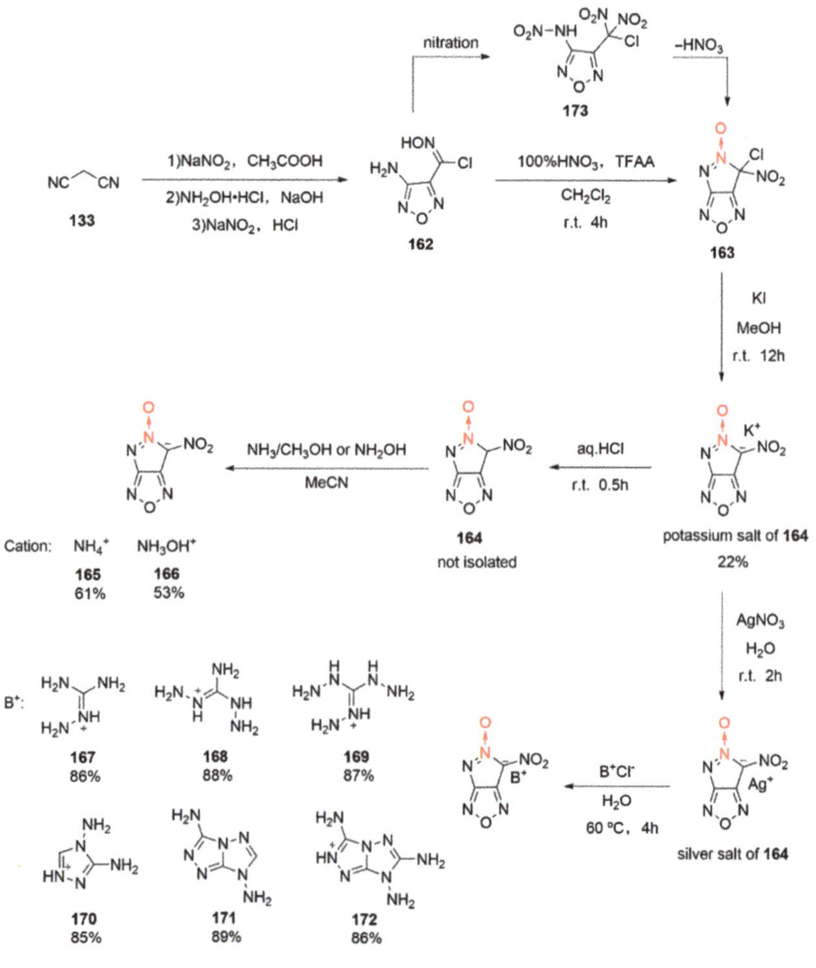

Scheme 31. Synthetic routes for energetic salts **165–172**.

2.2.3. Triazole-1-Oxides

The synthesis of 1,2,3-Triazolo [4,5-e]furazano [3,4-b]pyrazine-6-oxide(**177**) and its energetic ionic salts (**179–183**) was reported by Thottempudi et al. [64]. Using 3,4-diaminofurazan (**46**) as the starting material, diaminofurazano [3,4-b]pyrazine **175** was obtained by a three-step process of condensation-cyclization, chlorination, and ammoniation. Then, triazole-1-oxide (**176**) was synthetized using the nitration-cyclization of HNO_3/TFAA. Due to the acidity of compound **176**, a series of energetic ionic salts (**177–181**) were synthesized by the neutralization reaction with a yield ranging between 85% and 98%(Scheme 32). Compound **176** and ionic salts **177–181** display outstanding energy density and detonation properties (Table 3). Moreover, these compounds exhibit extremely low impact sensitivity, making them suitable for insensitive explosives.

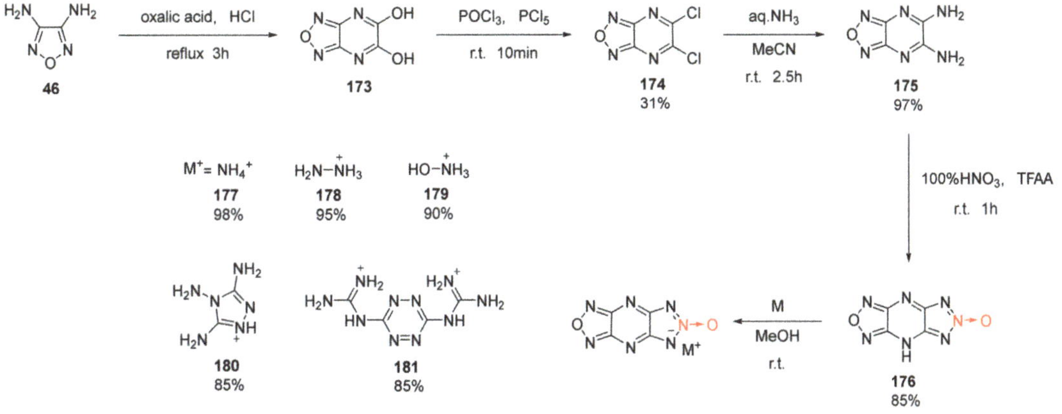

Scheme 32. Synthetic routes for compounds 177–181.

Table 3. Physical and detonation properties of compounds 177–181.

Compound	$\rho/\text{g}\cdot\text{cm}^{-3}$	$D/\text{m}\cdot\text{s}^{-1}$	P/GPa	IS/J	$T_d/^\circ\text{C}$
176	1.85	8532	32.4	32	281
177	1.73	8079	26.3	>40	270
178	1.76	8378	30.0	38	141
179	1.74	8518	30.3	35	157
180	1.70	7972	25.0	>40	274
181	1.69	7871	24.0	>40	301
TNT [63]	1.65	6881	19.5	15	295
PETN [63]	1.77	8564	31.3	5	150

Li Yanan et al. discussed the nitration-cyclization reaction mechanism of triazole-1-oxide 175 in 100% HNO_3/TFAA system [65] (Scheme 33). Two nitro groups in compound 175 were first converted into nitroamino groups through nitration to obtain intermediate 182. One nitroamino group in intermediate 182 was rearranged into a hydroxyloxyazo group, and the rearranged intermediate 182 was stripped of a molecule of HNO_3 to produce compound 177.

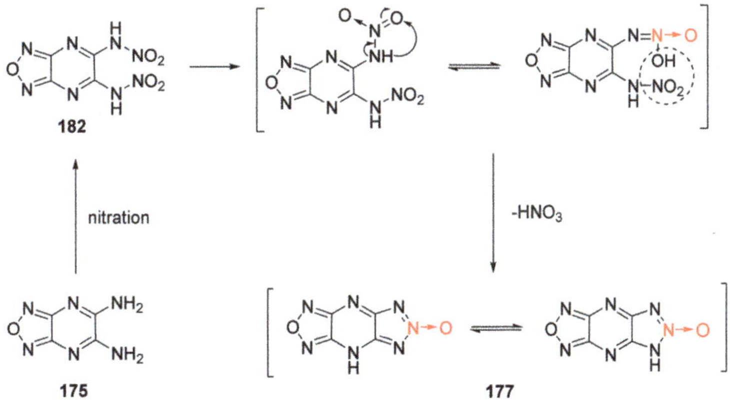

Scheme 33. Synthetic route for compound 177.

3. Conclusions

N-oxides can be synthesized by either direct N-oxidation or cyclization with N→O coordination bonds. As for the second approach, the N→O coordination bonds in energetic N-oxides could be introduced into the cyclization process, avoiding the pollution of the N-oxidation reaction, shortening the synthetic route, and improving efficiency. In this study, the synthetic methods towards two kinds of N-oxides (nine energetic N-oxides) are introduced in detail. The reaction mechanisms towards cyclization reaction are also discussed to clarify the formation of N→O coordination bonds. The physicochemical properties and detonation performances of typical N-oxides are also introduced, laying a foundation for future applications.

The synthesis of N-oxides by cyclization is one of the most important directions of energetic materials. At present, the industrialized manufacturing process of heat-resistant explosive LLM-105 has been found based on its synthetic method. With improved synthetic methods and technological advancements, new N-oxides with excellent performance are expected to emerge, and the manufacturing process for more N-oxide energetic materials will be improved to meet the need of mass production.

Author Contributions: W.S. completed the main content of the article; Z.X., L.Z., J.Z. and J.H. assisted with literature research and scheme drawing; W.P. and B.W. designed the main content and scope of the review. All authors have read and agreed to the published version of the manuscript.

Funding: This research was funded by the National Natural Science Foundation of China (Grant No. 22105155).

Conflicts of Interest: The authors declare no conflict of interest.

References

1. Gao, H.X.; Shreeve, J.M. Azole-based energetic salts. *Chem. Rev.* **2011**, *111*, 7377–7436. [CrossRef] [PubMed]
2. Badgujar, D.M.; Talawar, M.B.; Asthana, S.N.; Mahulikar, P.P. Advances in science and technology of modern energetic materials: An overview. *J. Hazard. Mater.* **2008**, *151*, 289–305. [CrossRef] [PubMed]
3. Wang, Y.; Liu, Y.; Song, S.W.; Yang, Z.J.; Qi, X.J.; Wang, K.C.; Liu, Y.; Zhang, Q.H.; Tian, Y. Accelerating the discovery of insensitive high-energy-density materials by a materials genome approach. *Nat. Commun.* **2018**, *9*, 2444. [CrossRef]
4. Tang, Y.X.; Kumar, D.; Shreeve, J.M. Balancing excellent performance and high thermal stability in a dinitropyrazole fused 1,2,3,4-tetrazine. *J. Am. Chem. Soc.* **2017**, *139*, 13684–13687. [CrossRef] [PubMed]
5. Zhai, L.J.; Bi, F.Q.; Huo, H.; Luo, Y.F.; Li, X.Z.; Chen, S.P.; Wang, B.Z. The ingenious synthesis of a nitro-free insensitive high-energy material featuring face-to-face and edge-to-face π-interactions. *Front. Chem.* **2019**, *7*, 559. [CrossRef] [PubMed]
6. Huo, H.; Zhang, J.L.; Dong, J.; Zhai, L.J.; Guo, T.; Wang, Z.J.; Bi, F.Q.; Wang, B.Z. A promising insensitive energetic material based on a fluorodinitromethyl explosophore group and 1,2,3,4-tetrahydro-1,3,5-triazine: Synthesis, crystal structure and performance. *RSC Adv.* **2020**, *10*, 11816–11822. [CrossRef]
7. Xu, Y.G.; Shen, C.; Lin, Q.H.; Wang, P.C.; Jiang, C.; Lu, M. 1-Nitro-2-trinitromethyl substituted imidazoles: A new family of high performance energetic materials. *J. Mater. Chem. A* **2020**, *4*, 17791–17800. [CrossRef]
8. Barton, L.M.; Edwards, J.T.; Johnson, E.C.; Bukowski, E.J.; Sausa, R.C.; Byrd, E.F.C.; Orlicki, J.A.; Sabatini, J.J.; Baran, P.S. Impact of stereo-and regiochemistry on energetic materials. *J. Am. Chem. Soc.* **2019**, *141*, 12531–12535. [CrossRef] [PubMed]
9. Ma, H.X.; Chen, X.; Zhang, C.; Zheng, W.W.; Tian, H.W.; Bai, Y. Reaserch on the S-Tetrazine-based Energetic Compounds. *Chin. J. Explos. Propellants* **2021**, *44*, 407–419. (In Chinese)
10. Han, Y.Z.; Yang, Y.Z.; Du, Z.M.; Zhang, Y.H.; Yao, Q. Progress of Study on Thermal Behaviors of Nitrogen-rich Compounds as Azole, Triazine and Furazan. *Chin. J. Explos. Propellants* **2016**, *39*, 1–11. (In Chinese)
11. Liu, Y.; Gong, L.S.; Yi, X.Y.; He, P.; Zhang, J.G. Tunable 1,2,3-triazole-N-oxides towards high energy density materials: Theoretical insight into structure-property correlations. *New J. Chem.* **2022**, *46*, 11741–11750. [CrossRef]
12. Tang, Y.X.; Zhang, J.H.; Mitchell, L.A.; Parrish, D.A.; Shreeve, J.M. Taming of 3, 4-Di(nitramino)furazan. *J. Am. Chem. Soc.* **2015**, *137*, 15984–15987. [CrossRef] [PubMed]
13. Chavez, D.; Klapötke, T.M.; Parrish, D.; Piercey, D.G.; Stierstorfer, J. The Synthesis and Energetic Properties of 3, 4-Bis(2,2,2-trinitroethylamino)furazan (BTNEDAF). *Propellants Explos. Pyrotech.* **2010**, *35*, 1–9. [CrossRef]
14. Li, J.; Ma, Q.; Tang, S.H.; Fan, G.J. Crystal Structure and Thermal Decomposition Properties of *N,N'*-Bis(2-fluoro-2,2'-dinitroethyl)-3,4'-dinitraminefurazan. *Chin. J. Energ. Mater.* **2019**, *27*, 41–46. (In Chinese)
15. Xiao, X.; Ge, Z.X.; Wang, W.; Liu, Q.; Su, H.P.; Li, T.Q.; Bi, F.Q.; Ji, X.T. Progress of 3-Azido-1,2,4-Triazole and Its Derivatives. *Chin. J. Energ. Mater.* **2014**, *22*, 100–107. (In Chinese)

16. Lai, Y.; Liu, Y.; Huang, W.; Zeng, Z.Z.; Yang, H.W.; Tang, Y.X. Synthesis and Characterization of Pyrazole- and Imidazole- Derived Energetic Compounds Featuring Ortho Azido/nitro Groups. *FirePhysChem* **2022**, *2*, 160–164. [CrossRef]

17. Wu, B.; Yang, L.F.; Zhai, D.D.; Ma, G.M.; Pei, C.H. Facile Synthesis of 4-Amino-3,5-dinitropyrazolated Energetic Derivatives via 4-Bromopyrazole and Their Performances. *FirePhysChem* **2021**, *1*, 76–82. [CrossRef]

18. Vishnevskiy, Y.V.; Tikhonov, D.S.; Schwabedissen, J.; Stammler, H.G.; Moll, R.; Krumm, B.; Klapötke, T.M.; Mitzel, N.W. Tetranitromethane: A nightmare of molecular flexibility in the gaseous and solid States. *Angew. Chem. Int. Edit.* **2017**, *56*, 9619–9623. [CrossRef]

19. Zhai, L.J.; Zhang, J.L.; Zhang, J.R.; Wu, M.J.; Bi, F.Q.; Wang, B.Z. Progress in Synthesis and Properties of High Energy Density Compounds Regulated by N-F Bond. *Chin. J. Org. Chem.* **2020**, *40*, 1484–1501. (In Chinese) [CrossRef]

20. Zhou, J.; Zhang, J.L.; Wang, B.Z.; Qiu, L.L.; Xu, R.Q.; Sheremete, A.B. Recent Synthetic Efforts towards High Energy Density Materials: How to Design High-Performance Energetic Structures? *FirePhysChem* **2022**, *2*, 83–139. [CrossRef]

21. Yuan, J.; Long, X.P.; Zhang, C.Y. Influence of N-Oxide Introduction on the Stability of Nitrogen-Rich Heteroaromatic Rings: A Quantum Chemical Study. *J. Phys. Chem. A* **2016**, *120*, 9446–9457. [CrossRef] [PubMed]

22. Politzer, P.; Lane, P.; Murray, J.S. Computational analysis of relative stabilities of polyazine N-oxides. *Struct. Chem.* **2013**, *24*, 1965–1974. [CrossRef]

23. Lai, W.P.; Lian, P.; Ge, Z.X.; Liu, Y.Z.; Yu, T.; Lv, J. Theoretical study of the effect of N-oxides on the performances of energetic compounds. *J. Mol. Model.* **2016**, *22*, 1–11. [CrossRef] [PubMed]

24. Liu, S.; Shi, W.; Wang, Y.; Zhang, Q.H. Research Progress of Nitrogen-rich Fused-ring N-oxides. *Chin. J. Energ. Mater.* **2021**, *29*, 567–578. (In Chinese)

25. Zhu, J.P.; Ren, J.; Han, X.Y.; Li, Y.X.; Wang, J.L.; Cao, R.L. Study on Structures and Detonation Performance for Polynitropyrazine-N-oxides by Density Functional Theory. *Chin. J. Explos. Propllants* **2010**, *33*, 47–51. (In Chinese)

26. Churakov, A.M.; Ioffe, S.L.; Strelenko, Y.A.; Tartakovskii, V.A. 1,2,3,4-tetrazine 1,3-dioxides-a new class of heterocyclic compounds. *Bull. Acad. Sci. USSR Div. Chem. Sci.* **1990**, *39*, 639–640. [CrossRef]

27. Bi, F.Q.; Wang, B.Z.; Li, X.Z.; Fan, X.Z.; Xu, C.; Ge, Z.X. Progress in the Energetic Materials Based on 1,2,3,4-Tetrazine 1,3-Dioxid. *Chin. J. Energ. Mater.* **2012**, *20*, 630–637. (In Chinese)

28. Klapötke, T.M.; Piercey, D.G.; Stierstorfer, J.; Weyrauther, M. The Synthesis and Energetic Properties of 5,7-Dinitrobenzo-1,2,3,4-tetrazine-1,3-dioxide (DNBTDO). *Propellants Explos. Pyrotech.* **2012**, *37*, 527–535. [CrossRef]

29. Christe, K.O.; Dixon, D.A.; Vasiliu, M.; Wagner, R.I.; Haiges, R.; Boatz, J.A.; Ammon, H.L. Are DTTO and iso-DTTO Worthwhile Targets for Synthesis? *Propellants Explos. Pyrotech.* **2015**, *40*, 463–468. [CrossRef]

30. Khakimov, D.V.; Dzyabchenko, A.V.; Pivina, T.S. Computer simulation of the crystal structure of tetrazino-tetrazine tetraoxide (TTTO) isomers with one and two independent molecules in the unit cell. *Russ. Chem. Bull.* **2020**, *69*, 212–217. [CrossRef]

31. Mendoza-Cortes, J.L.; An, Q.; Goddard, W.A., III; Ye, C.; Zybin, S. Prediction of the crystal packing of di-tetrazine-tetroxide (DTTO) energetic material. *J. Comput. Chem.* **2016**, *37*, 163–167. [CrossRef] [PubMed]

32. Song, X.L.; Li, J.C.; Hou, H.; Wang, B.S. Extensive theoretical studies of a new energetic material: Tetrazino-tetrazine-tetraoxide (TTTO). *J. Comput. Chem.* **2009**, *30*, 1816–1820. [CrossRef]

33. Klenov, M.S.; Guskov, A.A.; Anikin, O.V.; Churakov, A.M.; Strelenko, Y.A.; Fedyanin, I.V.; Lyssenko, K.A.; Tartakovsky, V.A. Synthesis of Tetrazino-tetrazine 1,3,6,8-Tetraoxide (TTTO). *Angew. Chem. Int. Ed.* **2016**, *55*, 11472–11475. [CrossRef] [PubMed]

34. Frumkin, A.E.; Churakov, A.M.; Strelenko, Y.A.; Tartakovsky, V.A. Synthesis of 1,2,3,4-tetrazino[5, 6-g]benzo-1,2,3,4-tetrazine-1,3,7,9-tetraoxides. *Russ. Chem. Bull.* **2006**, *55*, 1654–1658. [CrossRef]

35. Frumkin, A.E.; Churakov, A.M.; Strelenko, Y.A.; Kachala, V.V.; Tartakovsky, V.A. Synthesis of 1,2,3,4-Tetrazino[5, 6-f]benzo-1,2,3,4-tetrazine-1,3,7,9-Tetra-N-oxides. *Org. Lett.* **1999**, *1*, 721–724. [CrossRef]

36. Churakov, A.M.; Ioffe, S.L.; Tartakovsky, V.A. Synthesis of [1,2,5]Oxadiazolo [3,4-e][1,2,3,4]tetrazine-4,6-Di-N-oxide. *Mendeleev Commun.* **1995**, *6*, 227–228. [CrossRef]

37. Dong, L.L.; Zhang, G.Q.; Chi, Y.; Fan, G.J.; He, L.; Tao, G.H.; Huang, M. Synthesis of Furazano[3,4-e]-1,2,3,4-tetrazine-1,3-dioxide. *Chin. J. Energ. Mater.* **2012**, *20*, 690–692. (In Chinese)

38. Zelenov, V.P.; Lobanova, A.A.; Sysolyatin, S.V.; Sevodina, N.V. New syntheses of [1,2,5]oxadiazolo[3,4-e][1,2,3,4]tetrazine 4,6-dioxide. *Russ. J. Org. Chem.* **2013**, *49*, 455–465. [CrossRef]

39. Li, X.Z.; Wang, B.Z.; Li, H.; Li, Y.A.; Bi, F.Q.; Huo, H.H.; Fan, X.Z. Novel Synthetic Route and Characterization of [1,2,5]oxadiazolo-[3,4-e][1,2,3,4]tetrazine 4,6-Di-N-oxide (FTDO). *Chin. J. Org. Chem.* **2012**, *32*, 1975–1980. [CrossRef]

40. Voronin, A.A.; Zelenov, V.P.; Churakov, A.M.; Strelenko, Y.A.; Tartakovsky, V.A. Alkylation of 1-hydroxy-1H-[1,2,3]triazolo[4,5-e][1,2,3,4]tetrazine 5,7-dioxide. *Russ. Chem. Bull.* **2014**, *63*, 475–479. [CrossRef]

41. Voronin, A.A.; Zelenov, V.P.; Churakov, A.M.; Strelenko, Y.A.; Fedyanin, I.V.; Tartakovsky, V.A. Synthesis of 1,2,3,4-tetrazine 1,3-dioxides annulated with 1,2,3-triazoles and 1,2,3-triazole 1-oxides. *Tetrahedron* **2014**, *70*, 3018–3022. [CrossRef]

42. Wang, T.Y.; Zheng, C.M.; Liu, Y.; Gong, X.D.; Xia, M.Z. Theoretical studies of the structure, stability, and detonation properties of vicinal-tetrazine 1,3-dioxide annulated with a five-membered heterocycle. 1. Annulation with a triazole ring. *J. Mol. Model.* **2015**, *21*, 1–9.

43. Shvets, A.O.; Konnov, A.A.; Klenov, M.S.; Churakov, A.M.; Strelenko, Y.A.; Tartakovsky, V.A. Synthesis of 2-(6-nitrobenzofuroxan-4-yl)-2H-[1,2,3]triazolo-[4,5-e][1,2,3,4]tetrazine 4,6-dioxide. *Russ. Chem. Bull.* **2020**, *69*, 739–741. [CrossRef]

44. Bian, C.M.; Dong, X.; Zhang, X.H.; Zhou, Z.M.; Zhang, M.; Li, C. The unique synthesis and energetic properties of a novel fused heterocycle: 7-nitro-4-oxo-4, 8-dihydro-[1,2,4]triazolo[5,1-d][1,2,3,5]tetrazine 2-oxide and its energetic salts. *J. Mater. Chem. A* **2015**, *3*, 3594–3601. [CrossRef]

45. Zhou, Z.M.; Bian, C.M. Energetic ionic salts of 7-nitryl-4-ketone-4,8-dihydro-[1,2,4]triazole[5,1-d][1,2,3,5]tetrazine-2-oxide and preparation method of energetic ionic salt. CN Patent 104447762, 25 March 2015. (In Chinese)

46. Zhao, B.J.; Wang, P.P.; Fu, W.; Li, C.; Zhou, Z.M. High Density of a New Fused Heterocycle: 7,8-Dinitro-4-oxo-4,6-dihydropyrazolo[5,1-d][1,2,3,5]tetrazine 2-oxide and Its Energetic Salts. *ChemistrySelect* **2018**, *3*, 4797–4803. [CrossRef]

47. Lei, C.J.; Cheng, G.B.; Yi, Z.X.; Zhang, Q.H.; Yang, H.W. A facile strategy for synthesizing promising pyrazole-fused energetic compounds. *Chem. Eng. J.* **2021**, *416*, 129190. [CrossRef]

48. Tang, Y.X.; Imler, G.H.; Parrish, D.A.; Shreeve, J.M. Energetic and fluorescent azole-fused 4-amino-1,2,3-triazine-3-N-oxides. *ACS Appl. Energy Mater.* **2019**, *2*, 8871–8877. [CrossRef]

49. Deng, M.C.; Feng, Y.G.; Zhang, W.Q.; Qi, X.J.; Zhang, Q.H. A green metal-free fused-ring initiating substance. *Nat. Commun.* **2019**, *10*, 1339. [CrossRef]

50. Deng, M.C.; Feng, Y.A.; Zhang, Q.H. Green environmental protection type primary explosive and preparation method thereof. CN Patent 108752349, 11 June 2018. (In Chinese)

51. Ogurtsov, V.A.; Dorovatovskii, P.V.; Zubavichus, Y.V.; Khrustalev, V.N.; Fakhrutdinov, A.N.; Zlotin, S.G.; Rakitin, O.A. [1,2,5]Oxadiazolo[3,4-d] pyridazine 1,5,6-trioxides: Efficient synthesis via the reaction of 3,4-bis (hydroxyimino) methyl)-1,2,5-oxadiazole 2-oxides with a mixture of concentrated nitric and trifluoroacetic acids and structural characterization. *Tetrahedron Lett.* **2018**, *59*, 3143–3146. [CrossRef]

52. Liu, Y.G.; Huang, Z.; Yu, X.J. Progress of research of new insensitive energetic material LLM-105. *Expl. Shock Wave* **2004**, *24*, 465–469. (In Chinese)

53. Jing, S.M.; Liu, Y.C.; Yuan, J.M. New Synthesis of LLM-105 and Experimental Study of Replacement Process. *Initiat. Pyrotech.* **2012**, *2*, 34–36. (In Chinese)

54. Zhao, X.F.; Liu, Z.L.; Yao, Q.Z.; Chen, J.; Dong, Y. Synthesis of 2,6-Diamino-3,5-dinitropyrazine-1-oxide. *Chin. J. Explos. Propellants* **2012**, *35*, 15–17. (In Chinese) [CrossRef]

55. Wang, J.M.; Du, Y. Optimization of Synthetic Method of 2,6-Diamino-pyrazine-1-oxide. *Chin. J. Explos. Propellants* **2017**, *40*, 25–27. (In Chinese)

56. Li, Y.H.; Wang, Y.; Liao, L.Y.; Guo, Z.C.; Chen, L.P.; Wu, W.Q. Thermal Risk Analysis Based on Reaction Mechanism: Application to the 2,6-Diaminopyrazine-1-oxide Synthesis Process. *Org. Process Res. Dev.* **2021**, *25*, 849–857. [CrossRef]

57. Luo, Y.F.; Ma, L.; Wang, B.Z.; Zhou, Y.S.; Huo, H.; Jia, S.Y. Synthesis and Characterization of 3,3'-Dicyano-4,4'-azofuroxan. *Chin. J. Energ. Mater.* **2010**, *18*, 538–540. (In Chinese)

58. He, J.X.; Lu, Y.H.; Lei, Q.; Cao, Y.L. Synthesis and Properties of High Energetic Compound 3,3'-Dinitro-4,4'-azofuroxan. *Chin. J. Energ. Mater.* **2011**, *34*, 9–12. (In Chinese)

59. Ma, C.M.; Pan, Y.; Jiang, J.C.; Liu, Z.L.; Yao, Q.Z. Synthesis and thermal behavior of a fused, tricyclic pyridine-based energetic material: 4-amino-5-nitro-[1,2,5]oxadiazolo[3,4-e]tetrazolo[1,5-a]pyridine-3-oxide. *New J. Chem.* **2018**, *42*, 11259–11263. [CrossRef]

60. He, C.L.; Gao, H.X.; Imler, G.H.; Parrish, D.A.; Shreeve, J.M. Boosting energetic performance by trimerizing furoxan. *J. Mater. Chem. A* **2018**, *6*, 9391–9396. [CrossRef]

61. Sun, K.L.; Wu, N.; Yang, H.; Yang, X.F.; Li, H. Independent Component Analysis Combined with On-line Infrared Spectroscopy for Researching the Synthesis Reaction Mechanism of 3,4'-Bis(4'-aminofurazano-3')furoxan. *Chem. J. Chin. Univ.* **2014**, *35*, 244–249. (In Chinese)

62. Li, Y.; Huang, H.F.; Shi, Y.M.; Yang, J.; Pan, R.M.; Lin, X.Y. Potassium nitraminofurazan derivatives: Potential green primary explosives with high energy and comparable low friction sensitivities. *Chem. Eur. J.* **2017**, *23*, 7353–7360. [CrossRef]

63. Tang, Y.X.; He, C.L.; Shreeve, J.M. A furazan-fused pyrazole N-oxide via unusual cyclization. *J. Mater. Chem. A* **2017**, *5*, 4314–4319. [CrossRef]

64. Thottempudi, V.; Yin, P.; Zhang, J.H.; Parrish, D.A.; Shreeve, J.M. 1,2,3-Triazolo [4,5,-e]furazano[3,4-b]pyrazine 6-Oxide—A Fused Heterocycle with a Roving Hydrogen Forms a New Class of Insensitive Energetic Materials. *Chem. Eur. J.* **2014**, *20*, 542–548. [CrossRef]

65. Li, Y.N.; Hu, J.J.; Chen, T.; Wang, B.; Chang, P.; Wang, B.Z. Synthesis and Properties of 4-Amino-1,2,3-triazolo[4,5-e]furazano[3,4-b]pyrazine 6-oxide. *Chin. J. Energ. Mater.* **2020**, *28*, 664–669. (In Chinese)

Article

Ignition Growth Characteristics of JEOL Explosive during Cook-Off Tests

Xinyu Wang [1], Chunlan Jiang [1], Zaicheng Wang [1,*], Wenxing Lei [2] and Yuande Fang [1]

1   State Key Laboratory of Explosion Science and Technology, Beijing Institute of Technology, Beijing 100081, China
2   Shanxi Jiangyang Chemical Co., Ltd., Taiyuan 030041, China
*   Correspondence: wangskyshark@bit.edu.cn

Abstract: In order to study the reaction growth process of insensitive JEOL explosive after ignition under cook-off, a series of cook-off tests were carried out on JEOL explosive using a self-designed small cook-off bomb system. A thermocouple was used to measure the internal temperature of the explosive, and a camera recorded macro images of the cook-off process. The temperature change law of JEOL explosive before and after ignition under different heating rates and the smoke ejection caused by the reaction in the slit were studied. The research results showed that the ignition time decreased as the heating rate increased, while the ignition temperature was not sensitive to the heating rate. When the heating rate was faster, the internal temperature gradient of the explosive was larger, and the ignition point appeared at the highest temperature position. As the heating rate decreased, the internal temperature gradient of the explosive decreased, the ignition point appeared random, and multiple ignition points appeared at the same time. The growth process of the ignition point could be divided into severe thermal decomposition, slow combustion, and violent combustion stages. When the heating rate reduced from 7 to 1 °C/min, the burning rate obviously increased.

Keywords: insensitive munition; JEOL; cook-off; ignition characteristic

Citation: Wang, X.; Jiang, C.; Wang, Z.; Lei, W.; Fang, Y. Ignition Growth Characteristics of JEOL Explosive during Cook-Off Tests. *Crystals* **2022**, *12*, 1375. https://doi.org/10.3390/cryst12101375

Academic Editors: Rui Liu, Yushi Wen and Weiqiang Pang

Received: 28 August 2022
Accepted: 24 September 2022
Published: 28 September 2022

Publisher's Note: MDPI stays neutral with regard to jurisdictional claims in published maps and institutional affiliations.

1. Introduction

Ammunition will go through stages of transportation, storage, maintenance, and launch during its entire life cycle and may experience unexpected stimulation at all stages, such as vibration during transportation; drop during hoisting and transportation; and impact caused by bullets or fragments, fire, heat, etc. These stimuli can cause burning, explosion, or even detonation of ammunition, resulting in loss of personnel and equipment and weakening combat effectiveness. Among the abovementioned types of stimulation, exposure to fire is one of the most important factors affecting the safety of ammunition [1,2]. When ammunition is exposed to fire, two situations are possible: the ammunition is completely immersed in flames or a fire occurs in the nearby environment and high-temperature air heats the ammunition convectively. In the first case, the heating rate of the projectile is extremely fast, and the ammunition reacts very rapidly. In the second case, the heating rate of the projectile is relatively slow. After a long period of heat accumulation, the ammunition reacts more violently than when it heats up quickly. Therefore, studying the reaction mechanism and structural response of ammunition under different heating and thermal stimuli has important guiding significance for the safety design, manufacture, and use of ammunition.

The response of explosives is completely different under the conditions of fast cook-off and slow cook-off. Under fast cook-off, only the outer wall reaches the reaction temperature when the explosive reacts, so the reaction intensity is weaker than slow cook-off [3–5]. Therefore, the response of explosives under slow cook-off is a key concern for researchers. HMX-based plastic-bonded explosives have been discussed by many researchers. PBX9501 (95% HMX, 5% binder system) is a high-density pressed explosive [6,7]. Under cook-off, this type of explosive generally does not experience violent reactions above combustion. However, if the explosive is under strong restraint conditions, this kind of explosive may experience deflagration-to-detonation transition (DDT) [8–16]. Similarly, other HMX-based plastic-bonded explosives, such as LX-14 (95.5 wt% HMX, 4.5 wt% Estane 5702), LX-10 (94.5 wt% HMX, 5.5 wt% Viton A), and PBX9012 (90 wt% HMX, 10 wt% Viton A), are likely to go through DDT [17,18]. Because HMX will undergo a phase change from β to δ at 160–180 °C, the internal temperature of the PBX explosive will have a plateau during the firing cook-off [19–21]. With increase in insensitive components (such as TATB) in the PBX, the ignition temperature of the explosive rises, and the explosive reaction severity decreases. When reaching a certain temperature, the melt cast explosive will melt, so the response of this type of explosive under slow cook-off is also different from that of pressed explosives. Comp-B (63 wt% RDX, 36 wt% TNT, 1 wt% wax) is a widely used melt cast explosive [22,23]. Under cook-off, as the temperature rises, the inside of Comp-B begins to melt, which will affect the temperature distribution inside the explosive, and the internal temperature of the internal explosive measured by the thermocouple will also have a plateau. Due to the sedimentation of the solid, the ignition position of the melt-cast explosive will also appear in the lower part of the explosive. The thermal safety of solid propellants has also been extensively studied [24–30]. With different energetic subjects, the propellant will have completely different responses at the same heating rate [31]. When the heating rate is 1 K/min, the ignition time of insensitive propellant (AP/HTPE/Al (70/12/18)) is 87% longer than that of traditional propellant (AP/HTPB/Al (70/12/18)).

Regarding the standard cook-off bomb structure with a cylindrical charge and a shell, researcher have previously obtained data including the internal temperature distribution of the explosive, the ignition time, the ignition temperature, the shell flying speed, and the degree of violent reaction through experiments [10–19]. However, there is little information on the location of the ignition point and subsequent growth. The ignition position characterizes the combustion reaction spreading from there to the surrounding area, and it grows into a more violent reaction as the internal pressure of the ammunition increases. Therefore, the ignition position and the growth of the ignition point determine the final degree of the combustion reaction to a large extent. Determining the location of the ignition point and the growth mechanism and process of the ignition point is of great help to understand the combustion process of explosives after ignition under cook off and can improve the thermal safety of explosives and the viability of ammunition. It is of great significance to study the basic properties of energetic materials and the safe production of ammunition.

JEOL explosive (the composition and mechanical sensitivity as well as detonation parameters are listed in Table 1) is a new type of high-energy insensitive explosive. Although active metal aluminum is added to the explosive, the mechanical sensitivity of the explosive is at a very low level due to the stability of NTO. It has the advantages of high density, detonation heat, and detonation velocity as well as good thermal stability. Its main component, 3-nitro-1,2,4-triazol-5-one (NTO), is a triazolone type of nitrogen heterocyclic compound. Its theoretical density is as high as 1.93 g/cm^3, and its detonation performance is close to that of RDX. It has been used in the formulation of insensitive ammunition of multinational armies. Research on the thermal safety of JEOL explosive under cook-off conditions has important guiding significance for the design of insensitive ammunition. However, the ignition characteristics of JEOL explosive in cook-off are rarely reported, and the reaction growth process and mechanism after ignition are still unclear. Therefore, in this study, a small cook-off bomb was used to conduct an experimental study on JEOL explosive. According to the temperature jump time, jump amplitude, and temperature change law at different positions, the ignition position of JEOL explosive under cook off, the rule of ignition point position, ignition characteristics, and subsequent growth process were obtained.

Table 1. Composition/mechanical sensitivity/detonation parameters of JEOL.

Composition			
32 wt% 1,3,5,7-tetranitro-1,3,5,7-tetrazocane (HMX)	32 wt% 5-nitro-1,2-dihydro-3*H*-1,2,4-triazin-3-one (NTO)	28 wt% Al	8 wt% binder system (hydrocarbon polymer as binder, dinitro compounds as plasticizer, wax as insensitive agent)
Shock sensitivity [1]	Friction sensitivity [2]	Heat of detonation (kJ/kg)	Velocity of detonation(m/s)
18%	15%	6790	7821

[1] Drop weight: 10.00 ± 0.01 kg; drop high: 250 ± 1 mm; dosage: 50 ± 1 mg. [2] Swing angle: 90 ± 10°; dosage: 50 ± 1 mg.

2. Materials and Methods

In this paper, JEOL explosive was prepared by the solution water suspension process. The specific process was as follows. A certain amount of NTO, HMX, and Al was weighed according to the formula ratio and placed into a glass beaker. Saturated absolute ethanol solution of NTO was added, a few drops of dispersant were dipped into the beaker with a stirring bar, and a suspension was obtained. The suspension was fully stirred for even dispersion. The beaker was placed in the thermostatic water bath, and heating was commenced. The temperature was raised to 50 °C, and the binder solution was added slowly at a certain stirring speed. After the addition, the temperature was kept constant for 20 min. The temperature of the incubator was adjusted to 60 °C to disperse the solvent. After the solvent was dispersed, the suspension was cooled to 2 °C and filtered, washed, and dried to obtain the explosive molding powder. The obtained molding powder of JEOL explosive is shown in Figure 1. The molding powder was press packed, and the grain is shown in the figure.

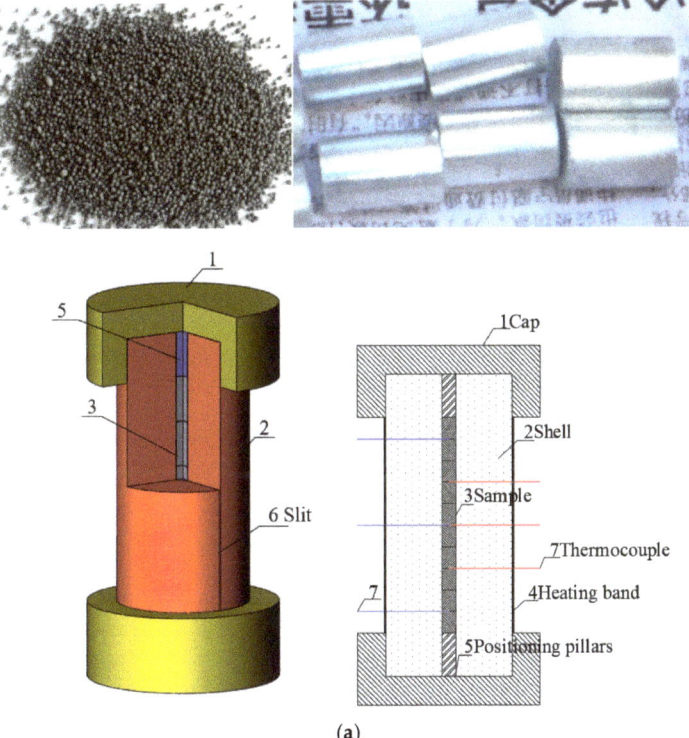

(a)

Figure 1. *Cont.*

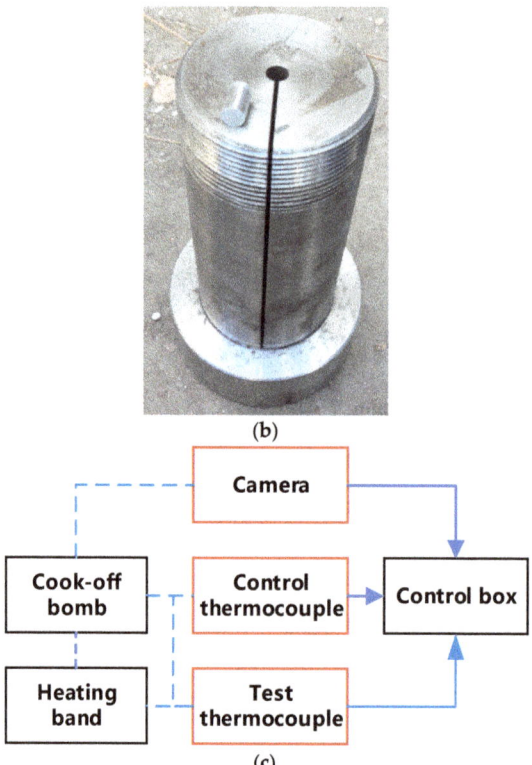

(b)

Camera

Cook-off
bomb

Control
thermocouple

Control box

Heating
band

Test
thermocouple

(c)

Figure 1. Cook-off experimental system. (**a**) Cross-section of cook-off bomb. (**b**) Sample and the shell of cook-off bomb. (**c**) Test flow chart.

There are two phenomena of sublimation and thermal decomposition of NTO, which is the main component of JEOL, in the process of heating up. Sublimation occurs first at low pressure and low temperature. At this time, the activation energy of explosives is small. Research has shown that the thermal decomposition of NTO occurs in the range of 200–260 °C. At this time, the explosive has a high activation energy under the combined action of sublimation and thermal decomposition, so NTO has high thermal safety. According to the adiabatic decomposition process of JEOL explosive, the explosive also has two exothermic stages, which indicates that its thermal safety is affected by the NTO composition.

The small cook-off bomb used in the present study is presented in Figure 1. The length of the shell was 210 mm, the inner diameter was 10 mm, and the outer diameter was 90 mm. In order to obtain temperature data after igniting the explosive, there were 2 mm slits in the axial direction to ensure that the cook-off bomb does not have a more violent reaction than combustion, which would destroy the thermocouple. The explosive charge was 10 mm in diameter and 15 mm in length. The sample used in each test consisted of 10 such charges stacked on top of each other for the total charge length of 150 mm. The steel positioning pillars with length of 30 mm were placed on the upper and lower sides so that the whole explosive was placed in the middle of the slit. When the explosive starts to react, the slit ensures that the gas produced in the cook-off bomb can be discharged in time and the internal pressure of the device will not be too high. When the cook-off bomb is in a completely airtight and strongly constrained state, the explosive is very likely to experience DDT, and it is impossible to obtain temperature data after the explosive reaction. The outer wall of the shell was wrapped with a heating band, and the asbestos was wrapped around the cook-off bomb for heat preservation. A thermocouple was placed between the samples, and the temperature of the explosive was recorded over time. The sampling frequency was 5 Hz. In order to eliminate the influence of the position of the thermocouple on the reaction, two thermocouple arrangements were used. The first was placing a thermocouple in the center of the charge and placing the other two thermocouples at a distance of 45 mm from the top and bottom of the charge, as shown by the red horizontal line in Figure 1a. The second was placing a thermocouple in the center of the charge and placing the other two thermocouples 15 mm away from the top and bottom of the charge, as shown by the blue horizontal line in Figure 1a. When the cook-off bomb was placed vertically, the thermocouple near the base was the lower thermocouple, and the middle and upper thermocouples were from the bottom to the top. A thermocouple was also placed between the heating band and the shell to record the temperature change of the shell. When the explosive reacted, the ignition sequence was judged according to the sudden jump in temperature, and the image was recorded by a camera. The effect of different heating rates on ignition position and ignition point growth was studied, where shot 1 was rapid heating and shots 2 and 3 were slow heating. The specific shots are shown in Table 2.

Table 2. Test conditions.

Shot	Heating Rate (°C/min)	Position of Thermocouple
1	7	Red
2	1	Red
3	1 [1]	Blue

[1] Shot 3: heated up to 50 °C at 3 °C/min, kept for 20 min, then heated up at 1 °C/min until the reaction.

3. Results and Discussion

3.1. Experimental Phenomenon

After heating for a certain period of time, strong white smoke was sprayed from the slit in all three shots, as shown in Figure 2. There was no explosion sound, indicating that the explosive did not react violently. After the reaction, the cap and the shell remained intact, as shown in Figure 3. It was judged that the reaction degree of the explosives was combustion of all three shots. The temperature–time curves of the three shots obtained in the experiment are shown in Figure 4. It can be seen from Figure 4 that, except for the warm-up stage and temperature jumps, the overall temperature curve showed a linear upward trend. When there was a sudden jump in temperature as measured by the thermocouple, it was considered that the explosive had ignited there. The test completely recorded the temperature data of the combustion stage after ignition.

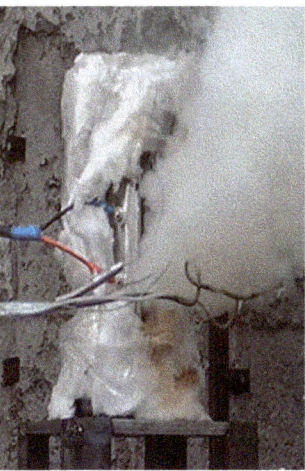

Figure 2. Jetting smoke.

Figure 3. Cook-off bomb after tests.

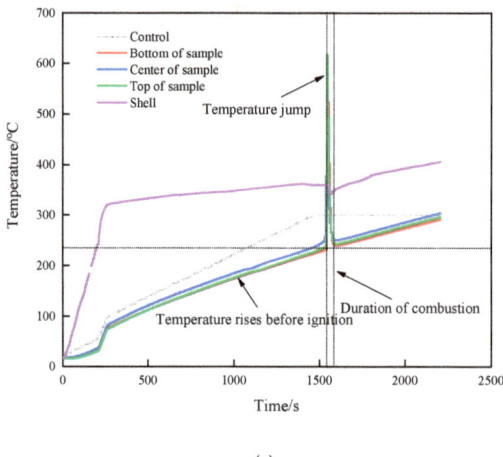

(a)

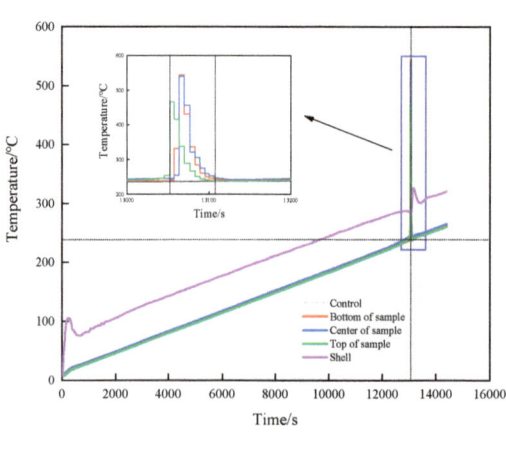

(b)

Figure 4. *Cont.*

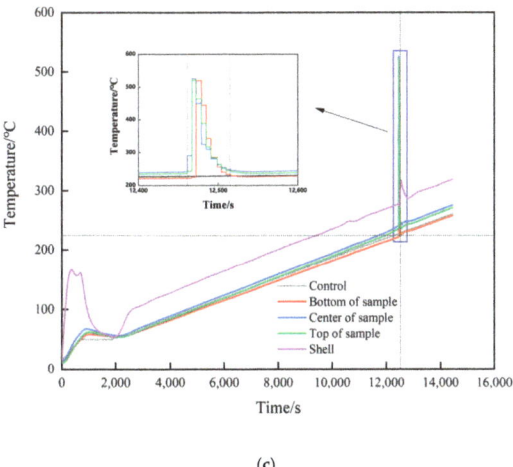

(c)

Figure 4. Temperature–time curve. (**a**) 7 °C/min. (**b**) 1 °C/min (no temperature preservation). (**c**) 1 °C/min (temperature preservation for 20 min).

3.2. Analysis of Ignition Time and Ignition Temperature

The changes in internal temperature of the explosive with time is shown in Figure 4. The start of heating was recorded as zero time, and the ignition time was recorded when there was a significant jump in temperature as measured by the thermocouple. The rising trend of the temperature curve under the three shots was similar, but with the change in the heating rate, the ignition time was obviously different. Shot 1 heated up rapidly at a rate of 7 °C/min, and the ignition time was 1531 s. The ignition time was 13,046 s when the heating rate was 1 °C/min, and the ignition time was 12,461 s when the heat was preserved for 20 min. The difference between the latter two shots was 585 s, which only accounted for 5% of the entire ignition time. The ignition time of the two could be considered as similar. The increase in the heating rate meant the heat transferred from the outside to the inside of the explosive per unit time increased, and the heating rate of the explosive itself increased accordingly. The self-heating reaction of the explosive followed the Arrhenius law, which led to an increase in the thermal decomposition rate of the explosive. With the combined action of heat and external heat, a local hot spot was formed inside the explosive and ignition occurred. Compared to the experiment of Hobbs et al., the experimental device used in this study was in an open state. Because the thermal decomposition reaction of explosives is related to environmental pressure, the higher the environmental pressure, the higher the thermal decomposition rate of explosives. Therefore, the ignition time of the explosive obtained in this study was relatively increased compared to the cook-off test under closed conditions.

When ignition occurred in shot 1, the ignition temperature was 249.4 °C. As the heating rate decreased, the ignition temperatures of shots 2 and 3 were 245.9 and 236.5 °C, respectively. The ignition temperature of the three shots witnessed little change, the heating rate had no obvious influence on the ignition temperature, and the ignition temperature change caused by the heating rate change was almost negligible. The ignition temperature under heat preservation condition was 9.4 °C lower than that without heat preservation, which was 3.9% of the total temperature rise of the explosive before ignition. The ignition temperature of shots 2 and 3 could be considered as similar. Analysis shows that because the diameter of the explosive used in this study was small, under the condition of low heating rate, the internal temperature of the explosive will be uniform without the insulation stage. Therefore, the temperature field inside the explosive in shots 2 and 3 is the same from the macro perspective. The ignition temperature of a certain energetic material at different heating rates can be roughly estimated based on the ignition temperature at a known heating rate. Analysis shows that whether the explosive ignites or not is determined by the balance between the heat generated by the explosive itself and the heat lost to the outside world. Under the test conditions in this study, the explosive structure was the same, the external environment temperature was the same, and the heat loss could be considered to be the same. Therefore, the ignition of explosives was determined by the heat production. The heat production of explosives is determined by the Arrhenius' law. In this study, the physicochemical parameters of JEOL explosive did not change significantly with the increase in temperature, so whether the explosives ignited could be determined by the temperature of the explosives. Therefore, the ignition temperature of JEOL explosive was the same at different heating rates. The relationship between ignition time, ignition temperature, and heating rate is shown in Figure 5a.

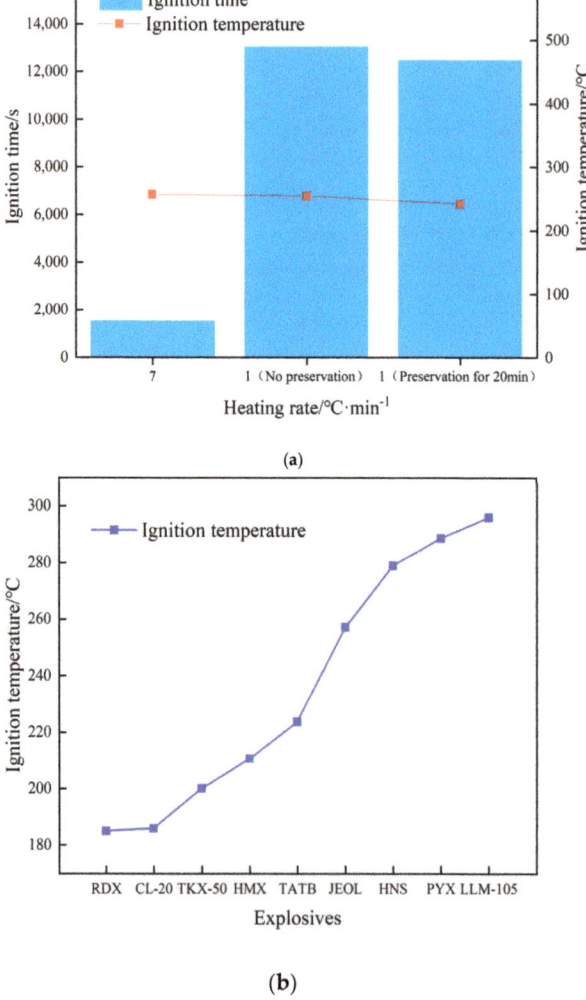

Figure 5. Ignition time/ignition temperature–heating rate curve. (**a**) Ignition time/ignition temperature–heating rate curve. (**b**) Ignition temperature of different explosives.

Figure 5b shows the comparison between the ignition temperatures of JEOL and eight other commonly used explosives [12,13,15,18,19]. It can be seen from the figure that the ignition temperature of JEOL is lower than that of heat-resistant explosives HNS, PYX, and LLM-105 but higher than that of high-energy explosives, such as RDX, CL-20, and HMX, indicating that JEOL has high thermal safety.

3.3. Analysis of Ignition Growth and Combustion Process

During the cook-off process, when the explosive did not ignite, the temperature measured by the thermocouple increased linearly. After ignition, the temperature at the ignition point suddenly jumped. Through the temperature data measured by the thermocouple, the ignition point position, the temperature rise of the ignition point, and the combustion duration was obtained. The macro image of the cook-off bomb taken by the camera was used to analyze the state of the explosive in the combustion stage.

3.3.1. Ignition Point Position

According to the temperature jump, the specific location of the ignition point was obtained. The temperature rise at the ignition point is shown in Table 3. The distance between the thermocouples in shot 1 was 30 mm; the center temperature measured before ignition was 249.4 °C; the upper and lower thermocouple temperatures were 230.2 and 236.4 °C, respectively; and the temperature gradients were 19.2 and 13 °C, respectively. The temperature of the center was higher than the two ends because the temperature of the external environment in the test was only 10 °C, and the cook-off bomb was placed on a metal base. Although the thermal insulation asbestos was wrapped, the upper end cover, lower end cover, and metal shelf of the cook-off bomb had the effect of increasing the heat dissipation area, causing the temperature at both ends to be lower than the center. At the next acquisition time, 0.2 s later, the temperature measured by the middle thermocouple was 257.2 °C; the upper and lower thermocouple temperatures were 230.9 and 237.6 °C, respectively; and the temperature increments were 7.8, 0.7, and 0.2 °C, respectively. The increase in the middle was much larger than the changes in the upper and lower positions and the temperature increase preset by the program, so it was judged that the initial ignition position was at the center of the charge axis.

Table 3. Response characteristics of cook-off.

Position	Shot 1			Shot 2			Shot 3		
	Temperature before Ignition/°C	Initial Ignition Temperature/°C	Temperature Variation/°C	Temperature before Ignition/°C	Initial Ignition Temperature/°C	Temperature Variation/°C	Temperature before Ignition/°C	Initial Ignition Temperature/°C	Temperature Variation/°C
Top	236.4	237.6	1.2	245.9	254.6	8.7	236.5	243.2	6.7
Center	249.4	257.2	7.8	245.9	246.1	0.2	242.2	291.2	49
Bottom	230.2	230.9	0.7	238.6	238.7	0.1	221.8	222	0.2

As the heating rate decreased, the temperature distribution inside the explosive tended to be uniform. For shot 2, the temperatures measured from the top to bottom thermocouples before ignition were 245.9, 245.9, and 238.6 °C. As can be seen, the upper part and the middle part of the explosive had the same temperature. However, due to the heat dissipation effect of the lower end cover and the base, the temperature of the lower part was significantly lower than that of the middle and upper part. The maximum temperature difference between the thermocouples was only 7.3 °C, which was much smaller than the shot of 7 K/min. This conformed to the law that the slower the heating rate, the more uniform the temperature distribution inside the explosive. After ignition, the upper temperature increment was 8.7 °C. According to the temperature change, it was judged that the initial ignition position was in the upper part.

In shot 3, the position of the thermocouples was changed, and the distance between the thermocouples was 45 mm. The temperatures before ignition measured from top to bottom were 236.5, 242.2, and 221.8 °C, respectively. As can be seen, even though the temperature was kept for 20 min, due to the heat dissipation effect of the upper and lower end caps, the farther away from the center of the explosive, the lower was the temperature. At this time, the temperature increments measured at the upper, middle, and lower positions were 6.7, 49, and 0.2 °C, respectively. Both the upper part and the center had a large temperature rise exceeding the preset heating rate, and it was judged that the upper part and the center had ignition points at the same time. Combining the cook-off test of the three shots, it was judged that the initial position of ignition occurred at the position with the highest temperature inside the explosive. When there is a significant temperature gradient, the ignition point only appears in a single position, and there is no situation where multiple ignition points appear at the same time. Analysis shows that when the temperature gradient in the explosive tends to be the same, the whole explosive will approach the critical state with the increase in temperature. The structure of the press-packed explosive itself is uneven, the mixture of the explosive and the additive is not uniform when making the molding powder, and the air gap may be formed between particles during press-packed molding. At the same time, the state of the explosive will also change during the cook-off process, and the explosive volume will expand. At 180 °C, the HMX grain will change phase, and a large number of microcracks will be generated in JEOL explosive. Debonding between particles will also occur. Due to the heterogeneity of the explosive structure in the micro and macro aspects, the conduction of heat in the explosive will be uneven, and the accumulation of heat will also be uneven. Therefore, the occurrence of the ignition position will be random according to the explosive structure. When the temperature gradient is small and tends to be uniform, the position of the ignition point has a certain randomness and multiple initial ignition points may appear at the same time.

3.3.2. Ignition Point Growth Process

The temperature changes of various parts of shot 1 after ignition are shown in Figure 6. The central thermocouple held for 6 s after the temperature jumped to 257.2 °C. Then, the temperature suddenly changed to 378.2 °C, and the change was as high as 121 °C. The temperature lasted for 5.8 s, and the temperatures at the upper, middle, and lower locations suddenly changed to 616.4, 476.4, and 445 °C at the same time. Combined with the camera recording (see Figure 7), after the initial ignition point appeared, a very small amount of intermittent white smoke was injected into the slit, which was almost undetectable. When the temperature of the ignition point changed again, a small amount of continuous white smoke was emitted from the slit. When the temperature of the thermocouple underwent a sudden jump at the same time, a large amount of continuous white smoke was ejected from the slit. Combining temperature and image data, it was judged that the initial ignition point had a small sudden temperature jump, which was the violent thermal decomposition process of the explosive. After that, the ignition point continued to grow, and the secondary temperature had a large sudden jump, which was the slow combustion process of the explosive. The third temperature jump was a further increase in the flame. Analysis shows that the main reason for this phenomenon is that NTO, the main component of JEOL explosive, has two exothermic stages under adiabatic conditions. Among them, the first, exothermic stage occurs at 194.9 °C and the second, exothermic stage occurs at 220 °C. The heat release and heat release rate in the first stage are very small, and the second stage is the main stage of NTO heat release. The thermal decomposition temperature of HMX, another component of JEOL explosive, is 176 °C, indicating that adding NTO to HMX will greatly improve the thermal safety of HMX itself. We divided the ignition process of explosives into three stages: rapid thermal decomposition, slow combustion, and violent combustion.

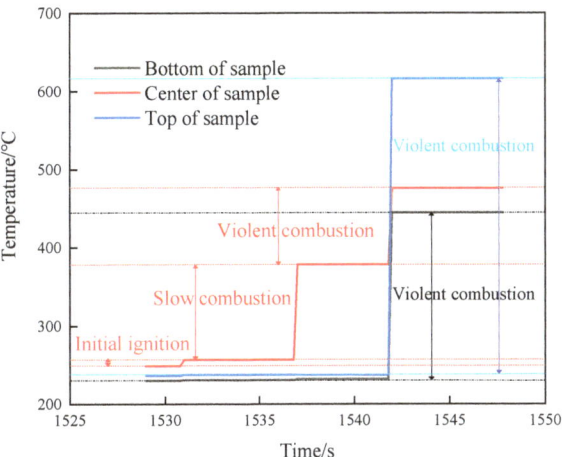

Figure 6. Temperature–time curve of shot 1.

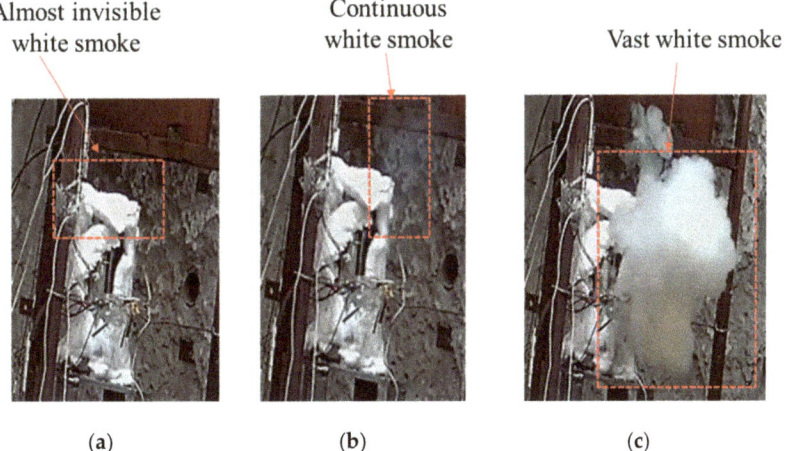

Figure 7. Response of cook-off bomb of shot 1. (**a**) Initial ignition. (**b**) Secondary jump. (**c**) Third jump.

The temperature changes of all parts of shot 2 after ignition are shown in Figure 8. The initial ignition point appeared in the upper part and lasted for 7 s, and then the temperature jumped from 211.7 to 466.3 °C, which lasted for 5s. After that, the temperature in the center and the lower part jumped suddenly, and the temperature increments were 8.8 and 91 °C, respectively. This temperature distribution lasted for 6.2 s, and the central and lower temperatures rose to 540.2 and 543.9 °C, respectively. Combined with the camera image (see Figure 9), after the initial ignition point appeared, the camera did not capture white smoke. When the upper temperature jumped for the second time, it was obvious that white smoke was ejected from the upper half of the slit, and there was no white smoke generated in the lower half. When the temperature of the middle and lower parts suddenly jumped, a large amount of white smoke was sprayed from the middle and lower parts of the slit. Combining the temperature change of this shot, it was found that the internal temperature of the explosive tended to be uniform under the condition of low heating rate, but due to the appearance of the initial ignition point, the explosive entered the first exothermic stage. Even with a temperature rise of only 8.7 °C, the heat released by the explosive was enough to destroy the heat balance in the explosive system. The appearance of the initial ignition point resulted in a temperature rise of more than 200 °C afterwards. However, the area with the same temperature as before did not have the effect of the initial ignition point. Here, the explosive did not enter the first exothermic stage, and the linear heating rate was maintained at the preset rate until the ignition point appeared again in the middle and lower parts.

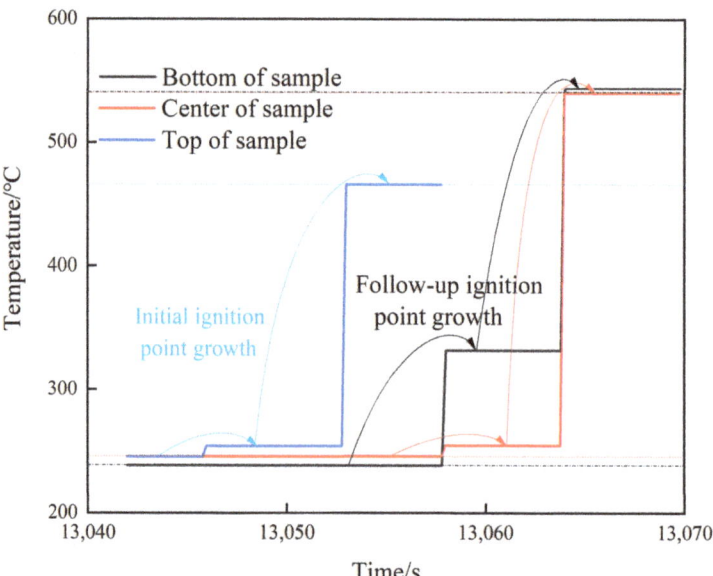

Figure 8. Temperature–time curve of shot 2.

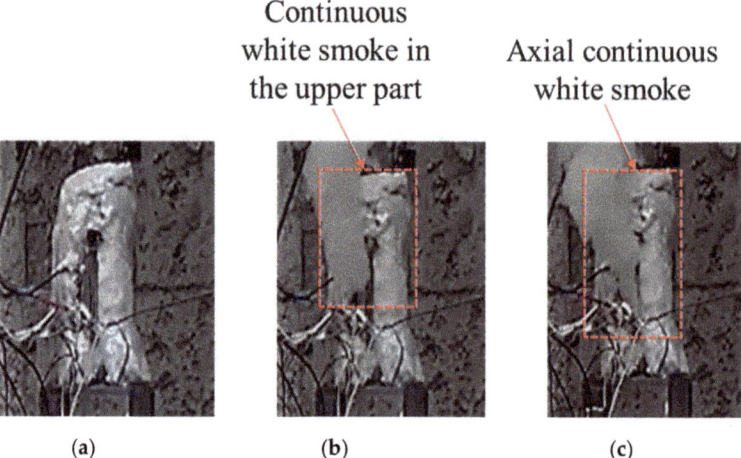

Figure 9. Response of cook-off bomb of shot 2. (**a**) Initial ignition. (**b**) Secondary jump. (**c**) Third jump.

The temperature changes at various locations in shot 3 after ignition are shown in Figure 10. The initial ignition position appeared at the center and upper part at the same time. After 6.2 s, further temperature jumps occurred at both locations with temperature rises of 234.4 and 277.8 °C, respectively. A temperature rise of 294.4 °C appeared in the lower part after 4.6 s. Combined with the camera image (see Figure 11), first, a small amount of intermittent white smoke was sprayed from the upper part of the slit. Then, a large amount of continuous white smoke was obviously sprayed from top to bottom, and the outer flame of the lower part lagged behind the upper middle part. Finally, the spraying was completed. Through shots 2 and 3, it can be seen that, compared to rapid heating, under the condition of slow heating, after the initial ignition point appears, it will directly grow into a violent combustion stage. There is no slow combustion stage, the growth is faster, and the temperature rise rate is higher. Analysis shows that when the heating rate is low, the temperature of JEOL explosive is relatively uniform and the internal energy is high. After the ignition of the explosive enters the first exothermic stage, the superposition of the released heat and the internal energy of the explosive makes the explosive quickly enter the second exothermic stage, and the explosive quickly releases a large amount of heat and enters the stage of intense combustion.

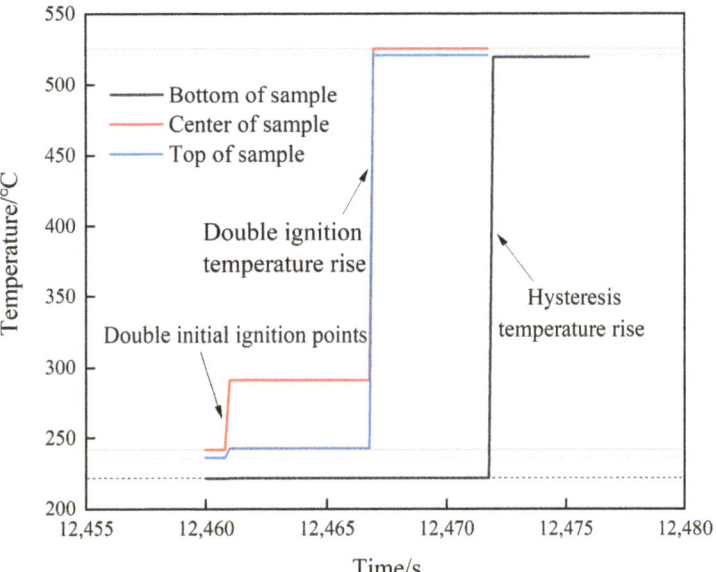

Figure 10. Temperature–time curve of shot 3.

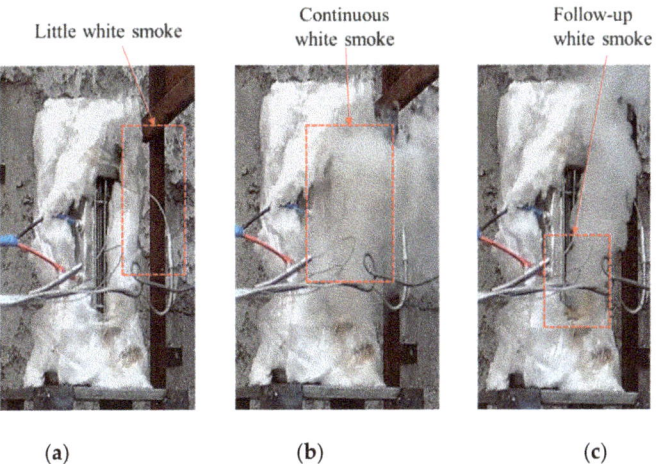

Figure 11. Response of cook-off bomb of shot 3. (**a**) Initial ignition. (**b**) Secondary jump. (**c**) Hysteretic jump.

3.3.3. Combustion Process

Combining the three shots, the temperature jump can be divided into two forms. One is a multistep jump that finally stabilizes at the combustion temperature. According to the ignition temperature and smoke phenomenon, this situation is judged to be the growth process of ignition. The other is a single-step jump that reaches the combustion temperature, as shown in Figure 12. The situation is the temperature rise caused by flame propagation. Based on this judgment, we further analyzed the temperature changes under the three shots.

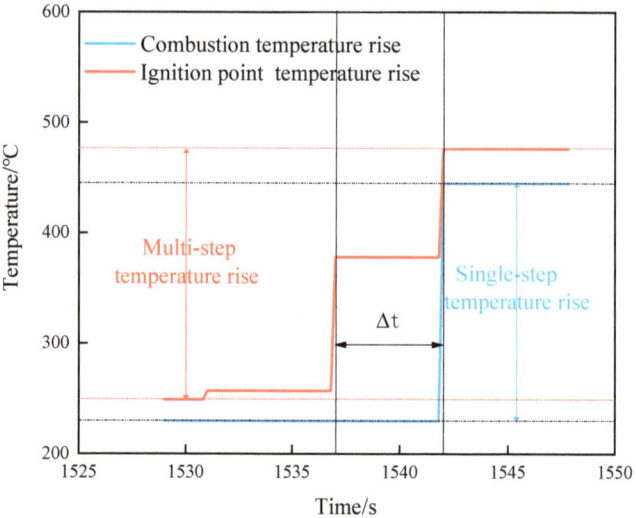

Figure 12. Temperature rise curve.

The only ignition point appeared in the center of operating shot 1, and then it steadily grew into combustion. The flame propagated from the center to the two ends along the axis, and the upper and lower parts jumped directly from the preignition temperature to the combustion temperature.

The ignition point first appeared in the upper part in shot 2, and then there were multistep temperature jumps in the center and lower part at the same time. This showed that the temperature change in the middle and lower parts was not caused by flame propagation. The ignition point reappeared when the internal temperature field was uniform.

In shot 3, the initial ignition point appeared at the upper and center at the same time and then grew into combustion together. The temperature in the lower part jumped directly to the combustion temperature in one step, indicating that the temperature rise at this place was caused by flame propagation in the center.

The flame propagation speed was calculated according to the temperature change time of shots 1 and 3 and the distance between the thermocouples. The distance between the thermocouples in shot 1 was 30 mm, the time difference between the two temperature jumps was 5 s, and the combustion speed was calculated to be 6mm/s. The distance between the thermocouple in shot 3 was 45 mm, the difference between the two temperature jumps was 4.8 s, and the combustion speed was calculated to be 9.4 mm/s. Low heating rate increases the burning rate by 57% compared to high heating rate. It is believed that thermal damage will occur inside the explosive during the baking process, including porosity increase, cracks, holes, and debonding between the matrix and the binder. When the heating rate is high, the ignition time of the explosive is short, the thermal damage in the explosive is less, and it is not easy for the flame to enter the crack of the explosive when the explosive is burning. The combustion mode is heat conduction combustion. When the heating rate is low, the ignition time of the explosive is long, and macro cracks are easily generated in the explosive (crack width: 200–500 μm). This causes the flame to easily enter the crack of the explosive under the pressure of the combustion product. At this time, the combustion area of the explosive increases rapidly, and the combustion mode changes from heat conduction combustion to convection combustion, resulting in a rapid increase in the combustion rate of the explosive [32,33].

4. Conclusions

In this study, a small cook-off bomb test system was used to study the ignition characteristics and subsequent reaction growth process of JEOL explosive under cook-off conditions. The main conclusions are as follows.

As the heating rate increases, the ignition time of JEOL explosive is significantly shortened, but the ignition temperature does not change significantly.

When the heating rate is high, the internal temperature gradient of JEOL explosive is large, and the ignition point only appears at the highest temperature. With the decrease in heating rate, the internal temperature distribution of the explosive tends to be uniform, the ignition point appears random, and there may be multiple ignition points at the same time.

The reaction growth process of JEOL explosive after ignition can be divided into three stages: violent thermal decomposition, slow combustion, and violent combustion. When the internal temperature of the explosive is uniform, there will be a further temperature jump at the position where the initial ignition point appears, and the other positions will continue to maintain linear heating. Under the condition of slow heating, after the ignition point appears, it is easier to directly grow into violent combustion.

After JEOL explosive is ignited, there are two forms of temperature jump inside the explosive. One is a multistep temperature rise, which characterizes the growth process of the explosive ignition point, while the other is a single step temperature rise, which characterizes the flame propagation process in the explosive. The burning rate at low heating rate is much greater than that at high heating rate.

Although this study took aluminum-containing pressed explosives as the discussion object, the final conclusion was derived from the explosive hot spot ignition mechanism, the Arrhenius chemical reaction mechanism, and the explosive combustion law. Therefore, the conclusion of this article should be applicable to other pressed and cast explosives.

Author Contributions: Conceptualization, X.W. and C.J.; methodology, X.W.; validation, X.W., C.J. and W.L.; formal analysis, Y.F.; investigation, W.L.; resources, C.J. and Z.W.; data curation, X.W.; writing—original draft preparation, X.W.; writing—review and editing, X.W.; visualization, C.J.; supervision, C.J.; project administration, C.J.; funding acquisition, Z.W. All authors have read and agreed to the published version of the manuscript.

Funding: This research received no external funding.

Institutional Review Board Statement: Not applicable.

Informed Consent Statement: Not applicable.

Conflicts of Interest: The authors declare no conflict of interest.

References

1. Li, G.; Zhao, J.; Du, C. *Safety Assessment and Technology of Conventional Missile Ammunition*; China Astronautic Publishing House: Beijing, China, 2015; pp. 1–12.
2. Baalisampang, T.; Abbassi, R.; Garaniya, V.; Khan, F.; Dadashzadeh, M. Review and Analysis of Fire and Explosion Accidents in Maritime Transportation. *Ocean Eng.* **2018**, *158*, 350–366. [CrossRef]
3. Gross, M.L.; Meredith, K.V.; Beckstead, M.W. Fast Cook-off Modeling of HMX. *Combust. Flame* **2015**, *162*, 3307–3315. [CrossRef]
4. Zhu, M.; Wang, S.; Huang, H.; Huang, G.; Wu, F.; Sun, S.; Li, B.; Xu, Z. Numerical and Experimental Study on the Response Characteristics of Warhead in the Fast Cook-off Process. *Defence Technol.* **2021**, *17*, 1444–1452. [CrossRef]
5. Wang, P.; Cheng, L.; Feng, C. Numerical simulation of cook-off for explosive at different heating rates. *Chin. J. Energetic Mater.* **2009**, *17*, 46–49.
6. Berghout, H.L.; Son, S.F.; Skidmore, C.B.; Idar, D.J.; Asay, B.W. Combustion of Damaged PBX 9501 Explosive. *Thermochim. Acta* **2002**, *384*, 261–277. [CrossRef]

7. Berghout, H.L.; Son, S.F.; Hill, L.G.; Asay, B.W. Flame Spread through Cracks of PBX 9501 (a Composite Octahydro-1,3,5,7-Tetranitro-1,3,5,7-Tetrazocine-Based Explosive). *J. Appl. Phys.* **2006**, *99*, 114901. [CrossRef]
8. Herrmann, M.; Engel, W.; Eisenreich, N. Thermal Expansion, Transitions, Sensitivities and Burning Rates of HMX. *Propellants Explos. Pyrotech.* **1992**, *17*, 190–195. [CrossRef]
9. Baer, M.R.; Gross, R.J.; Nunziato, J.W.; Igel, E.A. An Experimental and Theoretical Study of Deflagration-to-Detonation Transition (DDT) in the Granular Explosive, CP. *Combust. Flame* **1986**, *65*, 15–30. [CrossRef]
10. Griffiths, N.; Groocock, J.M. 814. The Burning to Detonation of Solid Explosives. *J. Chem. Soc.* **1960**, 4154–4162. [CrossRef]
11. Burnham, A.K.; Weese, R.K. Kinetics of Thermal Degradation of Explosive Binders Viton A, Estane, and Kel-F. *Thermochim. Acta* **2005**, *426*, 85–92. [CrossRef]
12. Garcia, F.; Vandersall, K.S.; Forbes, J.W.; Tarver, C.M.; Greenwood, D. Thermal Cook-Off Experiments of the HMX Based High Explosive LX-04 to Characterize Violence with Varying Confinement. *AIP Conf. Proc.* **2006**, *845*, 1061–1064.
13. Parker, G.R.; Heatwole, E.M.; Holmes, M.D.; Asay, B.W.; Dickson, P.M.; McAfee, J.M. Deflagration-to-Detonation Transition in Hot HMX and HMX-Based Polymer-Bonded Explosives. *Combust. Flame* **2020**, *215*, 295–308. [CrossRef]
14. Parker, G.R.; Dickerson, P.M.; Asay, B.W.; Mc Afee, J.M. *DDT of Hot, Thermally Damaged PBX 9501 in Heavy Confinement*; LA-UR-10-01356; LA-UR-10-1356; Los Alamos National Lab. (LANL): Los Alamos, NM, USA, 2010.
15. Price, D.; Bernecker, R.R. Effect of Initial Particle Size on the DDT of Pressed Solid Explosives. *Propellants Explos. Pyrotech.* **1981**, *6*, 5–10. [CrossRef]
16. Bernecker, R.R.; Price, D. Studies in the Transition from Deflagration to Detonation in Granular Explosives—II. Transitional Characteristics and Mechanisms Observed in 91/9 RDX/Wax. *Combust. Flame* **1974**, *22*, 119–129. [CrossRef]
17. Hsu, P.C.; DeHaven, M.; McClelland, M.; Maienschein, J.L. Thermal Damage on LX-04 Mock Material and Gas Permeability Assessment. *Propellants Explos. Pyrotech.* **2006**, *31*, 56–60. [CrossRef]
18. Tringe, J.W.; Glascoe, E.A.; McClelland, M.A.; Greenwood, D.; Chambers, R.D.; Springer, H.K.; Levie, H.W. Pre-Ignition Confinement and Deflagration Violence in LX-10 and PBX 9501. *J. Appl. Phys.* **2014**, *116*, 054903. [CrossRef]
19. Tarver, C.M.; Tran, T.D. Thermal Decomposition Models for HMX-Based Plastic Bonded Explosives. *Combust. Flame* **2004**, *137*, 50–62. [CrossRef]
20. Chidester, S.K.; Tarver, C.M.; Green, L.G.; Urtiew, P.A. On the violence of thermal explosion in solid explosives. *Combust. Flame* **1997**, *110*, 264–280. [CrossRef]
21. Glascoe, E.; Springer, H.K.; Tringe, J.W.; Maienschein, J.L. A comparison of deflagration rates at elevated pressures and temperatures with thermal explosion results. In Proceedings of the 17th Biennial International Conference of the APS Topical Group on Shock Compression of Condensed Matter, Chicago, IL, USA, 26 June–1 July 2011; 555–558.
22. Zhou, J.; Zhi, X.; Wang, S.; Hao, C. Rheological properties of Composition B in slow cook-off process. *Explosion Shock Waves* **2020**, *40*, 36–44.
23. Hobbs, M.L.; Kaneshige, M.J.; Erikson, W.W.; Brown, J.A.; Anderson, M.U.; Todd, S.N.; Moore, D.G. Cookoff experiments of a melt cast explosive (Comp-B3). *Combust. Flame* **2020**, *213*, 268–278. [CrossRef]
24. Kou, Y.; Chen, L.; Lu, J.; Geng, D.; Chen, W.; Wu, J. Assessing the thermal safety of solid propellant charges based on slow cook-off tests and numerical simulations. *Combust. Flame* **2021**, *228*, 154–162. [CrossRef]
25. Ye, Q.; Yu, Y. Numerical Simulation of Cook-off Characteristics for AP/HTPB. *Defence Technol.* **2018**, *14*, 451–456. [CrossRef]
26. Yang, H.-W.; Yu, Y.-G.; Ye, R.; Xue, X.-C.; Li, W.-F. Cook-off Test and Numerical Simulation of AP/HTPB Composite Solid Propellant. *J. Loss Prev. Process Ind.* **2016**, *40*, 1–9. [CrossRef]
27. Ye, Q.; Yu, Y. Numerical Analysis of Cook-off Behavior of Cluster Tubular Double-Based Propellant. *Appl. Therm. Eng.* **2020**, *181*, 115972. [CrossRef]
28. Liu, L.; Li, F.; Tan, L.; Ming, L.; Yi, Y. Effects of Nanometer Ni, Cu, Al and NiCu Powders on the Thermal Decomposition of Ammonium Perchlorate. *Propellants Explos. Pyrotech.* **2004**, *29*, 34–38. [CrossRef]
29. Ye, Q.; Yu, Y.; Li, W. Study on Cook-off Behavior of HTPE Propellant in Solid Rocket Motor. *Appl. Therm. Eng.* **2020**, *167*, 114798. [CrossRef]
30. Ho, S.Y. Thermomechanical Properties of Rocket Propellants and Correlation with Cookoff Behaviour. *Propellants Explos. Pyrotech.* **1995**, *20*, 206–214. [CrossRef]
31. Essel, J.T.; Nelson, A.P.; Smilowitz, L.B.; Henson, B.F.; Merriman, L.R.; Turnbaugh, D.; Gray, C.; Shermer, K.B. Investigating the Effect of Chemical Ingredient Modifications on the Slow Cook-off Violence of Ammonium Perchlorate Solid Propellants on the Laboratory Scale. *J. Energetic Mater.* **2020**, *38*, 127–141. [CrossRef]
32. Dickson, P.M.; Asay, B.W.; Henson, B.F.; Smilowitz, L.B. Thermal cook–off response of confined PBX 9501. *Proc. R. Soc. A Math. Phys. Eng. Sci.* **2004**, *460*, 3447–3455. [CrossRef]
33. Hu, H.; Fu, H.; Li, T.; Shang, H.; Wen, S. Progress in experimental studies on the evolution behaviors of non-shock initiation reaction in low porosity pressed explosive with confinement. *Explosion Shock Waves* **2021**, *40*, 011401.

 *crystals*

Article

A Novel Understanding of the Thermal Reaction Behavior and Mechanism of Ni/Al Energetic Structural Materials

Kunyu Wang, Peng Deng, Rui Liu, Chao Ge, Haifu Wang and Pengwan Chen *

State Key Laboratory of Explosion Science and Technology, Beijing Institute of Technology, Beijing 100081, China
* Correspondence: pwchen@bit.edu.cn

Abstract: Ni/Al energetic structural materials have attracted much attention due to their high energy release, but understanding their thermal reaction behavior and mechanism in order to guide their practical application is still a challenge. We reported a novel understanding of the thermal reaction behavior and mechanism of Ni/Al energetic structural materials in the inert atmosphere. The reaction kinetic model of Ni/Al energetic structural materials with Ni:Al molar ratios was obtained. The effect of the Ni:Al molar ratios on their thermal reactions was discussed based on the products of a Ni/Al thermal reaction. Moreover, depending on the melting point of Al, the thermal reaction stages were divided into two stages: the hard contact stage and soft contact stage. The liquid Al was adsorbed on the surface of Ni with high contact areas, leading in an aggravated thermal reaction of Ni/Al.

Keywords: Ni/Al energetic structural materials; thermal reaction; reaction kinetic model; two reaction stages; reaction mechanism

Citation: Wang, K.; Deng, P.; Liu, R.; Ge, C.; Wang, H.; Chen, P. A Novel Understanding of the Thermal Reaction Behavior and Mechanism of Ni/Al Energetic Structural Materials. *Crystals* **2022**, *12*, 1632. https://doi.org/10.3390/cryst12111632

Academic Editor: Evgeniy N. Mokhov

Received: 24 October 2022
Accepted: 7 November 2022
Published: 13 November 2022

Publisher's Note: MDPI stays neutral with regard to jurisdictional claims in published maps and institutional affiliations.

1. Introduction

All-metal energetic structural materials, such as Al/Ti, Al/Zr, Ni/Al, and so on, have received more and more attention due to their good strength and energy-releasing properties [1–5]. Among them, Ni/Al was considered as a promising material for further application in the defense industry, such as in the fields of fragments and shaped charges, because of its higher energy density (1507.7 J/g at the equal molar ratio), higher strength properties, and faster energy-releasing capacities. Its energy release, which originates from an intermetallic reaction, has received much attention in recent decades. However, its reaction behavior and mechanism have not been explained clearly, which has limited its application.

Currently, for Ni/Al energetic structural materials, a lot of works mainly focused on its macroscopic reaction. For example, Vandersall and Thadhani [6] reported that the shock response of Ni/Al energetic structural material was divided into two categories: shock-assisted chemical reaction and shock-induced chemical reaction. Song and Thadhani [7] proposed the thermodynamic calculation model for the shock reaction, based on the effects of the reaction energy release and the formation of products on the equation of state. Bennett and Horie [8] improved the thermodynamic reaction model to reduce the errors and ambiguities of existing Hugoniot calculations. Zhang et al. [9] also built the thermal chemical model of shock-induced chemical reaction. The reaction efficiency was evaluated by combining shock kinetics and chemical reaction kinetics. These works could be used to describe the macroscopic response of the Ni/Al energetic structural material. However, they did not illustrate the microscopic reaction mechanism in detail.

Essentially, the energy release of all-metal energetic structural materials depends on the chemical reaction process [10–12]. The critical parameters of the chemical reaction are determined through the impact-induced energy release test instead of the direct measurement [13]. This method strongly depends on the shock compression theory with the chemical reaction. Due to some assumption, it is difficult to widely use the reaction model for another type of energy release tests.

Generally, the thermal analysis test, referring to the differential scanning calorimeter, has been widely used to understand the chemical reaction of energetic materials [14,15]. To directly determine the activation energy and the pre-exponential factor in the chemical reaction equation of energetic materials, the Kissinger method [16], Flynn–Wall–Ozawa method [17], and Satava–Sestak method [18] were used. Moreover, for a complex chemical reaction, the classical differential methods and kinetic integration methods were built to analyze the thermal decomposition mechanism function and kinetic parameters [19,20]. These works showed a good analysis result, and further revealed that classical thermal analysis methods could be used to analyze the reaction kinetic parameters of energetic structural materials.

A few works on dynamic thermal analysis refer to the basic thermal reaction parameters of all-metal energetic structural materials. It was found that a simple analysis strategy was not used to match the whole thermal reaction process of Al-based energetic structural materials, especially for Ni/Al. According to the Ni–Al binary phase diagram, the Ni–Al eutectic temperature is higher than the melting point of Al. During the thermal reaction process, the state change of Al from solid to a liquid state occurred in a thermal reaction. Resulting from the state change of Al in the Al-based energetic structural materials, the existing reaction model mismatches the kinetic result of thermal reaction. The traditional analysis strategy only considers the solid-solid reaction in the thermal process, but ignores the influence of the state change of Al on the thermal reaction between Al and Ni. This causes the misunderstanding of the thermal reaction process. Therefore, it is urgent and important to study the thermal reaction behavior and mechanism of Ni/Al energetic structural materials depending on the state change of Al.

Herein, we studied the thermal reaction behavior and mechanism of Ni/Al energetic structural materials with the state of Al at two different reaction stages, depending on the melting point of Al. By fitting the differential scanning calorimeter (DSC) curves of Ni/Al energetic structural materials with different Ni:Al molar ratios, the hard contact stage and soft contact stage were distinguished. Their reaction kinetic models were obtained and the thermal reaction parameters, referring to activation energy ϵ, pre-exponential factors (A), and reaction function (f), were calibrated at different reaction stages. Reaction products of Ni/Al thermal reaction was used to analyze the effect of the Ni:Al molar ratios on their thermal reaction. Furthermore, the thermal reaction mechanism of Ni/Al energetic structural material was provided, based on two different reaction stages. This work offered a new way to understand thermal reaction behavior and mechanism of Ni/Al energetic structural materials under the different temperature stages.

2. Experimental

2.1. Materials

Different Ni and Al molar ratios will lead to different chemical reactions. Theoretically, when the molar ratio of Ni and Al is 1:1, 1:3, and 3:1, the corresponding apparent reaction is shown in the Equations (1)–(3), respectively [21].

$$Al + Ni \rightarrow AlNi \ - 1381.3 \, J/g \tag{1}$$

$$3Al + Ni \rightarrow Al_3Ni \ - 1078.24 \, J/g \tag{2}$$

$$Al + 3Ni \rightarrow AlNi_3 \ - 753.4 \, J/g \tag{3}$$

Three types of samples Ni/Al with different molar ratios were prepared. The molar ratio was set Ni/Al = 1:1, Ni/Al = 1:3, and Ni/Al = 3:1, respectively. The components Ni and Al with the particle size of 20 μm, and 25 μm, respectively, were used. Powders Ni and Al were purchased from Shanghai ST-Nano Sci & Tech., Co., Ltd., Shanghai, China. Alcohol was provided by Chengdu Kelong Chem., Tech., Co., Ltd., Chengdu, China.

The components Ni and Al were mixed with the different molar ratios. Then, the mixed powders were prepared by the milling technique for 5 h in the alcoholic environment. Finally, the powder was obtained with drying treatments.

2.2. Thermal Analysis and Characterization

The thermal reaction behavior of the mixed powder samples was recorded in an argon atmosphere by a STA449F3 differential scanning calorimeter (Netzsch, Bavaria, Gremany). The mass of each sample tested was 20 mg. The test conditions were alumina crucible with cover, 20 mL/min of protective gas, and 60 mL/min of purging gas. The DSC curves with the range from the room temperature to 1200 K were collected to study their thermal reaction processes under the different heating rates (5 K/min, 10 K/min, 15 K/min, and 20 K/min).

The structure of the reaction products from the thermal reaction of Ni/Al samples were characterized by X-ray diffraction (XRD). XRD patterns from 5–90 degrees were carried out via a D8 Advances XRD apparatus with the voltage of 40 kV and the current of 40 mA.

Generally, the chemical reaction kinetics equation is used to quantify the reaction behavior. According to the DSC test, the parameters of the equation can be obtained [22]. Assume that the reaction of Ni/Al samples follows the Equations (4) and (5), which builds the relation between the reaction degree and the temperature. Actually, the two equations are equivalent as the differential and integral forms of non-isothermal systems for calculating the thermodynamic parameters.

$$\frac{d\alpha}{dT} = \frac{A}{\beta} e^{\left(-\frac{E}{RT}\right)} f(\alpha) \tag{4}$$

$$\int_0^\alpha \frac{d\alpha}{f(\alpha)} = \int_0^T \left(\frac{A}{\beta}\right) e^{-\frac{E}{RT}} dT \tag{5}$$

where α is the reaction degree, T is the absolute temperature with the unit K, A is the pre-exponential factor with the unit min-1, β is the heating rate with the unit K/min, E is the apparent activation energy with the unit J/mol, R is the universal gas constant 8.31 J/(mol·K), and $f(\alpha)$ is the reaction function. According to the DSC curve, the reaction degree α means the ratio of the area enclosed by the curve at some temperature to the whole area enclosed by the whole DSC curve.

Considering the results of thermal analysis based on multiple heating rates are more accurate [23], the Ozawa method is used in the current work to determine the parameters in Equation (5). Firstly, define $u = \frac{E}{RT}$, the equation can be rewritten as

$$\int_0^T e^{-\frac{E}{RT}} dT = \int_{-\infty}^u -\frac{E}{R} \frac{e^{-u}}{u^2} du \tag{6}$$

Substitute Equation (6) into Equation (5), and define $g(\alpha) = \int_0^\alpha \frac{d\alpha}{f(\alpha)}$ and $P(u) = \int_{-\infty}^u -\frac{e^{-u}}{u^2} du$. Then, Equation (5) can be rewritten as

$$g(\alpha) = \frac{AE}{\beta R} P(u) \tag{7}$$

The Doyle approximation [24] is used to estimate $P(u)$,

$$lgP(u) = 2.315 - 0.4567u \tag{8}$$

Take the logarithm of Equation (4) combined with Equation (5), and Equation (6) can be rewritten

$$lg\beta = \left[lg\frac{AE}{Rg(\alpha)} - 2.315\right] - 0.4567\frac{E}{RT_\alpha} \tag{9}$$

In order to obtain the parameters in the equation, the least square method is used to fit the straight line as Equation (9), where $\frac{1}{T_\alpha}$ is the abscissa and $lg\beta$ is the ordinate. Four DSC curves, under different heating rate conditions, can determine four data points to be fitted. The apparent activation energy can be obtained according to the slope of the fitting line.

It should be stressed that in order to obtain E, α must be chosen to be 1. In addition, the pre-exponential factor A and the reaction function f are coupled in the vertical intercept.

Further, in order to determine the reaction function f, the master curve method will be used [25–27]. Firstly, to calculate $P(u)$, the reaction degree value α is chosen from 0.1–0.9, and the corresponding temperature T_α is obtained based on the DSC curve. Next, to calculate $g(\alpha)$, the form of $f(\alpha)$ need to be chosen. Generally, the reaction function has different forms, such as the nth-order reaction model, Avrami–Erofeev reaction model, and so on, and it depends on the type of materials. In the current work, the Avrami–Erofeev reaction function was chosen based on the reaction characteristics of Ni/Al energetic structural materials. It can be written as Equation (10), where n is the parameter related to the reaction mechanism.

$$f = n(1-\alpha)[-ln(1-\alpha)]^{\frac{n-1}{n}} \tag{10}$$

Based on Equation (7), the parameter of the reaction function is determined by using the master curve method. Considering the two-stage reaction, taking α_c as the transition point, Equation (7) can be rewritten as,

$$g(\alpha_c) = \frac{AE}{\beta R}P(u_c) \tag{11}$$

Divide Equation (7) by Equation (11),

$$P(u)/P(u_c) = g(\alpha)/g(\alpha_c) \tag{12}$$

According to the Equation (9), choose an appropriate Avrami–Erofeev reaction function parameter n and reaction degree α until the two reaction curves $(P(u)/P(u_c) - T_\alpha$ and $g(\alpha)/g(\alpha_c) - T_\alpha)$ have the highest correlation, and so the best reaction function f could be determined. Generally, $P(u)/P(u_c) - T_\alpha$ should be called the test reaction curve, and $g(\alpha)/g(\alpha_c) - T_\alpha$ should be called the standard reaction curve. The processing is conducted for the different heating rate conditions.

3. Results and Discussion

3.1. DSC Analysis

Figure 1 shows the morphology of the Ni/Al energetic structural material mixed powders with different molar ratios of 1:1, 1:3, and 3:1. It can be seen that by the mixed and ball milling technique, the Ni and Al particles were randomly dispersed, where the bright particle was Ni and the dark particle was Al.

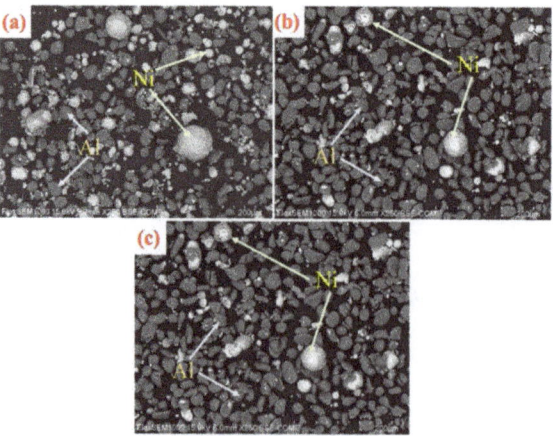

Figure 1. SEM images of (**a**) Ni/Al = 1:1, (**b**) Ni/Al = 1:3, and (**c**) Ni/Al = 3:1.

The DSC curves of Ni/Al energetic structural materials with the molar ratios 1:1, 1:3, and 3:1 at different heating rates are shown in Figure 2. For Ni/Al = 1:1, only one exothermic peak occurred during the thermal reaction process from room temperature to 1200 K, which were located at the range from ~870 K to ~950 K. In the DSC curves of Ni/Al = 1:3, two peaks, referring to an exothermic peak at ~900 K and endothermic peak at ~1150 K, appeared in Figure 2b. The exothermic peak represented the thermal reaction of Ni/Al, which was consistent with that of Ni/Al = 1:1 in Figure 2a. The endothermic peak was attributed to the melting process of NiAl₃, which was further discussed in XRD results. n addition, the DSC curves of Ni/Al = 3:1 are shown in Figure 2c. The thermal reaction processes between Ni and Al were seen at the exothermic peak.

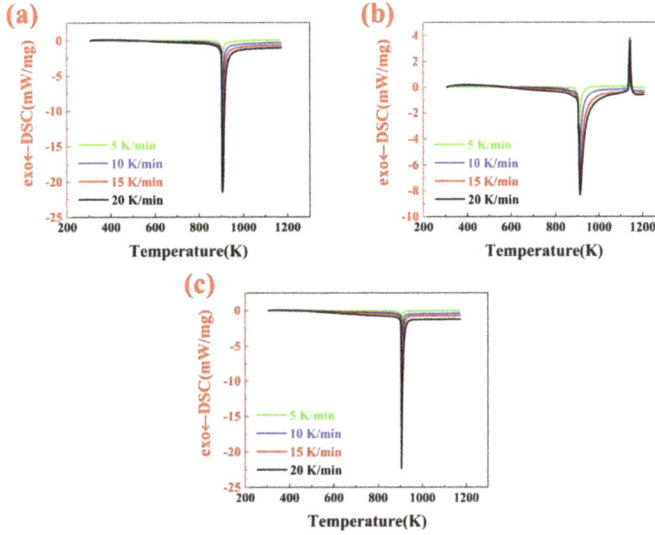

Figure 2. DSC curves of Ni/Al energetic structural materials: (**a**) Ni/Al = 1:1, (**b**) Ni/Al = 1:3, and (**c**) Ni/Al = 3:1.

From the DSC curve in Figure 2, the starting reaction temperature T_s and the reaction end temperature T_e were collected. For the endothermic process of Ni/Al = 1:3 at ~1175 K, the start melting temperature T_{s1} and the end melting temperature T_{e1} were also shown. The value of the heat release H was determined by the integral heat flow over time on the DSC curve. The analysis data is listed in the Table 1.

Table 1. Thermal reaction parameters of Ni/Al samples.

Heating Rate (K/min).		5	10	15	20
Ni/Al = 1:1	T_s (K)	904.05	903.45	902.85	902.75
	T_e (K)	912.04	923.11	934.37	948.17
	H (J/g)	840.70	840.30	839.40	840.90
Ni/Al = 1:3	T_s (K)	905.35	905.45	906.25	906.75
	T_e (K)	925.42	943.02	965.92	977.31
	H (J/g)	764.00	762.40	765.40	764.00
	T_{s1} (K)	1127.05	1127.05	1126.15	1126.35
	T_{e1} (K)	1152.95	1158.05	1165.85	1170.05
Ni/Al = 3:1	T_s (K)	905.75	905.15	904.75	903.85
	T_e (K)	913.11	923.73	932.55	942.83
	H (J/g)	463.70	464.00	464.00	464.10

Based on the thermal reaction characteristics of the Ni/Al samples, it could be found that T_s had no obvious changes, but T_e had increased obviously, as the heating rate increased.

With a higher heating rate, the peak value of Ni/Al samples was higher, and the reaction was faster. As the typical DSC curves of Ni/Al = 1:1, T_e had increased from 912.04 K to 948.17 K. The case of Ni/Al = 3:1 had a similar observation, where T_e increased from 913.11 K to 942.83 K. However, as the typical DSC curves of Ni/Al = 1:3, T_e had increased from 925.42 K to 977.31 K. The heat release H with different molar ratios Ni/Al = 1:1, Ni/Al = 1:3, and Ni/Al = 3:1 were about 840 J/g, 764 J/g, and 464 J/g, respectively. For all cases, the heat release H was almost constant as the heating rate increased.

3.2. Reaction Products Analysis

In order to determine the composition of the thermal reaction products of Ni/Al samples, the residue after DSC testing was collected for XRD analysis. The phase structure of the residue is shown in Figure 3. For the sample Ni/Al = 1:1, the main reaction products were Al_3Ni_2 and AlNi, as shown in Figure 3a. As the Al contents increased, the thermal reaction products of Ni/Al = 1:3 become complicated (shown in Figure 3b), including different Ni/Al intermetallic compounds, such as Al_3Ni, Al_4Ni_3, Al_3Ni_2, Ni_5Al_3, AlNi, and so on. As the Al contents decreased, the reaction products of Ni/Al = 3:1 (shown in Figure 3c) led to the XRD peaks of Al_4Ni_3, $AlNi_3$, Ni_5Al_3, AlNi and Al_3Ni_2. It could be found that the actual reaction products of Ni/Al powders with different molar ratios were different from the theoretical products, which indicated that the complex and incomplete reaction processes resulted in the diversity of products.

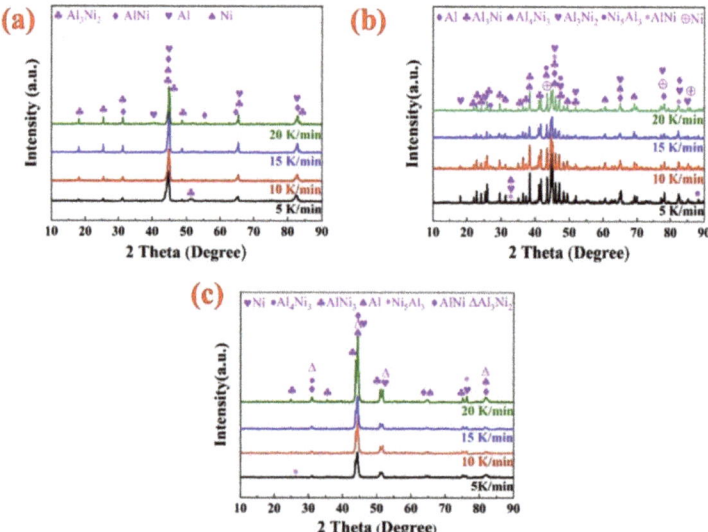

Figure 3. XRD results: (**a**) Ni/Al = 1:1, (**b**) Ni/Al = 1:3, and (**c**) Ni/Al = 3:1.

Generally speaking, when the heating temperature was lower than the Al melting point temperature, the reaction between Ni and Al took place in a solid–solid contact mode and the main product Al_3Ni was first formed [28]. When the heating temperature reached the temperature of the melting point of Al, Al and Al_3Ni would form a eutectic liquid phase. The liquid spread to the surface of Ni powders under the action of capillarity, which accelerated the liquid–solid contact with Ni particles. Ni would react with Al_3Ni in liquid phase to form Al_3Ni_2. Further, the formation of Al_3Ni_2 layer gradually covered the Ni powder and separated Ni from the liquid phase. Moreover, the ongoing formation of Al_3Ni_2 could only depend on the diffusion of atoms. At the same time, Al_3Ni_2 would continue to dissolve into the liquid phase side, and gradually form an enrichment layer. When the Al_3Ni_2 layer increased to a certain thickness, Al_3Ni_2 and Ni would form AlNi [29]. When the sample was heated to the reverse peritectic reaction temperature around 1130 K, the reverse peritectic reaction of Al$_3$Ni occurred, which corresponded to the endothermic process of Ni/Al = 1:3 in Figure 2b. Considering the low quantity of Al in Ni/Al = 1:1 and Ni/Al = 3:1, the product of Al$_3$Ni was low, which was not found by XRD. In the high content of Ni in Ni/Al = 3:1, AlNi$_3$ formed due to the diffusion reaction between AlNi and Ni [30].

3.3. The Reaction Kinetics Analysis

The kinetic parameters referring to the apparent activation energy E and the pre-exponential factor A were calculated by the Ozawa method described in the Section 2.2. The reaction function f was also obtained by the master curve method derived from the temperature integral described in the Section 2.2.

In order to calculate the apparent activation energy E, the end reaction temperature T_e under different heating rates were required. It should be explained that T_e corresponded to $\alpha = 1$. The temperature data of Ni/Al samples at four different heating rates of 5 K/min, 10 K/min, 15 K/min, and 20 K/min were listed in Table 1. According to the chemical reaction kinetics equations described in Section 2.2, T_e corresponding to β of each sample was taken out to calculate $lg\beta$ and $1/T_e$. The scatter plot shows the abscissa $1/T_e$ and the ordinate $lg\beta$. The apparent activation energy E was obtained by the least square fitting method. The results of the linear fitting and apparent activation energy E are shown in Figure 4.

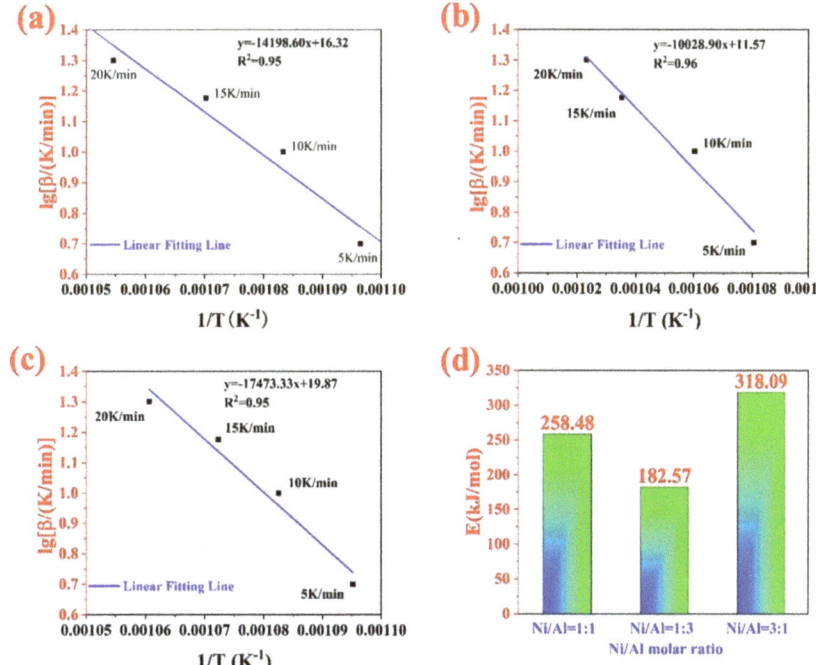

Figure 4. Linear fitting of samples at different heating rates (**a**) Ni/Al = 1:1, (**b**) Ni/Al = 1:3, and (**c**) Ni/Al = 3:1; (**d**) the apparent activation energy E of the samples.

Figure 4d showed that the apparent activation energy E of Ni:Al = 1:1, 1:3, and 3:1 are 258.48 kJ/mol, 182.57 kJ/mol, and 318.09 kJ/mol, respectively. When the contents of Al increased, the apparent activation energy E of Ni/Al materials reduced, resulting from the higher activity of Al than Ni. Moreover, when the melting of Al occurred, the liquid phase of Al increased the contact surface of Ni particles [31]. The higher quantity of Al would benefit from promoting the thermal reaction.

The Avrami–Erofeev reaction function f was generally used for energetic structural materials [6]. However, it was not a single reaction process; the segment fitting method was used here. The parameters in the reaction function were optimized for different segments. According to the theoretical calculation of the master curve method in Section 2.2, T_α was defined as the transition temperature of segmented reaction curves. The pre-exponential factor A was calculated for the different reaction stages at different heating rates based on Ozawa method.

Figure 5a–c show the fitting results of the test reaction curve at the heating rate 10 K/min. The two reaction curves revealed a good fitting effect. For the other heating rate conditions, it had a similar trend. In order to explore the mechanism on the occurrence of transition temperature, the transition data points of the samples are plotted in Figure 5d. It could be found that the transition temperature points were distributed in the Ni/Al liquid eutectic temperature range, which indicated that the transition from solid state to liquid state of Al was the critical factor, although the discrepancy was presented due to the reaction hysteresis at the high heating rates for Ni/Al = 1:3.

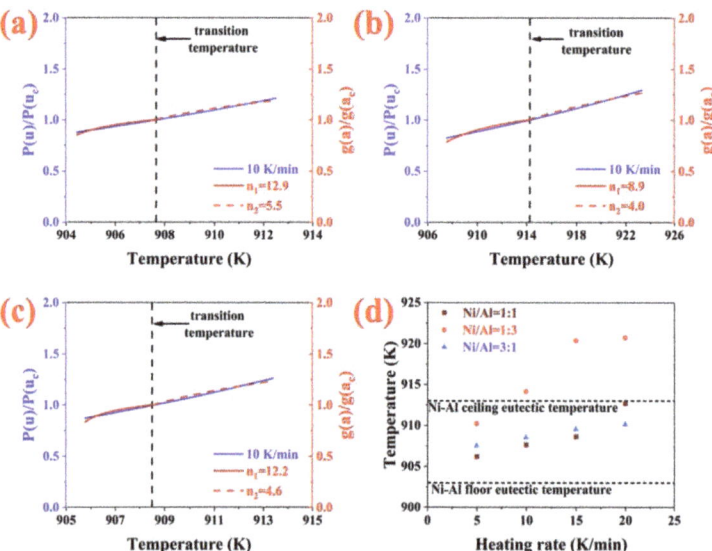

Figure 5. The comparison between the reaction curves and the test reaction curve at 10 K/min (**a**) Ni/Al = 1:1; (**b**) Ni/Al = 1:3; (**c**) Ni/Al = 3:1; (**d**) the transition temperature T_α.

Table 2 gives the reaction function parameter n and pre-exponential factor A of all the samples. For the Ni/Al powder samples with the same molar ratio, the different reaction parameters under different heating rates are collected in Table 2. It could be found that as the heating rate increased, both of the parameters n_1 and n_2 decreased. This is because the increase of the heating rate brought in the temperature accumulation of the sample, which included exothermic reaction. The weakened constraint between lattice atoms originated from the overheating effect, which was good for promoting the reaction process.

The reaction process was described by the Avrami–Erofee reaction function with two sets of parameters and was divided into reaction stage I and II, according to the transition temperature. The schematic diagram of Ni–Al reaction mechanism with the two-stage reaction is shown in Figure 6. In the reaction stage I, both Ni particles and Al particles were solid, and the contact was similar to the point contact. This stage was considered as the hard contact stage. The reaction to generate Al_3Ni only occurred at the contact reaction zones. Therefore, the solid phase reaction was limited. Once the reaction temperature had been heated over the melting point, the solid state of Al started to transfer into the liquid state. The reaction entered the reaction stage II, where the liquid Al had a soft contact stage with Ni particles. With a higher reaction temperature, the solid–liquid reaction between Ni and Al occurred at the surface of Ni particles. In this stage, the reaction rate become faster. In addition, the eutectic liquid would also exist at the reaction zones in the reaction stage II, as shown in Figure 6. The soft contact stage was also beneficial for promoting the thermal reaction of Ni/Al materials [32].

Table 2. The reaction function parameter n and pre-exponential factor A of all the samples.

Sample	Stage	n/A	Heating Rate (K/min)			
			5	10	15	20
Ni/Al = 1:1	I	n_1	27.6	12.9	10.3	5.5
		A_1 ($\times 10^{14}$ min^{-1})	1.3639	1.3590	1.3567	1.3504
	II	n_2	12.2	5.5	4.8	2.2
		A_2 ($\times 10^{14}$ min^{-1})	1.3585	1.3467	1.3436	1.3389
Ni/Al = 1:3	I	n_1	14.5	8.9	4.9	2.9
		A_1 ($\times 10^{9}$ min^{-1})	3.4174	3.4045	3.3773	3.1903
	II	n_2	6.5	4.0	2.1	1.9
		A_2 ($\times 10^{9}$ min^{-1})	3.3921	3.3638	3.2979	2.8624
Ni/Al = 3:1	I	n_1	23.5	12.2	8.2	6.3
		A_1 ($\times 10^{17}$ min^{-1})	3.9351	3.9216	3.9079	3.8954
	II	n_2	10.1	4.6	3.1	2.4
		A_2 ($\times 10^{17}$ min^{-1})	3.9157	3.8754	3.8400	3.8085

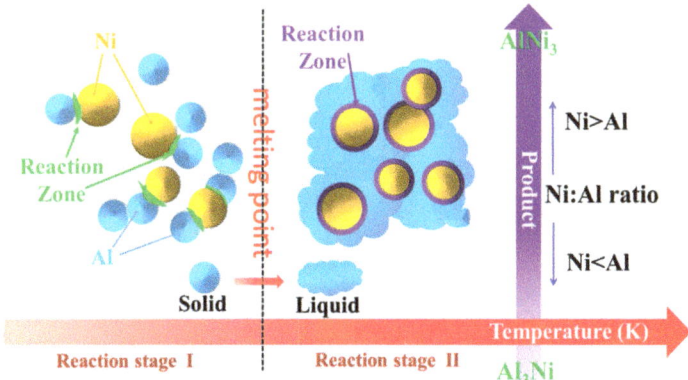

Figure 6. Schematic diagram of the reaction mechanism of Ni/Al materials.

4. Conclusions

In summary, a novel understanding of the thermal reaction behavior and mechanism of Ni/Al energetic structural materials was demonstrated. Depended on the melting point of Al, the thermal reaction stages of Ni/Al were divided into two stages: the hard contact stage and the soft contact stage. The thermal reaction behavior of Ni/Al energetic structural material powder was studied based on the DSC test, XRD characterization, and chemical reaction kinetics analysis. The reaction kinetic parameters and specific reaction mechanism were determined to describe the reaction process for Ni/Al powder. The parameters were used to determine the difficulty and mode of reaction. For the specific kinetic parameters, as the ratio of Al increased, the apparent activation energy of the material significantly reduced. Otherwise, as the ratio of Ni increased, the apparent activation energy increased. It could be found that the exothermic reaction function between Al and Ni was described by Avrami–Erofee segment reaction function. The transition of Al from solid to liquid was the critical factor affecting the establishment of segment reaction function. In addition, the thermal reaction mechanism of Ni/Al energetic structural material was provided based on the hard contact stage and soft contact stage. This work offered a new idea to understand the thermal reaction behavior and mechanism of Ni/Al energetic structural materials under different temperature stages.

Author Contributions: Conceptualization, P.C., H.W. and K.W.; resources, R.L.; writing—original draft preparation, K.W. and C.G.; writing—review and editing, R.L. and P.D.; supervision, P.C. All authors have read and agreed to the published version of the manuscript.

Funding: This work was supported by the National Natural Science Foundation of China No. 12132003 and 2019-JCJQ-ZD-011-00, State Key Laboratory of Explosion Science and Technology No. QNKT20-07 and Beijing Institute of Technology Research Fund Program for Young Scholars.

Data Availability Statement: The data that support the findings of this study are available from the corresponding author upon reasonable request.

Conflicts of Interest: The authors declare no conflict of interest.

References

1. Millar, D.; Marshall, W.; Oswald, I.; Pulham, C. High-pressure structural studies of energetic materials. *Crystallogr. Rev.* **2010**, *16*, 115–132. [CrossRef]
2. Hastings, D.; Dreizin, E. Reactive Structural Materials: Preparation and Characterization. *Adv. Eng. Mater.* **2018**, *20*, 1700631. [CrossRef]
3. Seropyan, S.; Saikov, I.; Andreev, D.; Saikova, G.; Alymov, M. Reactive Ni–Al-Based Materials: Strength and Combustion Behavior. *Metals* **2021**, *11*, 949. [CrossRef]
4. Wu, J.; Wang, H.; Fang, X.; Li, Y.; Mao, Y.; Yang, L.; Yin, Q.; Wu, S.; Yao, M.; Song, J. Investigation on the Thermal Behavior, Mechanical Properties and Reaction Characteristics of Al-PTFE Composites Enhanced by Ni Particle. *Materials* **2018**, *11*, 1741. [CrossRef]
5. Grapes, M.; Weihs, T. Exploring the reaction mechanism in self-propagating Al/Ni multilayers by adding inert material. *Combust. Flame* **2016**, *172*, 105–115. [CrossRef]
6. Vandersall, K.; Thadhani, N. Investigation of "shock-induced" and "shock-assisted" chemical reactions in Mo + 2Sipowder mixtures. *Metall. Mater. Trans. A* **2003**, *34*, 15–23. [CrossRef]
7. Song, I.; Thadhani, N. Shock-induced chemical-reactions and synthesis of nickel aluminides. *Metall. Mater. Trans. A* **1002**, *23*, 41–48. [CrossRef]
8. Bennett, L.; Horie, Y. Shock-induced inorganic reactions and condensed phase detonations. *Shock Waves* **1994**, *4*, 127–136. [CrossRef]
9. Zhang, X.; Shi, A.; Zhang, J.; Qiao, L.; He, Y.; Guan, Z. Thermochemical modeling of temperature controlled shock-induced chemical reactions in multifunctional energetic structural materials under shock compression. *J. Appl. Phys.* **2012**, *111*, 123501. [CrossRef]
10. Mason, B.; Groven, L.; Son, S. The role of microstructure refinement on the impact ignition and combustion behavior of mechanically activated Ni/Al reactive composites. *J. Appl. Phys.* **2013**, *114*, 113501. [CrossRef]
11. Wang, H.; Zheng, Y.; Yu, Q.; Liu, Z.; Yu, W. Impact-induced initiation and energy release behavior of reactive materials. *J. Appl. Phys.* **2011**, *110*, 074904.
12. Xiong, W.; Zhang, X.; Tan, M.; Liu, C.; Wu, X. The Energy Release Characteristics of Shock-Induced Chemical Reaction of Al/Ni Composites. *J. Phys. Chem. C* **2016**, *120*, 24551–24559. [CrossRef]
13. White, J.; Reeves, R.; Son, S.; Mukasyan, A. Thermal Explosion in Al-Ni System: Influence of Mechanical Activation. *J. Phys. Chem. A* **2009**, *113*, 13541–13547. [CrossRef]
14. Deng, P.; Chen, P.; Fang, H.; Liu, R.; Guo, X. The combustion behavior of boron particles by using molecular perovskite energetic materials as high-energy oxidants. *Combust. Flame* **2022**, *241*, 112118. [CrossRef]
15. Izato, Y.; Koshi, M.; Miyake, A.; Habu, H. Kinetics analysis of thermal decomposition of ammonium dinitramide (ADN). *J. Therm. Anal. Calorim.* **2017**, *127*, 255–264. [CrossRef]
16. Huang, C.; Mei, X.; Cheng, Y.; Li, Y.; Zhu, X. A model-free method for evaluating theoretical error of Kissinger equation. *J. Therm. Anal. Calorim.* **2014**, *116*, 1153–1157. [CrossRef]
17. Koga, N. Ozawa's kinetic method for analyzing thermoanalytical curves History and theoretical fundamentals. *J. Therm. Anal. Calorim.* **2013**, *113*, 1527–1541. [CrossRef]
18. Mian, I.; Li, X.; Jian, Y.; Dacres, O.; Zhong, M.; Liu, J.; Ma, F.; Rahman, N. Kinetic study of biomass pellet pyrolysis by using distributed activation energy model and Coats Redfern methods and their comparison. *Bioresour. Technol.* **2019**, *294*, 122099. [CrossRef]
19. Ren, N.; Wang, F.; Zhang, J.; Zheng, X. Progress in Thermal Analysis Kinetics. *Acta Phys.-Chim. Sin.* **2020**, *36*, 1905062.

20. Meng, F.; Zhou, Y.; Liu, J.; Wu, J.; Wang, G.; Li, R.; Zhang, Y. Thermal decomposition behaviors and kinetics of carrageenan-poly vinyl alcohol bio-composite film. *Carbohyd. Polym.* **2018**, *201*, 96–104. [CrossRef]
21. Hiramoto, M.; Okinaka, N.; Akiyama, T. Self-propagating high-temperature synthesis of nonstoichiometric wustite. *J. Alloys Compd.* **2012**, *520*, 59–64. [CrossRef]
22. Szterner, P.; Legendre, B.; Sghaier, M. Thermodynamic properties of polymorphic forms of theophylline. Part I: DSC, TG, X-ray study. *J. Therm. Anal. Calorim.* **2010**, *99*, 325–335. [CrossRef]
23. Vyazovkin, S.; Burnham, A.; Criado, J.; Perez-Maqueda, L.; Popescu, C.; Sbirrazzuoli, N. ICTAC Kinetics Committee recommendations for performing kinetic computations on thermal analysis data. *Thermochim. Acta* **2011**, *520*, 1–19. [CrossRef]
24. Cai, J.; Liu, R.; Shen, F. Improved version of Doyle integral method for nonisothermal kinetics of solid-state reactions. *J. Math. Chem.* **2008**, *43*, 1127–1133. [CrossRef]
25. Perez-Maqueda, L.; Criado, J.; Gotor, F.; Malek, J. Advantages of combined kinetic analysis of experimental data obtained under any heating profile. *J. Phys. Chem. A* **2022**, *106*, 2862–2868. [CrossRef]
26. Gotor, F.J.; Criado, J.M.; Malek, J.; Koga, N. Kinetic analysis of solid-state reactions: The universality of master plots for analyzing isothermal and nonisothermal experiments. *J. Phys. Chem. A* **2000**, *104*, 10777–10782. [CrossRef]
27. Mamleev, V.; Bourbigot, S.; Le Bras, M.; Duquesne, S.; Sestak, J. Modelling of nonisothermal kinetics in thermogravimetry. *Phys. Chem. Chem. Phys.* **2000**, *2*, 4708–4716. [CrossRef]
28. Philpot, K.; Munir, Z.; Holt, J. An investigation of the synthesis of nickel aluminides through gasless combustion. *J. Mater. Sci.* **1987**, *22*, 159–169. [CrossRef]
29. Bouche, K.; Barbier, F.; Coulet, A. Intermetallic compound layer growth between solid iron and molten aluminium. *Mat. Sci. Eng. A-Struct.* **1998**, *249*, 167–175. [CrossRef]
30. Morsi, K.; Moussa, S.; Wall, J. Reactive extrusion and high-temperature oxidation of Ni3Al. *J. Mater. Sci.* **2006**, *41*, 1265–1268. [CrossRef]
31. Chen, L.; Song, W.; Lv, J.; Wang, L.; Xie, C. Effect of heating rates on TG-DTA results of aluminum nanopowders prepared by laser heating evaporation. *J. Therm. Anal. Calorim.* **2009**, *96*, 141–145. [CrossRef]
32. Yu, L.; Meyers, M. Shock synthesis and synthesis-assisted shock consolidation of suicides. *J. Mater. Sci.* **1991**, *26*, 601–611. [CrossRef]

Article

Characterization and Analysis of Micromechanical Properties on DNTF and CL-20 Explosive Crystals

Hai Nan [1,2], Yiju Zhu [2], Guotao Niu [2], Xuanjun Wang [1,*], Peipei Sun [2], Fan Jiang [2] and Yufan Bu [2]

1   High-Tech Institute of Xi'an, Xi'an 710025, China
2   Xi'an Modern Chemical Research Institute, Xi'an 710065, China
*   Correspondence: wangxj503@sina.com

Abstract: To study the crystal mechanical properties of 3,4-dinitrofurazanofuroxan (DNTF) and hexanitrohexaazaisowurtzitane (CL-20) deeply, the crystals of DNTF and CL-20 were prepared by the solvent evaporation method. The crystal micromechanical loading procedure was characterized by the nanoindentation method, and then obtained the mechanical parameters. In addition, the crystal fracture behaviors were investigated with scanning probe microscopy (SPM). The results show that the hardness for DNTF and CL-20 was 0.57 GPa and 0.84 GPa, and the elastic modulus was 10.34 GPa and 20.30 GPa, respectively. CL-20 obviously exhibits a higher hardness, elastic modulus and local energy-dissipation and a smaller elastic recovery ability of crystals than those of DNTF. CL-20 crystals are more prone to cracking and have a lower fracture toughness value than DNTF. Compared to DNTF crystals, CL-20 is a kind of brittle material with higher modulus, hardness and sensitivity than that of DNTF, making the ignition response more likely to happen.

Keywords: DNTF; CL-20; nanoindentation; explosive crystals; micromechanical properties

1. Introduction

  An energetic crystal is a key composition for the explosive formulation designation and its application. In particular, its sensitivity has an important impact on the safety performance of explosive mixtures. Material mechanical performance plays a crucial role on the response behavior of crystals under external mechanical load (such as the impact, friction, impact, etc.), which could result in the formation of "hot spots", and also relate to impact sensitivity. It is reported that for most chemical compounds, sensitivity increases with an energy content rise, although this is not a strict rule [1]. It is of great significance to fully grasp the micromechanical properties of crystals for further understanding the safety properties of materials and analyzing the response mechanism.

  The traditional mechanical test method is only suitable for samples with a large size, and struggles to tests smaller ones, especially in the nanometer dimension. Additionally, this problem can be resolved effectively by nanoindentation technology. As a new testing method invented in the early 1990s, nanoindentation technology has been extensively applied to all kinds of materials in the nano/micro dimension, such as ceramics, metals, alloys, energetic materials, etc. [2–6]. Nanoindentation technology uses a computer-controlled load to push a rigid indenter of a specific shape into the surface of the material being tested. At the same time, a high-resolution displacement sensor is used to collect the depth of pressure on the surface of the measured material, and the load–displacement curve of the material surface is obtained. This can effectively measure some mechanical behaviors of materials at the micro/nanoscale, such as hardness, elastic modulus, fracture toughness, strain hardening effect, creep behavior, etc. [7]. Nanoindentation technology is becoming the primary choice for the mechanical property testing of micro/nanoscale materials and structures due to its advantages of simple test operation, high measurement efficiency and wide application range [8]. At present, the research on nanoindentation mainly focuses on

Citation: Nan, H.; Zhu, Y.; Niu, G.; Wang, X.; Sun, P.; Jiang, F.; Bu, Y. Characterization and Analysis of Micromechanical Properties on DNTF and CL-20 Explosive Crystals. *Crystals* **2023**, *13*, 35. https://doi.org/10.3390/cryst13010035

Academic Editors: Rui Liu, Yushi Wen and Weiqiang Pang

Received: 1 December 2022
Revised: 21 December 2022
Accepted: 22 December 2022
Published: 25 December 2022

the scale effect of indentation experiments. Early researchers [9,10] found that indentation hardness increased with the decrease in indentation depth through experimental studies. Gerberich [11] studied and found the size effect of indenter shape on indentation hardness. Swadender et al. [12] found that the hardness value decreases with the decrease in the radius of the contact area. Scholars mainly focus on vertical loading and unloading, and study the scale effect of mechanical properties of materials by fitting load–displacement curve and hardness-displacement curve.

With the development of nanoindentation technology, it is gradually applied to the characterization of energetic materials. Ramos et al. analyzed the deformation mechanism of brittle material with the nanoindentation test of cyclotetramethylene tetranitramine (HMX) simulative material [13] and the surface testing about monocrystalline Cyclotrimethylene trinitramine (RDX) [14]. Hudson et al. obtained the micromechanical properties of the different crystal RDX, which demonstrated a potential relationship with the degree of crystal internal defects [15]. Mathew and Sewell [16] studied the crystal micromechanical properties of 1,3,5-triamino-2,4,6-trinitrobenzene (TATB) and carried out its molecular dynamic simulation. Matthew et al. [17] tested and analyzed the elastic and plastic characteristics for FOX-7, HMX and ADAAF. Zhai et al. explored the yield behavior of PETN and found that the indentation modulus decreases with the increase in indentation depth [18]. According to the investigation on the regular jump phenomenon of RDX (210) [19], Li et al. computed the yield stress and hardness values, and analyzed the elastic modulus of crystal β-HMX [20]. Zhu et al. [21] found that DNAN had worse ability to resist deformation than TNT, but more obvious slow recovery elasticity and stronger impact energy absorption ability. Meanwhile, they found that HATO was harder and more brittle compared with RDX when impacted by external shock [22]. Moreover, Ekaterina et al. found that surface dynamics influence a material's ability to dissipate excess energy, acting as a buffer to mechanical initiation [23]. For the materials with less hardness, such as picric acid and 3,4-dinitropyrazole, the surfaces could be rearranged in response to mechanical deformation. DNTF and CL-20 are typical highly energetic materials with superior crystal density and energy compared to RDX, HMX, TNT, which are crucial parameters to improve the explosive properties. However, few studies have focused on the micromechanical properties of DNTF and CL-20. This paper analyzed parameters such as the elastic modulus and hardness, characterized the break behavior and explored the relationship between crystal properties and sensitivity by means of nanoindentation technology.

2. Materials and Methods

2.1. Materials

DNTF and CL-20 crystals were prepared by volatilization using acetone as a solvent at room temperature, where the CL-20 was from Qingyang Special Chemical Industry Co., Ltd. (Qingyang, China), and the DNTF was synthesized by Xi'an Modern Chemistry Research Institute. In addition, the nanomechanical analyzer, TI950, made by the Hysitron company in America, was used to employ the nanoindentation experiments to obtain the mechanical characteristics of crystalline CL-20 and DNTF. The mechanical characteristics included the testing of material hardness, elastic modulus, and fracture toughness, where the indenters were both kinds of Berkovich and the parameters of scanning probe image were 2 μN contact force and 15 μm × 15 μm size, respectively.

2.2. Methods

In the process of nanoindentation testing, the indenter was pressed into the surface of the samples with a certain load, and when the load reached a designed value, the external force was unloaded.

During the test loading, the indenter displacement (H) and load (P) were recorded by means of the high-precision load–displacement testing technique. At the contact with the indenter, in the direction of the load, the material had a certain degree of elastic recovery. Figure 1 shows the typical curves of displacement and load in the process of loading and unloading. The key parameters included the maximum load (P_{max}), maximum displacement (h_{max}), the final indentation depth after complete unloading (h_f) and the top slope S of the unloading curve.

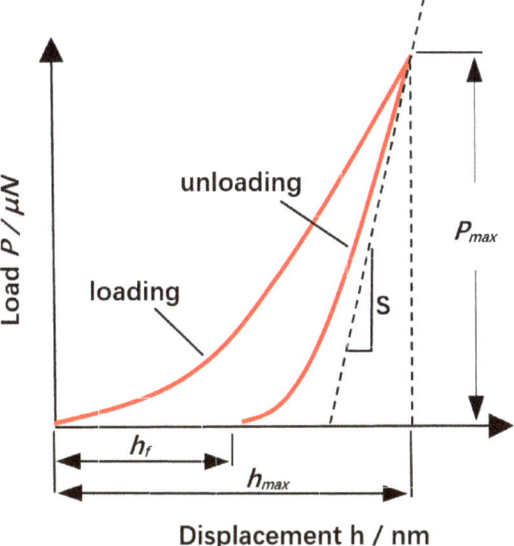

Figure 1. Typical crystal load–displacement curve.

In the experiments, the crystalline DNTF and CL-20 explosives were loaded with forces of 500 μN to 5000 μN with the same conditions of 5 s loading, 2 s pressure maintaining and 5 s unloading.

3. Results and Discussion

3.1. Indentation Curve of Crystalline Material

According to the nanoindentation testing, the curves of loading and unloading for DNTF and CL-20 were obtained as shown in Figure 2.

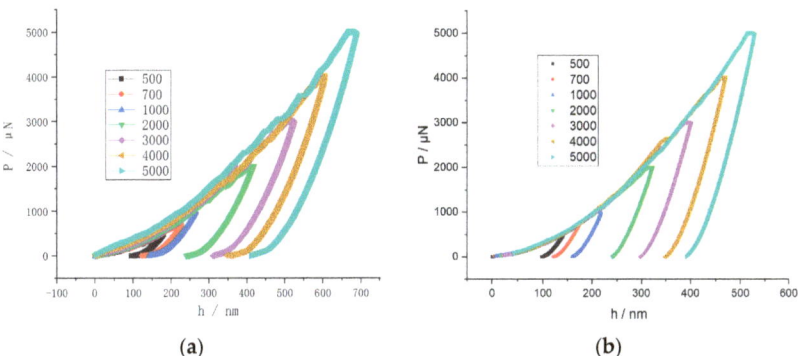

Figure 2. Quasi-static load–displacement curve for DNTF (**a**) and CL-20 (**b**).

The curves obtained exhibited similarity to the theory curve above. With the increase in loading, the maximum depth of the indenter (h_{max}) and the final indentation depth (h_f) increase continuously gradually. During the loading, the displacement on both crystals showed a sudden increase, which was mainly caused by internal microdefects such as microcracks, micropores, etc. When the contact surface of the indenter is close to the defect, the elasticity and hardness of the local material will decrease sharply, resulting in the sudden increase in the indenter displacement. Therefore, when the existing defects are sensed by the indenter, the indentation displacement increases. In order to clearly compare the loading characteristics of the two kinds of crystal mechanics, three loading–displacement curves at 1000, 3000 and 5000 μN were compared, and the results are shown in Figure 3.

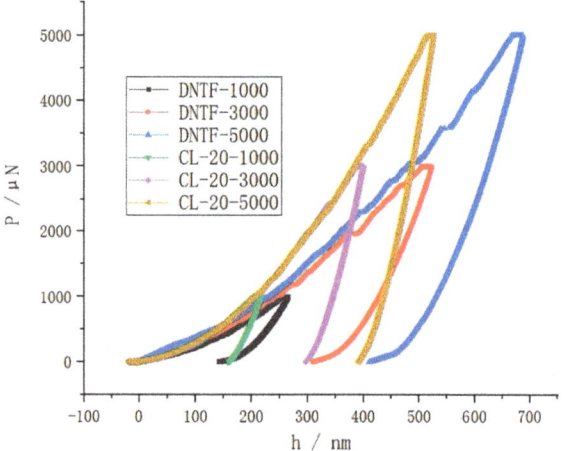

Figure 3. DNTF, CL-20 crystal load–displacement comparison curve.

It can be seen from Figure 3 that the two kinds of material showed different mechanical behaviors, in that the indentation depth on the crystal surface was quite different under the same load. Figure 4 shows that the maximum displacement (h_{max}) of DNTF was higher than that of CL-20 under the same load. In addition, when the pressure was completely unloaded, the final indentation depth h_{max} of the two crystals was basically the same. Therefore, the DNTF was more prone to deformation. Furthermore, the top slope S of the unloading curve was also named contact stiffness and increased with the load. The S_{CL-20} was obviously higher than S_{DNTF}, which indicated that CL-20 had a harder contact stiffness, as shown in Figure 5.

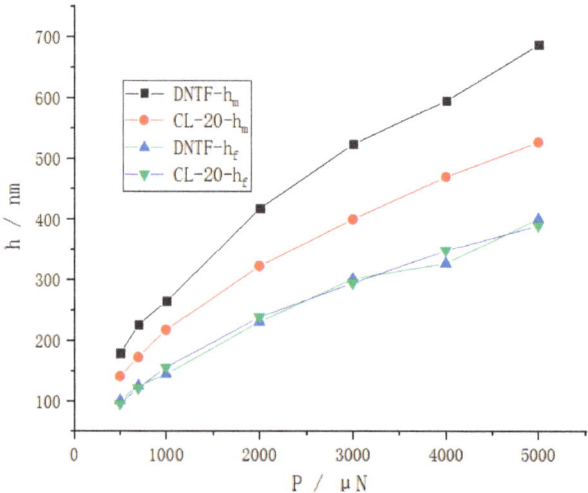

Figure 4. Load–displacement variation curve.

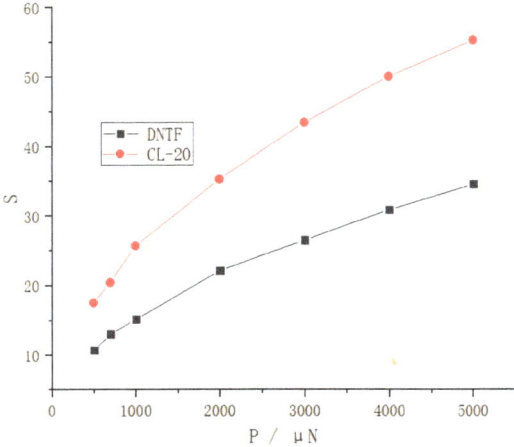

Figure 5. Load-contact stiffness variation curve.

3.2. Crystalline Elastic Modulus and Hardness

The hardness and elastic modulus of crystalline materials were calculated by the theory of Oliver–Pharr, and the formulas of hardness H and elastic modulus E are shown below:

$$H = \frac{P_{\max}}{A} \tag{1}$$

$$\frac{1}{E_r} = \frac{1 - v^2}{E} + \frac{1 - v_i^2}{E_i} \tag{2}$$

$$E_r = \frac{\sqrt{\pi}}{2\beta} \frac{S}{\sqrt{A}} \tag{3}$$

where E_r is the equivalent modulus. E_i is the modulus of indenter. A is the contact area. β is a constant related with indenter shape. v_i is the Poisson ratio of the indenter and v is the Poisson ratio of samples. According to the formulas above, the values of crystalline elastic modulus and hardness were obtained and the variation with load, as can be seen in Figures 6 and 7.

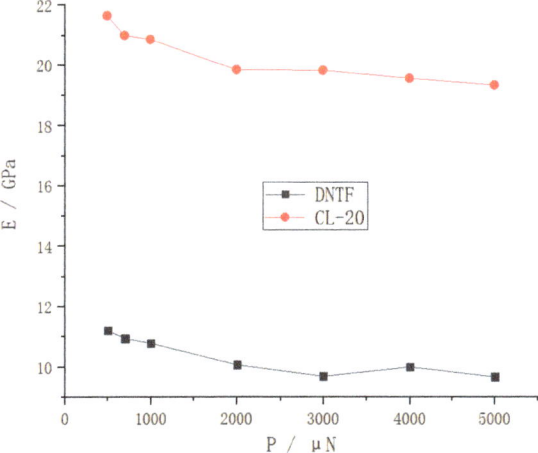

Figure 6. Load–crystal modulus of elasticity curve.

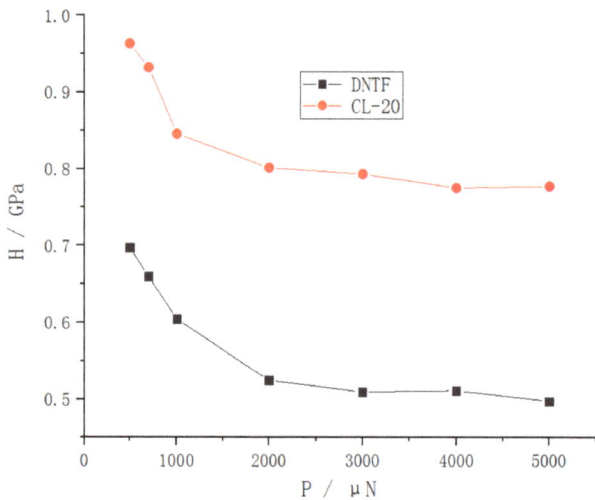

Figure 7. Load–crystal hardness curve.

As shown in Figures 6 and 7, with the increase in H_{max}, the hardness and elastic modulus of DNTF and CL-20 decreased first and then trended toward a fixed value. That is, when the compression depth is small, the mechanical parameters of the material are larger. As the depth of compression increases, the mechanical parameters of the material approach a constant value, which is called the "scale effect", and this effect is related to the plastic strain and the plastic strain gradient of the material [15]. The elastic modulus of the crystal is mainly determined by the strength of the intermolecular binding force. The stronger the intermolecular binding force is, the less easy it is to deform, and the higher the elastic modulus is. With the increase in indentation depth, the elastic modulus changes little but decreases slightly.

The average values of hardness for DNTF and CL-20 is 0.57 GPa and 0.84 GPa, and the elastic modulus is 10.34 GPa and 20.30 GPa, respectively. The average deviations of hardness for DNTF and CL-20 are 0.07 GPa and 0.06 GPa, and those of the elastic modulus are 0.54 GPa and 0.74 GPa, respectively. The elastic modulus and hardness of CL-20 are about 47% and 96% higher than those of DNTF, respectively, which indicates that CL-20 has a high stiffness and is difficult to deform. On the contrary, DNTF experiences easier indentation—namely, CL-20 is "hard" and DNTF is "soft".

In addition, the elastic modulus of a material is not directly proportional to its hardness. The elastic–plastic local deformation in the loading process determines the hardness of the material and the work conducted by external forces, and the elastic recovery in the unloading process reflects the local energy dissipation and elastic modulus of the material. Based on elastic contact theory, the relationship between the elastic modulus and hardness of solid materials depends on the energy dissipation capacity of materials. Additionally, the local energy dissipation R_S of the material is inversely proportional to the ratio of H/E [19]. The ratio of CL-20 and DNTF is calculated to be 0.041 and 0.055, respectively, so the local energy dissipation of CL-20 crystals is greater than that of DNTF, which will lead to a lower elastic recovery capacity around the indentation of CL-20 than that of DNTF.

The hardness and elastic modulus of crystals are closely related to the intermolecular binding force. Figures 8 and 9 show the molecular structure of CL-20 and DNTF.

CL-20 is a caged nitroamine explosive with molecular formula $C_6H_6O_{12}N_{12}$, and there are van der Waals forces and hydrogen bond interactions between molecules. Pampuram et al. [24] found a novel synthesis method of hexaazaisowurtzitane cages to access CL-20, where CL-20 with a yield of 25% was successfully obtained.

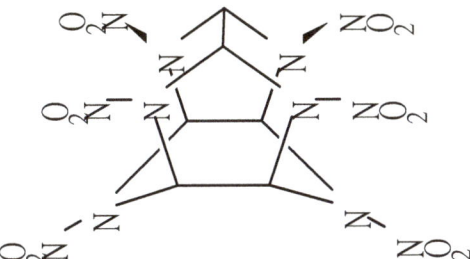

Figure 8. Schematic diagram of molecular structure for CL-20.

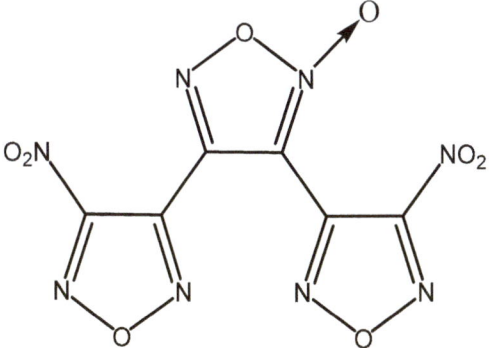

Figure 9. Schematic diagram of molecular structure for DNTF.

By contrast, DNTF is a typical furazan compound, which is composed of an oxidized furazan ring, a furazan ring, a nitro group and other groups. The molecular formula is $C_6O_8N_8$, and there is no hydrogen element in the molecule, so there is no hydrogen bond between molecules, which is mainly dominated by van der Waals forces. Therefore, the intermolecular binding force of DNTF is weaker than that of CL-20. Due to its strong intermolecular binding force, CL-20 is not easy to deform, resulting in its mechanical properties differing from those of DNTF.

3.3. Crystalline Elastic Property

In the testing of nanoindentation, pure elastic deformation is almost impossible. Due to the high local stress concentration, local plastic deformation inevitably occurs to some extent, so the variation of crystal depth mainly includes elastic and plastic deformation. In the process of pressing, the total work transforms to the sum of elastic and plastic work of materials. In addition, after unloading, only part of the elastic work is released. From the curve of loading and unloading, the total deformative work A_t and elastic work A_e are obtained, and accordingly, the plastic work A_p is calculated. The plasticity of crystalline materials can be expressed by dimensionless δ_A as follows [25–27]:

$$\delta_A = \frac{A_p}{A_t} = \frac{A_t - A_e}{A_t} \tag{4}$$

$$A_t = \int_0^{h_{max}} Pdh \tag{5}$$

$$A_e = \int_{h_p}^{h_{max}} Pdh \tag{6}$$

It can be seen in Table 1 that the total deformation work and elastic deformation work of the two materials increase with the increase in load. Additionally, under the same load, the total deformation work and elastic deformation work of DNTF crystal are significantly greater than that of CL-20. However, the values of δ_A of the two materials are basically the same, which shows that when the crystal is stimulated by external load, the ratio of elastic deformation to plastic deformation of DNTF and CL-20 remains unchanged, and most of the total deformation work is converted into plastic deformation work.

Table 1. Calculated elastic–plastic work of DNTF and CL-20.

$P_{max}/\mu N$		500	700	1000	2000	3000	4000	5000
	$A_t \times 10^{-10}/J$	0.38	0.67	1.1	3.52	6.65	9.95	14.32
DNTF	$A_e \times 10^{-10}/J$	0.15	0.24	0.43	1.2	2.21	3.45	4.72
	δ_A	0.61	0.64	0.61	0.66	0.67	0.65	0.67
	$A_t \times 10^{-10}/J$	0.31	0.49	0.89	2.55	4.67	7.41	9.95
CL-20	$A_e \times 10^{-10}/J$	0.09	0.14	0.24	0.67	1.25	1.94	2.67
	δ_A	0.71	0.7	0.73	0.74	0.73	0.74	0.73

Since the δ_A average value of DNTF (0.64) is smaller than that of CL-20 (0.73), it can be concluded that CL-20 crystals have a higher plastic deformation and transformation ability. However, the elastic transformation ability of DNTF crystal is stronger, and the elastic recovery ability of DNTF is higher than that of CL-20, which further reflects the characteristics of the large local energy dissipation of CL-20. This also means that under the same loading conditions, the structural integrity of CL-20 crystals is weaker than that of DNTF crystals, so it is more likely to be damaged under impact.

3.4. Fracture Toughness Property

Crack formation is an important form of crystal mechanics, and the fracture toughness (K_{IC}) parameter is generally used to quantify and measure the nanoindentation. K_{IC} reflects the energy required for crystal fracture, which means the ability of crystals to prevent crack propagation. In general, the higher the fracture toughness value of the material, the higher the critical stress required for crack instability propagation and the stronger the crack resistance. According to the theory of fracture mechanics and the analysis of the angular radial crack traces in the nanoindentation test, the mathematical relationship between the fracture toughness value and the indentation crack length is as follows:

$$K_{IC} = \alpha \sqrt{\frac{E}{H}} \left(\frac{P_m}{C^{3/2}} \right) \tag{7}$$

where P_m is the load, C is the radial crack length, and α is the empirical parameter related to the indenter (α of the cubic angle indenter is 0.036).

A cubic angle indenter was used to test the radial crack on the surface of DNTF and CL-20 crystals at 3000 μN, and the results are shown in Figure 10. It can be seen that obvious cracks appear in CL-20, while no cracks appear in DNTF. The radial cracks of DNTF and CL-20 and the average radial crack lengths were obtained under 4000 μN and 5000 μN loads by means of the same loading method. The elastic modulus and hardness obtained were substituted into Equation (7) to obtain the fracture toughness values under different loads, as shown in Table 2.

With the increase in loading from 3000 to 5000 μN, the crack length on the surface of the two crystals shows an increasing trend, and the crack length of CL-20 is more significant. With the increase in loading, the fracture toughness value of CL-20 decreases continuously, showing a typical material brittle fracture behavior. In addition, the fracture toughness value of CL-20 crystals is lower than that of DNTF crystals, and it is more prone to cracking than DNTF crystals. The experimental results show that CL-20 exhibits brittleness. Although the compressive strength is high under quasi-static conditions, the impact resistance is weak. In contrast, DNTF shows toughness—that is, under the same impact load, it will absorb more energy and undergo large deformation without sudden failure, and DNTF crystal has strong anti-failure ability.

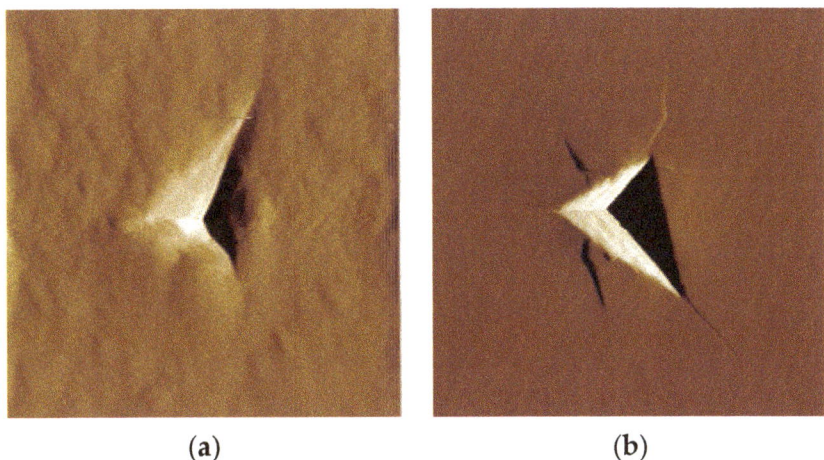

(a)   **(b)**

Figure 10. Crystal indentation crack at 3000 μN for (**a**) DNTF and (**b**) CL-20.

Table 2. Crack length and fracture toughness values at different loads.

$P_{max}/\mu N$	DNTF		CL-20	
	$L/\mu m$	K_{IC}/MPa	$L/\mu m$	K_{IC}/MPa
3000	–	–	1.89	92.36
4000	2.63	114.37	2.42	85.22
5000	3.25	114.80	3.37	64.50

3.5. Crystal Mechanics and Crystal Sensitivity

Under impact conditions, when materials with different elastic moduli deform, the higher the elastic modulus is, the higher the strain rate will be, and the greater the impact stress will be [28]. Therefore, under the dual action of stress and strain rate, stress concentration is more likely to lead to crystal fracture and the formation of "hot spots" for the high elastic modulus material. From the perspective of the mechanical properties of materials, by comparing the mechanical properties of CL-20 and DNTF crystals, it can be seen that CL-20 crystal has a high elastic modulus and hardness, and CL-20 is brittle and prone to cracking. Consequently, CL-20 is more likely to lead to an ignition response than DNTF. In addition, DNTF crystal is a typical high-energy melting and casting carrier. In addition to low modulus and hardness, it also has the property of endothermic melting, which is beneficial to inhibit the formation of "hot spots". Understandably, DNTF is less likely to react than CL-20.

As the typical high energy density materials, both DNTF and CL-20 have high energy and high sensitivity. According to the mechanical sensitivity test method of GJB772A-97, the impact sensitivity of CL-20 and DNTF is 100% and 88%, respectively [29,30], and the friction sensitivity is 100% and 84% [31,32], which means that DNTF mechanical sensitivity explosion probability is lower than CL-20. It can be concluded that although DNTF and CL-20 are both highly sensitive materials, DNTF is relatively safer than CL-20 according to the crystal mechanical properties and sensitivity performance data.

4. Conclusions

The average hardness of DNTF and CL-20 is 0.57 GPa and 0.84 GPa, and the average elastic modulus is 10.34 GPa and 20.30 GPa. The hardness, elastic modulus and local energy dissipation of CL-20 are significantly higher than those of DNTF.

Most of the pressing work of CL-20 crystal is converted to plastic work, and its elastic recovery ability is less than that of DNTF. The indentation morphology shows that CL-20 crystal is more prone to cracking than DNTF crystals, and the fracture toughness value is lower than that of DNTF crystals. Compared with DNTF crystals, CL-20 is a brittle crystal material with high modulus and high hardness.

Based on crystal micromechanics, CL-20 crystals have a lower sensitivity than DNTF and are more prone to an ignition response.

Author Contributions: Conceptualization, H.N., Y.Z. and X.W.; methodology, H.N., G.N. and Y.B.; formal analysis, H.N., Y.Z. and G.N.; investigation, G.N., P.S. and F.J.; writing—original draft preparation, H.N., G.N. and Y.B.; writing—review and editing, G.N. and Y.B.; supervision, H.N. and X.W. All authors have read and agreed to the published version of the manuscript.

Funding: This research received no external funding.

Institutional Review Board Statement: Not applicable.

Informed Consent Statement: Not applicable.

Data Availability Statement: Not applicable.

Conflicts of Interest: The authors declare no conflict of interest.

References

1. Muravyev, V.; Meerov, D.B.; Monogarov, K.A.; Melnikov, I.N.; Kosareva, E.K.; Fershtat, L.L.; Sheremetev, A.B.; Dalinger, I.L.; Fomenkov, I.V.; Pivkina, A.N. Sensitivity of Energetic Materials: Evidence of Thermodynamic Factor on a Large Array of CHNOFCl Compounds. *Chem. Eng. J.* **2021**, *421*, 129804. [CrossRef]
2. Li, Y.; Kong, J.-X.; Guo, W.-C. Current state and development trends of nano-indentation technology. *Mech. Sci. Technol. Aerosp. Eng.* **2017**, *36*, 469–474. [CrossRef]
3. Gong, J.-H. Theoretical foundation and data analyses of quasi-static nanoindentation. *J. Ceram.* **2021**, *42*, 181–245.
4. Liu, Y.; Chen, D.-F. Measurement of material mechanical properties using nanoindentation and finite element simulation. *J. Wuhan Univ. Technol. Transp. Sci. Eng.* **2003**, *27*, 690–693.
5. Ruestes, C.J.; Alhafez, I.A.; Urbassek, H.M. Atomistic Studies of Nanoindentation—A Review of Recent Advances. *Crystals* **2017**, *7*, 293. [CrossRef]
6. Milman, Y.V.; Chugunova, S.I.; Goncharova, I.V. Plasticity Characteristic Obtained by Indentation Technique for Crystalline and Noncrystalline Materials in the Wide Temperature Range. *High Temp. Mater. Process.* **2006**, *25*, 39–46. [CrossRef]
7. Poitrimolt, M.; Cheikh, M.; Bernhart, G.; Velay, V. Characterisation of the transverse mechanical properties of carbon/carbon composites by spherical indentation. *Carbon* **2014**, *66*, 234–245. [CrossRef]
8. Dong, H.E.; Zhu, J.C.; Lai, Z.H.; Yong, L.I.U.; Yang, X.W.; Nong, Z.S. Residual elastic stress-strain field and geometrically necessary dislocation density distribution around Nano-indentation in TA15 titanium alloy. *Trans. Nonferrous Met. Soc. China* **2013**, *23*, 7–13.
9. Iost, A.; Bigot, R. Indentation size effect: Reality or artifact. *Mater. Sci.* **1996**, *31*, 3573–3577. [CrossRef]
10. Shi, M.X.; Huang, Y.; Hwang, K.C. Fracture in a higher order elastic continuum. *Mech. Phys. Solids* **2000**, *48*, 2513–2538. [CrossRef]
11. Gerberich, W.W. Nanoindentation methods in interfacial fracture testing, Chapter 13. In *Comprehensive Structural Integrity*; Elsevier Ltd.: New York, NY, USA, 2003; pp. 453–494.
12. Swadener, J.G.; George, E.P.; Pharr, G.M. The correlation of the indentation size effect measured with indenter of various shapes. *Mech. Phys. Solids* **2002**, *50*, 681–694. [CrossRef]
13. Ramos, K.J.; Bahr, D.F. Mechanical behavior assessment of sucrose using nanoindentation. *J. Mater. Res.* **2007**, *22*, 2037–2045. [CrossRef]
14. Ramos, K.J.; Hooks, D.E.; Bahr, D.F. Direct observation of plasticity and quantitative hardness measurements in single crystal cyclotrimethylene trinitramine by nanoindentation. *Philos. Mag.* **2009**, *89*, 2381–2402. [CrossRef]
15. Hudson, R.J.; Zioupos, P.; Gill, P.P. Investigating the Mechanical Properties of RDX Crystals Using Nano-Indentation. *Propellants Explos. Pyrotech.* **2012**, *37*, 191–197. [CrossRef]
16. Mathew, N.; Sewell, T.D. Nanoindentation of the Triclinic Molecular Crystal 1,3,5-Triamino-2,4,6-Trinitrobenzene: A Molecular Dynamics Study. *J. Phys. Chem. C* **2016**, *120*, 8266–8277. [CrossRef]
17. Taw, M.R.; Yeager, J.D.; Hooks, D.E.; Carvajal, T.M.; Bahr, D.F. The mechanical properties of as-grown noncubic organic molecular crystals assessed by nanoindentation. *J. Mater. Res.* **2017**, *32*, 2728–2737. [CrossRef]
18. Zhai, M.; McKenna, G.B. Mechanical properties of pentaerythritol tetranitrate(PETN) single crystals from nano-indentation: Depth dependent response at the nano meter scale. *Cryst. Res. Technol.* **2016**, *51*, 414–427. [CrossRef]
19. Li, M.; Chen, T.; Pang, H.; Huang, M. Ruptures and mesoscale fracture behaviors of RDX crystals. *Chin. J. Energetic Mater.* **2013**, *21*, 200–204.

20. Li, M.; Tan, W.-J.; Kang, B.; Xu, R.-J.; Tang, W. The Elastic Modulus of b-HMX Crystals Determined by Nanoindentation. *Propellants Explos. Pyrotech.* **2010**, *35*, 379–383. [CrossRef]
21. Zhu, Y.J.; Tu, J.; Chang, H.; Su, P.F.; Chen, Z.Q.; Xu, M. Comparative study on micromechanical properties of DNAN and TNT crystals by nanoindentation. *Chin. J. Explos. Propellants* **2017**, *40*, 68–71, 84.
22. Zhu, Y.; Zhou, W.; Qu, C.; Li, X.; Xu, M.; Wang, M. Research on micromechanical properties of HATO and RDX crystals. *Initiat. Pyrotech.* **2021**, *000*, 41–44.
23. Kosareva, E.K.; Gainutdinov, R.V.; Michalchuk, A.A.; Ananyev, I.V.; Muravyev, N.V. Mechanical stimulation of energetic materials at the nanoscale. *Phys. Chem. Chem. Phys. PCCP* **2022**, *24*, 8890–8900. [CrossRef]
24. Aravindu, P.; Rani, K.D.; Shaik, A.M.; Kommu, N.; Rao, V.K. Synthesis of Novel Hexaazaisowurtzitane Cages to Access CL-20. *Asian J. Org. Chem.* **2022**, *11*, e202100680. [CrossRef]
25. Wen, M.P.; Xu, R.; Zhang, H.B.; Sun, J.; Yan, X.L.; Chi, Y. Modulus and hardness of TNT single crystal (100) plane by a nano indenter. *Chin. J. Energetic Mater.* **2014**, *22*, 430–432.
26. Wen, M.P.; Fu, T.; Tang, M.F.; Tan, K.Y.; Xu, R.; Chen, T.N. Linear correlation between micro-plastic properties of TNT, RDX and HMX explosives crystals and their corresponding impact sensitivities. *Chin. J. Energetic Mater.* **2020**, *28*, 1102–1108.
27. Bao, Y.; Wang, W.; Zhou, Y. Investigation of the relationship between elastic modulus and hardness based on depth-sensing indentation measurements. *Acta Mater.* **2004**, *52*, 5397–5404. [CrossRef]
28. Milman, Y.V. Plasticity characteristic obtained by indentation. *J. Phys. D Appl. Phys.* **2011**, *41*, 074013. [CrossRef]
29. Chen, C.-Y.; Li, B.-H.; Li, K.; Gao, L.L.; Wang, X.F.; Nan, H.; Ning, D. Relationship between ignition characteristics and mechanical properties of PBX under impact loading. *Chin. J. Explos. Propellants* **2018**, *41*, 369–374.
30. Zhang, C.; Zhang, K.; Zhang, J.-X. Research progress of CL-20 desensitization process. *Shanxi Chem. Ind.* **2021**, *41*, 35–38.
31. Hang, G.Y.; Yu, W.L.; Wang, T. Preparation and performance test of CL-20/RDX cocrystal explosive. *Chin. J. Explos. Propellants* **2021**, *44*, 484–488.
32. Gao, J.; Wang, H.; Luo, Y.M.; Wang, H.X.; Wang, W. Study on binary phase diagram of DNAN/DNTF mixed system and its mechanical sensitivity. *Chin. J. Explos. Propellants* **2020**, *43*, 213–218, 224.

Article

Preparation and Properties of RDX@FOX-7 Composites by Microfluidic Technology

Jin Yu, Hanyu Jiang, Siyu Xu, Heng Li, Yiping Wang, Ergang Yao, Qing Pei, Meng Li, Yang Zhang and Fengqi Zhao *

Science and Technology on Combustion and Explosion Laboratory, Xi'an Modern Chemistry Research Institute, Xi'an 710065, China
* Correspondence: npecc@163.com

Abstract: 1,3,5-trinitro-1,3,5-triazacyclohexane (RDX) is a type of high energy explosive, its application in weapon systems is limited by its high mechanical sensitivity. At the same time, 1,1-diamino-2,2-dinitroethylene (FOX-7) is a famous insensitive explosive. The preparation of RDX@FOX-7 composites can meet the requirements, high energy and low sensitivity, of the weapon systems. It is difficult for the reactor to achieve uniform quality of composite material, which affects its application performance. Based on the principle of solvent-anti-solvent, the recrystallization process was precisely controlled by microfluidic technology. The RDX@FOX-7 composites with different mass ratios were prepared. At the mass ratio of 10%, the RDX@FOX-7 composites are ellipsoid of about 15 μm with uniform distribution and quality. The advantages of microscale fabrication of composite materials were verified. The results of structure characterization showed that there is no new bond formation in RDX@FOX-7, but the distribution of two components on the surface of the composite was uniform. Based on the structure characterization, we established the structure model of RDX@RDX-7 and speculated the formation process of the composites in microscale. With the increase of FOX-7 mass ratios, the melting temperature of RDX was advanced, the thermal decomposition peak of RDX changed to double peaks, and the activation energy of RDX@FOX-7 composite decreased. These changes were more pronounced between 3 and 10% but not between 10 and 30%. The ignition delay time of RDX@FOX-7 was shorter than that of RDX and FOX-7. RDX@FOX-7 burned more completely than RDX indicating that FOX-7 can assist heat transfer and improve the combustion efficiency of RDX.

Keywords: composite explosives; microfluidic; laser ignition; RDX@FOX-7; thermal analysis; micro-Raman

Citation: Yu, J.; Jiang, H.; Xu, S.; Li, H.; Wang, Y.; Yao, E.; Pei, Q.; Li, M.; Zhang, Y.; Zhao, F. Preparation and Properties of RDX@FOX-7 Composites by Microfluidic Technology. *Crystals* **2023**, *13*, 167. https://doi.org/10.3390/cryst13020167

Academic Editors: Rui Liu, Yushi Wen, Weiqiang Pang and Robert F. Klie

Received: 15 December 2022
Revised: 11 January 2023
Accepted: 15 January 2023
Published: 18 January 2023

1. Introduction

With the development of modern weapon systems, ammunition requires not only higher energy levels and energy release efficiency [1] but also a higher security in the process of preparation, storage, transportation, and combat readiness [2]. Therefore, high energy insensitive explosives have become an important direction for modern ammunition development [3]. RDX is a type of high energy explosive with good stability, which is often used in propellant as a high energy additive. However, the high mechanical sensitivity limits its application in weapon system [4]. At present, the research on reducing the sensitivity of RDX, whether to improve the crystal quality or to cover the insensitive material, has been very thorough [4–8]. Shi [9] and Guo [10] prepared RDX-based composite energetic microspheres with narrow particle size distribution. The particle size changes from 50~200 μm to 0.5~7 μm, and dense coating by polymer coating, which improved the thermal safety and mechanical sensitivity of RDX.

Furthermore, it was expected that the energy properties of the explosives can be maintained or even improved while desensitizing treatment [11]. FOX-7 has stable molecular structure and excellent comprehensive properties, exists in α phase under normal temperature, and normal pressure [12–14]. Its crystal density is 1.878 g·cm^{-3}, detonation velocity is 8870 m·s^{-1}, detonation pressure is 34.0 GPa, detonation energy is high, and it is insensitive to impact, spark, and friction [15]. What matters is that it is compatible with RDX. Gao [16] and Wei [17] employed theoretical methods including molecular dynamics (MD) simulation and quantum-chemical DFT and MP2 calculation to estimate the properties of 2,4,6,8,10,12-hexanitrohexaazaisowurtzitane; 1,3,5,7-tetranitro-1,3,5,7-tetrazacyclooctane; and FOX-7 co-crystal, respectively. The result indicates that CL-20/FOX-7 co-crystal may have the high forming probability in a molar ratio of 1:1 and meet a given standard of low sensitive high energetic materials. There is no research on RDX@FOX-7 co-crystal. Tian [18] has prepared pressed explosives containing FOX-7 and RDX. It was found that FOX-7 can improve the rapid ignition growth of RDX. Zhou [19] used the principle of solvent–nonsolvent to cover FOX-7 on the surface of RDX. When the mass ratio of FOX-7 to RDX was 3:1, the impact sensitivity of RDX was reduced from 80 to 32%, achieving a positive impact on security performance. However, the distribution uniformity of RDX and FOX-7 is poor and unstable. This is due to that it is inevitable that there exists a large range of concentration gradient and temperature gradient in the reactor, which makes the reaction environment in different regions of the reactor vary greatly, resulting in different crystalline forms after recrystallization, polycrystalline particles and single crystal particles are often doped [20]. The material quality is not easy to control.

In recent years, microfluidic technology has gradually emerged in the field of synthesis and preparation of energetic materials. Microfluidic technology is to process or manipulate microfluids in micropipes [21], which can realize the rapid and uniform mixing of multifluids in microscale. It is widely used in the fields of fine chemicals preparation [22], nanofunctional materials preparation [23], and so on [24]. The recrystallization process can be controlled precisely in microscale [25], so it is suitable for preparing explosives with small particle size and narrow distribution. Wu [26] used this technique to obtain nano-RDX particles ranging from 150 to 900 nm. Based on these characteristics, microfluidic technology has unique advantages in the preparation of composite explosives. Zhou illustrated the key to the preparation of Pb·BaTNR co-crystal is to enable the Stephen acid group to be able to combine lead ions and ions at the same time, which can be achieved by providing limited ion concentration and reaction space through the microfluidic technology, but not in the reactor. Similarly, a series of composite explosives, such as HNS@NC [27], nAl@PVDF@CL-20 [28], combustion catalyst A@NC, Pb(N3)2@NC [29], Zr@NC [30], TATB/F2602 [31], and HMX/F2602 [23], were prepared by continuous phase extraction. All the components in these composite explosives are physically compound, and there is no new chemical bond to form. The quality of the composite explosive is controlled by concentration and flow rate [32,33]. Generally speaking, composite explosives have two forms of co-crystal and coating. Actually, there are many different combinations of components; however, no reports have explored this.

It is very valuable to realize the crystal quality treatment and coating desensitization in one step, through the integration of technology upgrades. In this paper, RDX@FOX-7 composite with uniform quality was prepared by microfluidic technology based on the solvent–nonsolvent method. The advantages of microfluidic control are demonstrated by the characterization of the morphology, structure, and uniformity of RDX@FOX -7, which is expected to provide new ideas for the preparation of high quality composite explosives. The thermal decomposition characteristics and laser ignition characteristics are also investigated to provide basic data for its application.

2. Experimental Details

2.1. Raw Materials

RDX was provided by Gansu Yinguang Chemical Industry Group Co.,Ltd. (Yinguang, China). FOX-7 was produced by Xi'an Modern Chemistry Research Institute. Dimethyl sulfoxide (DMSO) was purchased from Chengdu Cologne Chemical Co.,Ltd. (Chengdu, China). The deionized water was made in the laboratory.

2.2. Preparation of RDX@FOX-7 Composite

RDX was dissolved in DMSO with a concentration of 0.03 g/mol. FOX-7 was added in various mass percentages (3%, 10%, 30%) to the solution and stirred until completely dissolved as the solvent phase. Deionized water was selected as the non-solvent phase. The two-phase solution was injected into the microfluidic chip at the flow rates of 1 and 5 mL/min, respectively by the driving device. The samples were filtered out as suspension and washed several times with deionized water to remove the residual solvent, and RDX@FOX-7 composites were obtained by freeze drying method. The production flow chart is showed in Figure 1. The microreactor used in the microfluidic system is heart-shaped, with high mixing efficiency, and the depth is 700 μm. In order to avoid blocking, the pipe connected at the outlet has an inner diameter of 1 mm and a length of 0.3 m.

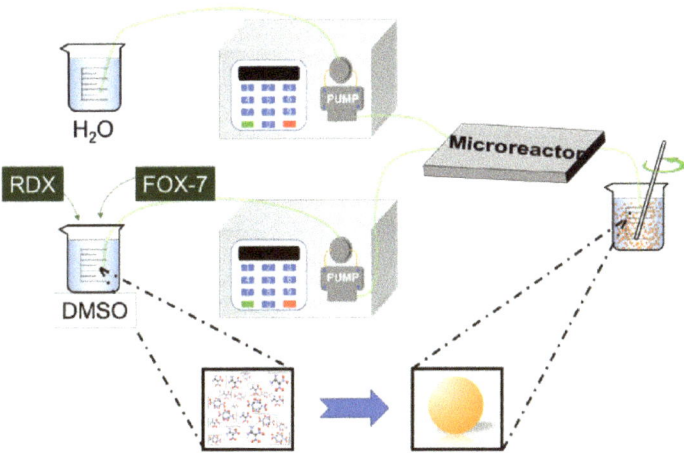

Figure 1. The flowchart for preparation of RDX@FOX-7 with microfluidic device.

2.3. Characterization Methods

The morphology of the sample was observed by scanning electron microscopy (SEM, FEI JSM-5800). The structure and composition were analyzed using X-ray diffraction (XRD, PANalytical Empyrean, Shanghai, China) and Fourier transform infrared spectroscopy (FT-IR, Tensor 27, Bruker, Germany). The quality of the composite was evaluated by a micro-Raman spectrometer (INVIAnvia type, spatial resolution is 1μm). In order to avoid the accumulation of thermal effects, it is necessary to test random sampling points, and the distance between two points should be greater than 1 mm. The effect of FOX-7 content on thermal decomposition characteristics of RDX was studied by differential scanning calorimeter (DSC, 200 F3, NETZSCH, Selb, Germany) with a heating rate of 5, 10, 15, 20 °C·min^{-1}, respectively (sample mass of (0.5 ± 0.2) mg, using an aluminum crucible with holes, normal pressure, and nitrogen flow rate of 50 mL/min). Kissinger, Ozawa, Friedman, and Starink methods were employed to obtain the activation energies of RDX@FOX-7 with different content (Equations (1)–(4)).

$$\ln\left(\frac{\beta}{T_P^2}\right) = \ln\left(\frac{AR}{E_k}\right) - \frac{E_k}{R}\frac{1}{T_P} \tag{1}$$

$$\lg\beta = \lg\left[\frac{AE_O}{RG(\alpha)}\right] - 2.315 - 0.4567\frac{E_O}{RT_P} \tag{2}$$

$$\ln\left(\beta\frac{d\alpha}{dT_P}\right) = \ln[Af(\alpha)] - \frac{E_F}{RT_P} \tag{3}$$

$$\ln\left(\frac{\beta}{T_p{}^{1.92}}\right) = \text{Const} - 1.0008\frac{E_S}{RT_p} \tag{4}$$

where E_k, E_O, E_F, and E_S are the activation energy with different content; T_p is thermal decomposition peak temperature; β is heating rate; R is universal gas constant; and A is pre-exponential factor.

3. Results and Discussion

3.1. Morphological and Structure Analysis

For explosives, morphological is very important for security performance. When stimulated by the outside, the irregular particles are easy to form hot spots at the corners. The spherical-like crystal has a smooth surface without edges and angles, less accumulation of hot spots, and mechanical sensitivity will be reduced. Smaller particle size also facilitates hot spot dispersion. Figure 2 shows electron microscopic images of RDX, FOX-7, and RDX@FOX-7. It is evident that the RDX crystal has the characteristics of smooth surface and large size. FOX-7 crystal is more irregular. The morphology and particle size distribution of RDX@FOX-7 are significantly different from those of the raw materials. When the mass ratio of FOX-7 is 3%, FOX-7 is adsorbed on the surface of the RDX particles in the form of small particles. This may be due to the low supersaturation of FOX-7 in the solution, which leads to the late crystallization point. When RDX reaches the equilibrium of dissolution and RDX crystal reaches the growth limit, FOX-7 begins to precipitate slowly, which results in the formation of such surface adsorption. When the FOX-7 mass ratio is 10%, it is much better, the particle size of RDX@FOX-7 was reduced obviously, about 15μm. On the other hand, RDX@FOX-7 is similar to RDX in shape of ellipsoid, and its surface is smooth without obvious edges and defects. RDX cannot be distinguished from FOX-7 by the SEM images, which indicates that the composite may be in the form of coating or co-crystal. When the mass ratio of FOX-7 was 30%, some larger particles appeared, which made the overall consistency of the composite worse. Therefore, it is concluded that 10% RDX@FOX-7 composite explosive has better comprehensive performance. Compared with the RDX@FOX-7 prepared in the reactor [19], which can be defined as the bonding of two component particles, the microfluidic technique has obvious advantages in morphology control.

Eutectic is a type of multi-component molecular crystal which is formed in the same lattice by non-covalent bond. When co-crystal is formed, the crystal structure and internal composition of each component changes radically. If the result is RDX@FOX-7 co-crystal, there must be a new bond formation. Thus, the composite was further characterized by XRD and FT-IR, and the results are showed in Figures 3 and 4. Peaks at 13.3°, 18.1°, 22.2° and 15.2°, 28.2° were used as markers for RDX and FOX-7, respectively. Many characteristic peaks of RDX can be observed in RDX@FOX-7 with different mass ratios, and the peaks are stronger. This is because the main component of the composite is RDX. In the spectra of RDX@FOX-7 with mass ratios of 3%, there were no FOX-7 characteristic peaks at 15.2° and 28.2°. The two peaks became stronger with the increase of mass ratio of FOX-7. The formation of RDX@FOX-7 composite can be determined. However, despite the relative intensity of the diffraction peak has changed, there is no new diffraction peak shows that the RDX@FOX-7 composite explosive does not form co-crystal. In the infrared spectrum, we can observe that the amino absorption peaks of RDX@FOX-7 at 3423, 3332, 3298 cm^{-1} have a blue shift of 2~3 cm^{-1} wave number compared to FOX-7, the C-H bond stretching vibration peak at 3074 and 3000 cm^{-1} has a red shift of 2~3 cm^{-1} wave number compared to RDX, and the nitro adsorption peak at 1594 cm^{-1} has shifted compared to FOX-7 and RDX. This may be because the amino of FOX-7 and the nitro of RDX formed a part of the hydrogen bond after forming the composition. The interaction between the molecules is weak, which can only lead to a certain degree of amino, nitro absorption peak shift, but in the infrared spectrum of the generation of new absorption peak. Therefore, we infer that the prepared sample is a type of composite explosive.

Figure 2. SEM images of samples: (**a**) raw RDX; (**b**) raw FOX-7; (**c–e**) RDX@FOX-7 with mass ratios of 10%, 3%, 30%.

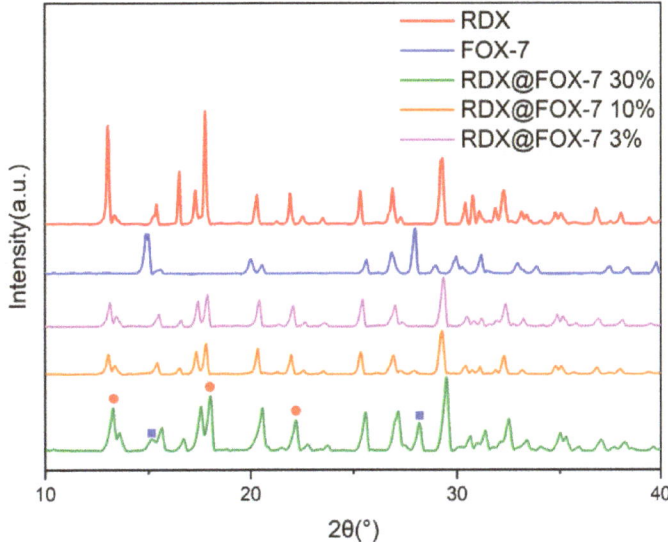

Figure 3. XRD patterns of the RDX, FOX-7, and RDX@FOX-7 composite.

Most composites have problems with uneven distribution These unevenly distributed composites are easy to be separated in use and cannot guarantee stable performance, which has a great impact on the reliability and service life of products. Raman spectroscopy can be used to identify the components distribution of composite by observing the characteristic peak of Raman characteristic peak. The crystal quality of RDX@FOX-7 was characterized by micro-Raman spectra at 10 different points, and the results are shown in Figure 5. It can be seen that the peak shape in the Raman spectra of each point has a good consistency. Seven important Raman peaks were observed. The characteristic peaks at 476 cm^{-1}, 1340 cm^{-1}, and 1548 cm^{-1} belong to the symmetric rocking vibration peak corresponding to N-H, the symmetric stretching vibration peak corresponding to C-NO$_2$, and the torsional vibration peak corresponding to NH$_2$ in FOX-7. Similarly, the characteristic peaks at 467 cm^{-1}, 923 cm^{-1}, 1274 cm^{-1}, 1311 cm^{-1}, 1388 cm^{-1}, and 1377 cm^{-1} belong to the tensile vibration of N-N, the rocking vibration of CH$_2$, the tensile vibration of N-NO$_2$ and the bending vibration outside the C-H plane in RDX, indicating that a type of composite explosive with excellent uniformity was achieved.

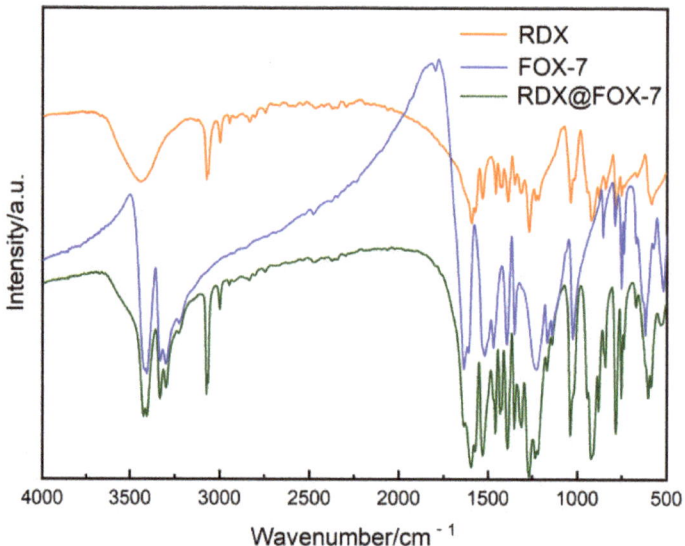

Figure 4. FT-IR spectra of the RDX, FOX-7, and RDX@FOX-7 composite with mass ratios of 10%.

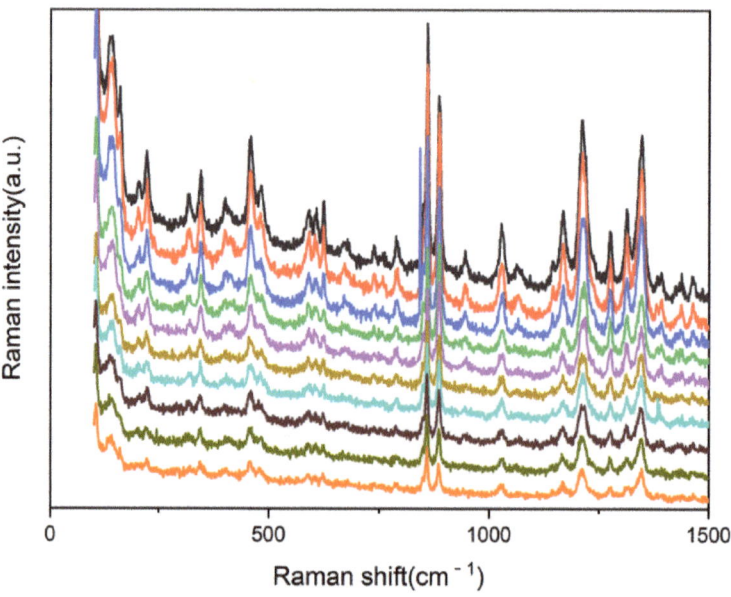

Figure 5. Micro-Raman spectra of RDX@FOX-7 at 10 different points.

Deeply, we modeled the structure of RDX@FOX-7 (shown in Figure 6). The reason for this effect lies in the formation of supersaturation, when two-phase solutions meet at the microscale, which causes the burst nucleation of crystals in a short time. Fluid mixing depends mainly on the diffusion of fluid molecules. The typical feature of microscale flow is that the Reynolds number (Re) is very low and the flow is mostly in the laminar flow region. It is very difficult to achieve efficient mixing without an external physical field. Therefore, we introduce disturbances through the structure of the microreactor to disrupt the stratified flow and enhance the mixing between the layers. Thus, the supersaturation of each point in the channel remains highly consistent and provides a stable driving force for crystal precipitation. In the pipe connected at the outlet, flow will maintain a uniform supersaturation, crystals continue to grow precipitation. Finally, the composite material with uniform distribution of two components is formed. Based on this, we believe that the uniformity of the composite material can be controlled by controlling the crystallization points of different components. When the crystallization points of two components are close, the uniform composite material such as RDX@FOX-7 is formed. When the process of two sets of crystals is independent in time, the composite material of the core shell structure can be obtained. Of course, we can also prepare more composite materials through other parameters, such as regulatory concentration and temperature, to meet the needs. It is interesting to be able to precisely control the internal structure of composites.

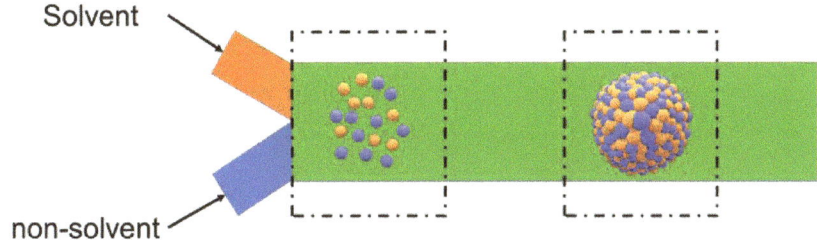

Figure 6. Structure model for RDX@FOX-7.

3.2. Thermal Behavior of RDX@FOX-7 Composite

Thermal decomposition is the initial stage of explosive combustion. The thermal decomposition characteristics of each component of composite explosive directly affect the combustion performance. The thermal decomposition properties of RDX@FOX-7 at different mass ratios were carried out by differential scanning calorimetry, as showed in Figure 7. The peak temperatures of thermal decomposition at various heating rates are listed in Table 1. Combined with DSC curve, the endothermic peak of $\alpha \rightarrow \beta$ and $\beta \rightarrow \gamma$ phase transition cannot be observed at about 116 °C and 145 °C when the content of FOX-7 is 3% and 10%. Thermal decomposition is the initial stage of explosive combustion. The thermal decomposition characteristics of each component of composite explosive directly affect the combustion performance. It can be seen that the melting endothermic peak of RDX is advanced with the increase of FOX-7 content, but the change is obvious between 3% and 10% of FOX-7 content, and it is advanced at about 4 °C. The effect of FOX-7 on the melting peak is not obvious, but the peak width and peak area increases, the heat needed for melting increases, and the thermal stability is better. At 10% of FOX-7, a sharp exothermic peak appears around 205 °C. This represents the first thermal decomposition of the FOX-7 with $\gamma \rightarrow \delta$ phase transition. In addition, with the increase of the heating rate, the peak temperature both of RDX@FOX-7 at different mass ratios moves to high temperature, the peak width becomes wider, the peak area increases, the heat release increases, and the reaction process becomes more complete. At the same heating rate, based on the two-step decomposition principle of FOX-7, with the content of FOX-7 increasing gradually, the heat release of the second decomposition increased, and the thermal decomposition of RDX gradually separated; the thermal decomposition peak of RDX@FOX-7 has a tendency to change from single peak to double peak.

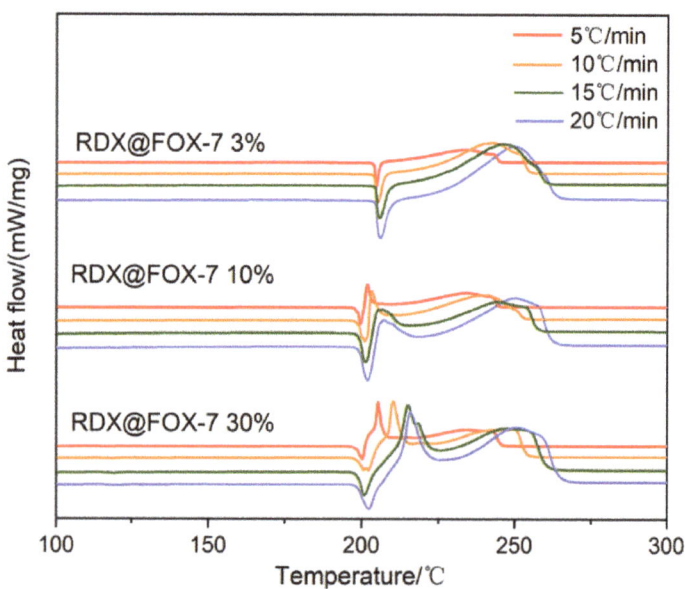

Figure 7. DSC curves of RDX@FOX-7 composite with different mass ratios at different heating rates.

Table 1. Thermal decomposition temperatures and kinetic parameters of RDX@FOX-7 with different mass ratios at different heating rates.

Sample	T_p/°C				E_O kJ/mol	E_K kJ/mol	E_F kJ/mol	E_S kJ/mol
	5 °C/min	10 °C/min	15 °C/min	20 °C/min				
RDX@FOX-7 3%	233.7	242.8	245.7	250.1	180.43	181.19	186.51	181.38
RDX@FOX-7 10%	234.2	239.7	244.3	250.5	178.36	179.00	185.40	179.19
RDX@FOX-7 30%	230.8	240.1	248.4	251.7	132.37	134.01	138.20	132.61

The thermal decomposition activation energy of RDX@FOX-7 samples were evaluated by the method of Kissinger, Flynn-Wall-Ozawa, Friedman, and Starink methods by using DSC data, and the results are showed in Table 1. The activation energy decreased with the increase of FOX-7 content from 3 to 10 %, which had little effect on the activation energy, about 1 KJ/mol. From 10 to 30 %, the activation energy decreased more obviously, about 40 KJ/mol. Combined with DSC data and activation energy calculation results, it can be fully demonstrated that FOX-7 has an obvious effect on improving the thermal stability of RDX. The effect of 10% addition of FOX-7 to RDX on the thermal stability was the best.

3.3. Ignition and Combustion Properties of RDX@FOX-7 Composite

Ignition is the beginning of combustion process, followed by pyrolysis reaction. Studying the ignition characteristics of RDX@FOX-7 can provide information support for its further application in the solid propellants. The ignition and combustion process of all samples were studied by using a laser ignition apparatus, which is shown in Figure 8. The radiation wavelength of the laser is 10.6 μm and the irradiation time is 5 ms. The time between the laser emission and the appearance of the flame is defined as the ignition delay time. The kinetics and mechanism of thermal decomposition affect the ignition delay time and flame structure, respectively. The change of ignition delay time with laser power density is shown in Figure 9. With the increase of laser power density, the ignition delay time first decreases rapidly and then tends to be constant. RDX@FOX-7 has a lower latency compared to RDX and FOX-7. This can be explained in relation to the results of thermal decomposition. The first thermal decomposition of the FOX-7 interacted with RDX, bringing the overall advance.

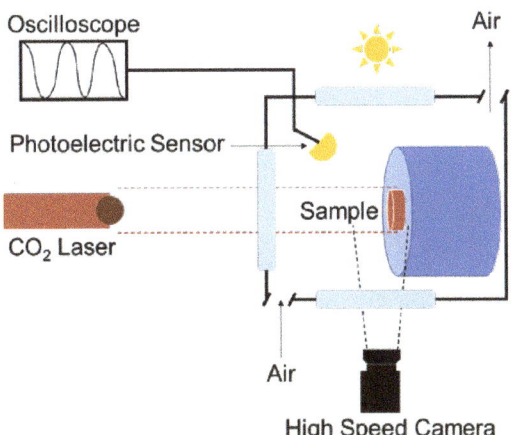

Figure 8. Schematic illustration of laser ignition apparatus.

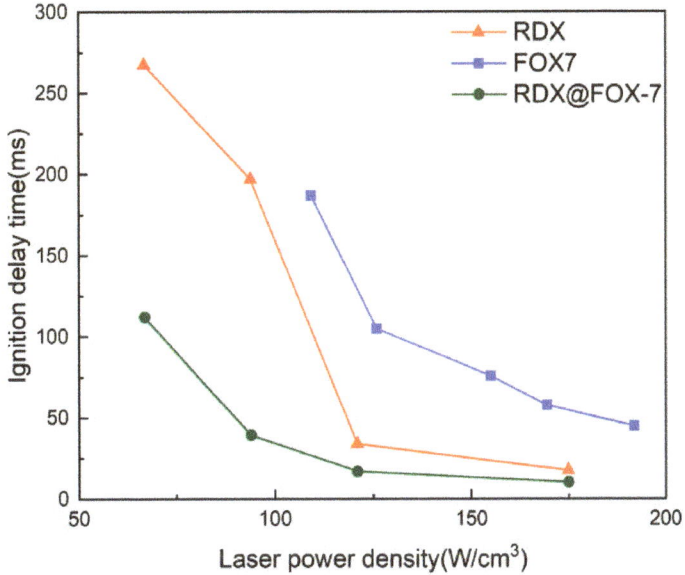

Figure 9. Ignition delay time of RDX, FOX-7 and RDX@FOX-7.

Figure 10 shows the combustion flame structures of RDX@FOX-7. When the laser power density is 169.5 W/cm^2, the ignited RDX powder has a dark red flame, and the inner layer is the bright core produced by the particle combustion. The further development of a more moderate flame has large gas clouds and more combustion of unreacted particles. It can be observed that there are melting surface and unreacted particles from the combustion residues of RDX shown in Figure 11a. Therefore, we infer that the accumulation of local heat due to the simultaneous melting and thermal decomposition of RDX will cause incomplete combustion. RDX@FOX-7 has a similar combustion flame evolution process. However, there are almost no unreacted particles, the complete combustion in the core of the flame lasts longer, the structure is stable, and the gas cloud area is smaller. Moreover, there are pores in the molten surface of the RDX@FOX-7 combustion residues, which is shown in Figure 11b, and the unreacted particles are obviously reduced. This may be due to the two thermal decompositions of FOX-7 experienced to spread the heat more evenly, accelerate the RDX melting heat of absorption, and, ultimately, make the combustion more complete. This is very critical for improving the application performance of RDX@FOX-7 in propellant. The unstable combustion of propellant is intermittent combustion, flameout during combustion, and incomplete chemical reaction. When the combustion of propellant is incomplete, it will destroy the internal ballistic performance of the engine and change the expected thrust plan. The addition of FOX-7 can ensure the energy output level to a certain extent, and it also has a high safety performance. In addition, as mentioned above, there is a very high application value for the auxiliary for RDX completely burning.

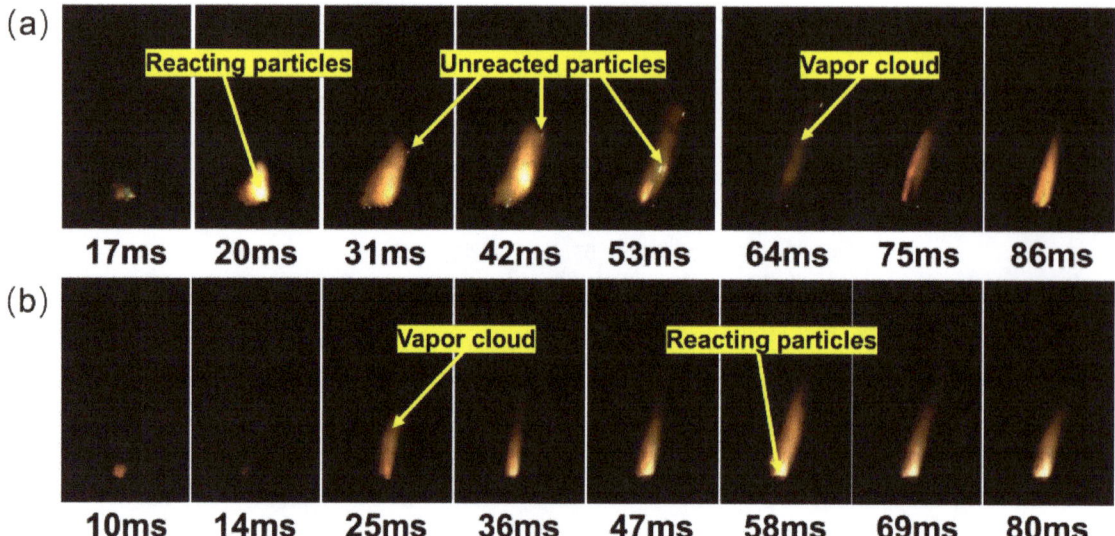

Figure 10. Sequential images for the combustion initial process of samples at the laser power density of 175 W/cm^2: (**a**) RDX, (**b**) RDX@FOX-7 composite.

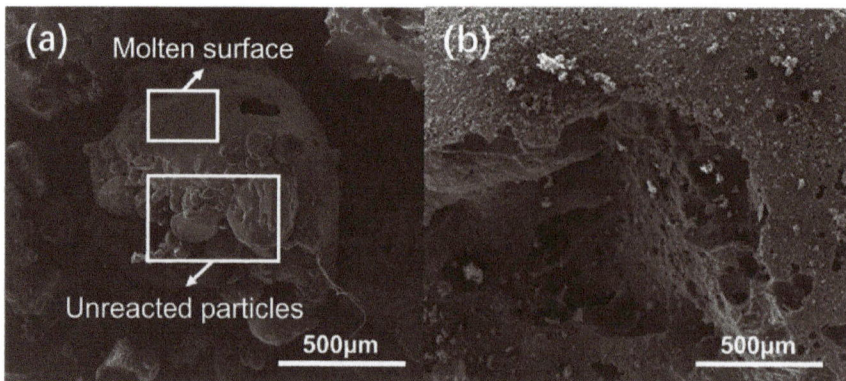

Figure 11. SEM images of the combustion residues of samples: (**a**) RDX, (**b**) RDX@FOX-7 composite.

4. Conclusions

(1) The RDX@FOX-7 composite with different mass ratios was successfully prepared by microfluidic technique. The structure was characterized by SEM, XRD, FT-IR, and micro-Raman. RDX@FOX-7 with 10% content has the best morphology, and the particle size is about 15 μm. The obtained composite explosive is of uniform distribution and uniform quality, and no co-crystal is formed.

(2) The advantages of microscale fabrication were verified. The structure model of RDX@FOX-7 is established. The formation process of the composites in microscale is predicted. It is considered that the uniformity of composites can be controlled by controlling the crystallization point of different components. In the future, more types of composites can be prepared by adjusting the concentration, temperature, and other parameters to meet the needs.

(3) With the increase of FOX-7 mass ratios, the melting temperature of RDX was advanced, the thermal decomposition peak of RDX changed to double peaks, and the activation energy of RDX@FOX-7 composite decreased. These changes were more pronounced between 3 and 10% but not between 10 and 30%.

(4) RDX@FOX-7 composite has a shorter ignition delay compared to RDX and FOX-7. FOX-7 can assist in heat transfer, making RDX@FOX-7 composite's combustion more complete.

In summary, the effect of a 10% addition of FOX-7 to RDX makes the overall comprehensive properties of the composites reach the best. This study provides a promising new method for the preparation of homogeneous and stable composites. Nevertheless, this study still has some limitations. For example, consider if the additional compressing tests would lead to specific reorientation of the crystals inside the composite, as well as a more comprehensive characterization of the applied properties of composite explosives. Next, we will also continue to study this in depth.

Author Contributions: S.X. and F.Z. contributed to the conception of the study; J.Y. and H.J. performed the experiments; Y.W., E.Y. and H.L. contributed significantly to analysis and manuscript preparation; J.Y. performed the data analysis and wrote the manuscript; Q.P., Y.Z. and M.L. helped perform the analysis with constructive discussions. All authors have read and agreed to the published version of the manuscript.

Funding: This work was financially supported by the National Natural Science Foundation of China (Funder: Jianhua Yi, Funding number: 22075226), (Funder: Yang Zhang, Funding number: 22205178).

Institutional Review Board Statement: Not applicable.

Informed Consent Statement: Informed consent was obtained from all subjects involved in the study.

Data Availability Statement: Not applicable.

Acknowledgments: The authors would like to gratefully thank Pang for his valuable suggestions to improve the quality of this article.

Conflicts of Interest: The authors declare no conflict of interest.

References

1. Jia, X.; Hou, C.; Wang, J.; Tan, Y.; Zhang, Y.; Li, C. Research Progress on the Desensitization Technology of Nitramine Explosives. *Chin. J. Explos. Propellants* **2018**, *41*, 326–333.
2. Liu, J.; Jiang, W.; Yang, Q.; Song, J.; Hao, G.-Z.; Li, F.-S. Study of nano-nitramine explosives: Preparation, sensitivity and application. *Def. Technol.* **2014**, *10*, 184–189. [CrossRef]
3. Sikder, A.; Sikder, N. A review of advanced high performance, insensitive and thermally stable energetic materials emerging for military and space applications. *J. Hazard. Mater.* **2004**, *112*, 1–15. [CrossRef] [PubMed]
4. Jia, X.; Hu, Y.; Xu, L.; Liu, X.; Ma, Y.; Fu, M.; Wang, J.; Xu, J. Preparation and Molecular Dynamics Simulation of RDX/MUF Nanocomposite Energetic Microspheres with Reduced Sensitivity. *Processes* **2019**, *7*, 692. [CrossRef]
5. Xu, W.; Deng, J.; Liang, X.; Wang, J.; Li, H.; Guo, F.; Li, Y.; Yan, T.; Wang, J. Comparison of the effects of several binders on the combination properties of cyclotrimethylene trinitramine (RDX). *Sci. Technol. Energ. Mater.* **2021**, *82*, 29–37.
6. Jia, X.; Cao, Q.; Guo, W.; Li, C.; Shen, J.; Geng, X.; Wang, J.; Hou, C. Synthesis, thermolysis, and solid spherical of RDX/PMMA energetic composite materials. *J. Mater. Sci. Mater. Electron.* **2019**, *30*, 20166–20173. [CrossRef]

7. Li, N.; Xiao, L.; Jian, X.; Xu, F.; Zhou, W. Coating of RDX with GAP-based energetic polyurethane elastomer. *J. Solid Rocket Technol.* **2012**, *35*, 212–215.
8. Zhang, J.; Zhang, J.; Wang, B.; Liu, S.; Zhang, T.; Zhou, D. Technology of Ultra-fine RDX Coating with Rapid Expansion of Supercritical Solution Method. *Energ. Mater.* **2011**, *19*, 147–151.
9. Shi, X.; Wang, J.; Li, X.; Wang, J. Preparation and characterization of RDX-based composite energetic microspheres. *Chin. J. Energ. Mater.* **2015**, *23*, 428–432.
10. Guo, C.; Jia, X.; Geng, X.; Wang, J.; Hou, C.; Tan, Y. Effect of high temperature water vapor on morphology and thermal decomposition properties of RDX/PMMA. *J. Solid Rocket Technol.* **2018**, *41*, 338–342.
11. Klaumünzer, M.; Pessina, F.; Spitzer, D. Indicating Inconsistency of Desensitizing High Explosives against Impact through Recrystallization at the Nanoscale. *J. Energ. Mater.* **2017**, *35*, 375–384. [CrossRef]
12. Li, X.; Sun, H.; Song, C.; Zhang, X.; Yang, W.; Wang, J. Effect of binder on properties of Cl-20/Fox-7 based PBX. *Chin. J. Explos. Propellant* **2020**, *43*, 51–56.
13. Lal, S.; Staples, R.J.; Jean'ne, M.S. FOX-7 based nitrogen rich green energetic salts: Synthesis, characterization, propulsive and detonation performance. *Chem. Eng. J.* **2023**, *452*, 139600. [CrossRef]
14. Fangjian, S.; Ting, W.; Yinhua, M.; Meiheng, L. Theoretical study on several important decomposition paths of FOX-7 and its derivatives. *Comput. Theor. Chem.* **2022**, *1217*, 113895.
15. Nazin, G.M.; Dubikhin, V.V.; Kazakov, A.I.; Nabatova, A.V.; Krisyuk, B.E.; Volkova, N.N.; Shastin, A.V. Kinetics of the Decomposition of 1,1-Diamino-2,2-Dinitroethylene (FOX-7). Part 4: Comparison of the Decomposition Reactions of FOX-7 and Its Diazacyclic Derivatives. *Russ. J. Phys. Chem. B* **2022**, *16*, 308–315. [CrossRef]
16. Gao, H.F.; Zhang, S.H.; Ren, F.D.; Liu, F.; Gou, R.J.; Ding, X. Theoretical insight into the co-crystal explosive of 2,4,6,8,10,12-hexanitrohexaazaisowurtzitane (CL-20)/1,1-diamino-2,2-dinitroethylene (FOX-7). *Comput. Mater. Sci.* **2015**, *107*, 33–41. [CrossRef]
17. Wei, Y.; Ren, F.; Shi, W.; Zhao, Q. Theoretical insight into the influences of molecular ratios on stabilities and mechanical properties, solvent effect of HMX/FOX-7 cocrystal explosive. *J. Energ. Mater.* **2016**, *34*, 426–439. [CrossRef]
18. Tian, X.; Huang, Y.; Wang, X.; Xu, H.; Li, W.; Zhao, K.; Yang, H.; Li, Y. Influence of Mixture Ratio of FOX-7 and RDX on Slow Cook-off and Shock Sensitivity of Pressed Explosives. *Explos. Mater.* **2019**, *48*, 38–41.
19. Zhou, C.; Chang, P.; Hu, L.; Li, X.; Xiong, C.; Wang, B. Preparation, Characterization and Properties of FOX-7/RDX Composite Explosive. *Chin. J. Explos. Propellant* **2020**, *43*, 669–673+680.
20. Zhou, N.; Zhu, P.; Rong, Y.; Xia, H.; Shen, R.; Ye, Y.; Lv, S. Microfluidic Synthesis of Size-Controlled and Morphologically Homogeneous Lead Trinitroresorcinate Produced by Segmented Flow. *Propellants Explos. Pyrotech.* **2016**, *41*, 899–905. [CrossRef]
21. Shi, J.; Zhao, S.; Jiang, H.; Xu, S.; Zhao, F.; Shen, R.; Ye, Y.; Zhu, P. Multi-size control of homogeneous explosives by coaxial microfluidics. *React. Chem. Eng.* **2021**, *6*, 2354–2363. [CrossRef]
22. Lu, J.; Wang, H.; Pan, J.; Fang, Q. Research progress of microfluidic technology in synthesis of micro/nano materials. *Acta Chim. Sinica* **2021**, *79*, 809–819. [CrossRef]
23. Zhou, J.; Wu, B.; Wang, M.; Liu, S.; Xie, Z.; An, C.; Wang, J. Accurate and efficient droplet microfluidic strategy for controlling the morphology of energetic microspheres. *J. Energ. Mater.* **2021**, 1–18. [CrossRef]
24. Yang, Z.; Zhu, P.; Zhang, Q.; Shi, J.; Wang, K.; Shen, R. Microcrystalline PETN Prepared Using Microfluidic Recrystallization Platform and Its Performance Characterization. *Propellants Explos. Pyrotech.* **2021**, *46*, 1097–1106. [CrossRef]
25. Yu, J.; Xu, S.Y.; Jiang, H.Y.; Zhao, F.Q. Application and Development Trend of Microfluidic Technology in Preparation of Energetic Materials. *Acta Pyrodynamites* **2021**, *6*, 2354–2363.
26. Wu, J.W.; Xia, H.M.; Zhang, Y.Y.; Zhao, S.F.; Zhu, P.; Wang, Z.P. An efficient micromixer combining oscillatory flow and divergent circular chambers. *Microsyst. Technol.* **2018**, *25*, 2741–2750. [CrossRef]
27. Han, R.; Chen, J.; Zhang, F.; Wang, Y.; Zhang, L.; Lu, F.; Wang, H.; Chu, E. Fabrication of microspherical Hexanitrostilbene (HNS) with droplet microfluidic technology. *Powder Technol.* **2021**, *379*, 184–190. [CrossRef]
28. Yazhi, C.; Qian, W.; Hui, R.; Tao, Y. Preparation and Characterization of nAl@PVDF@CL-20 Composite Energetic Particles Assembled via Microfluidic Method. *Chin. J. Energ. Mater.* **2022**, *30*, 1–9.
29. Ruishan, H.; Fang, Z.; Feipeng, L.; Yanlan, W.; Lei, Z.; Jianhua, C.; Rui, Z.; Haifu, W.; Enyi, C. Preparation and Modification Technology of Lead Azide Primary Explosive Based on Microfluidics. *Chin. J. Energ. Mater.* **2022**, *30*, 451–458.
30. Yipeng, F.; Jinyu, S.; Peng, Z.; Ruiqi, S.; Bin, Y.; Anmin, Y.; Enyi, C. Microscale Continuous Flow Preparation and Characterization of Ultra Zr@NC. *Chin. J. Energ. Mater.* **2022**, *30*, 417–423.
31. Jinqiang, Z.; Kai, L.; Yunyan, G.; Rui, Z.; Jiahui, S.; Bidong, W.; Chongwei, A.; Jingyu, W. Preparation of TATB-based PBX Composite Microspheres by Droplet Microfluidic Technology. *Chin. J. Energ. Mater.* **2022**, *30*, 439–445.
32. Li, L.; Ling, H.; Tao, J.; Pei, C.; Duan, X. Microchannel-confined crystallization: Shape-controlled continuous preparation of a high-quality CL-20/HMX cocrystal. *Cryst. Eng. Comm.* **2022**, *24*, 1523–1528. [CrossRef]
33. Li, L.; Yin, T.; Wu, B.; Duan, X.; Pei, C. Preparation of CL-20/HMX Co-crystal by Microchannel Crystallization Based on Solvent/non-solvent Method. *Chin. J. Energ. Mater.* **2021**, *29*, 62–69.

 *crystals*

Article

Micromagnetic Simulation of Increased Coercivity of (Sm, Zr)(Co, Fe, Cu)$_z$ Permanent Magnets

Mark V. Zheleznyi [1,2,3], Natalia B. Kolchugina [1,2,*], Vladislav L. Kurichenko [3], Nikolay A. Dormidontov [1,2], Pavel A. Prokofev [2], Yuriy V. Milov [1], Aleksandr S. Andreenko [1], Ivan A. Sipin [1], Andrey G. Dormidontov [1] and Anna S. Bakulina [2]

1   LLC Magnitoelectromechanics, ul. Tvardovskogo 8, build. 1, Moscow 123458, Russia
2   Baikov Institute of Metallurgy and Materials Science, Russian Academy of Sciences, Leninskii pr. 49, Moscow 119334, Russia
3   National University of Science and Technology MISiS, Leninskii pr. 4, Moscow 119991, Russia
*   Correspondence: nkolchugina@imet.ac.ru

Abstract: The finite element micromagnetic simulation is used to study the role of complex composition of 2:17R-cell boundaries in the realization of magnetization reversal processes of (Sm, Zr)(Co, Cu, Fe)$_z$ alloys intended for high-energy permanent magnets. A modified sandwich model is considered for the combinations of 2:7R/1:5H phase and 5:19R/1:5H phase layers as the 2:17R-cell boundaries in the alloy structure. The results of the simulation represented in the form of coercive force vs. total width of cell boundary showed the possibility of reaching the increased coercivity at the expense of 180°-domain wall pinning at the additional barriers within cell boundaries. The phase and structural states of the as-cast $Sm_{1-x}Zr_x(Co_{0.702}Cu_{0.088}Fe_{0.210})_z$ alloy sample with $x = 0.13$ and $z = 6.4$ are studied, and the presence of the above phases in the vicinity of the 1:5H phase was demonstrated.

Keywords: high-energy permanent magnets; (Sm, Zr)(Co; Cu; Fe)$_z$ alloys; as-cast state; coherent 1:5 phase; micromagnetic simulation; sandwich model; MuMax3

Citation: Zheleznyi, M.V.; Kolchugina, N.B.; Kurichenko, V.L.; Dormidontov, N.A.; Prokofev, P.A.; Milov, Y.V.; Andreenko, A.S.; Sipin, I.A.; Dormidontov, A.G.; Bakulina, A.S. Micromagnetic Simulation of Increased Coercivity of (Sm, Zr)(Co, Fe, Cu)$_z$ Permanent Magnets. *Crystals* **2023**, *13*, 177. https://doi.org/10.3390/cryst13020177

Academic Editor: Arcady Zhukov

Received: 8 December 2022
Revised: 12 January 2023
Accepted: 17 January 2023
Published: 19 January 2023

1. Introduction

Despite the wide application of Nd-Fe-B permanent magnets, sintered Sm-Co-based permanent magnets continue to be of interest for research and practice. The magnetism in intermetallic phases of rare-earth metals and transition metals, such as the high-performance magnet based on $SmCo_5$ and Sm_2Co_{17}, is a result of the synergy between the 4f RE electrons, which provide a high anisotropy due to spin-orbit coupling, and the 3d TM electrons, which have large magnetic moments and provide strong ferromagnetic exchange interactions, thus enabling long-range order [1]. The synergy between 3d and 4f electrons depends crucially on the local atomic environments, and thus, gives the variety of the magnetic properties of Sm-Co system compounds. The coercivity of samarium-based magnets originates from the Sm sublattice anisotropy, whereas the transition metals, such as Co, sublattice yields a high Curie temperature and thus stabilizes through inter-sublattice exchange.

The $Sm(Co, Cu, Fe, Zr)_z$ alloys, the microstructure of which is characterized by the presence of three constituents (rhombohedral 2:17R phase cells, coherent 1:5H phase boundaries of the cells, and coherent Z-phase (1:3R) lamellae), are widely used as permanent magnets with the high time and temperature stability because of their capacity to retain the high intrinsic coercivity $_IH_C$ due to the high magnetic anisotropy field of the 2:17R and 1:5H constituents and the high maximum energy product $(BH)_{max}$, which is related to the high remanence remaining at elevated temperatures [2]. The above microstructure determines the coercivity of the magnets, and therefore the improvement or justification of the microstructure can result in an increase in the coercivity, which is the issue of numerous investigations.

In [3], it was shown that depending on the heat treatment conditions, the cell boundaries can consist of a mixture of coherent phases (1:5H + 2:7R and/or 5:19R) due to the eutectoid decomposition occurred at a temperature of 800 °C. Data available in [4] deserve special attention; isotropic Sm–Co thin films comprising various Sm_xCo_y phases were prepared by triode sputtering of targets of variable compositions. The authors found that stacking faults appear within 1:5H phase grains, which correspond to local phase variants, including 2:7R, 5:19R and 1:3R; the domain wall pinning at grain boundaries and likely at 1:3R stacking faults is the main source of the enhanced coercive force. Such inhomogeneities considered in the film nanocrystalline material characterized by the high magnetocrystalline anisotropy result in the substantially increased coercive force.

In many studies, the effect of microchemistry on the increase in the coercivity of the Sm-Co magnets was considered both experimentally and by simulation. The authors of [3] used TEM-EDS analysis and showed that the Cu content at the cell boundary phase affects its pinning strength against magnetic domain wall motion. Depending on the Cu concentration and Cu distribution inside the cell boundary phase, the domain wall energy and magnetocrystalline anisotropy of the cell boundary phase change, and this affects its pinning strength. In [5], the mechanism of the increase in the coercivity of Sm_2Co_{17}-type permanent magnets is surveyed based on numerous literature data, and in particular, its dependence on the cooling rate from aging temperature of ~850 °C to 400 °C (slow cooling) and subsequent quenching to room temperature and Cu content is considered. Thus, the importance of the microchemistry of the cell boundary phase for the coercivity of these materials is highlighted. In [6], the cooling rate was related to the Cu concentration in the cell boundary phase. It was supposed that the higher Cu concentration in the cell boundary phase and its diffusive feature to the interface lead to a large difference between K_1 values of 2:17 cell phase and 1:5 cell boundary phase, causing an attractive domain wall pinning at the cell boundary and a higher coercivity. Thus, the importance of the difference of the K_1 value between the 2:17 cell phase and 1:5 cell boundary phase should be noted for the increase in the coercivity. In this case, the low coercivity of the sample quenched (rather than slowly cooled) from the aging temperature and was related to the smaller Cu content in the cell boundary phase, and therefore, a smaller K_1 difference between 2:17 cell and 1:5 cell boundary, which results in a repulsive barrier of the cell boundary for the domain wall pinning.

In [5], the increase in the Cu content, broadening the Cu distribution and sharp and smooth 1:5/2:17 interface structure, and Fe concentration were discussed in accordance with the cooling rate. This situation was simulated [5] in considering the Sm_2Co_{17}-type magnet, including the 2:17 cell phase, 1:5 cell boundary phase, and Z-phase; the $Sm(Co_{0.9}Cu_{0.1})_5$ phase as the cell boundary with Cu diffused region at the 1:5/2:17 interface and $Sm(Co_{0.8}Cu_{0.2})_5$ cell boundary phase free of any interface defects were considered. It was shown that the existence of Cu diffused interface reduces the coercivity. This fact disagrees with data of [6], according to which the Cu concentration in the $SmCo_5$ phase increases during slow cooling from 820 to 520 °C, having a wider concentration profile than that of Sm at cell boundaries. This experimental result explains the substantial increase in coercivity during the slow cooling process.

In [5], the lower coercivity was simulated by a new model related to the microchemistry of cell-boundary phases, i.e., the reduction of K_1 value of the cell boundary phase is due to the enrichment in Fe and diffusive feature of Cu. This results in a gradual increase in K_1 from the interface to the center of the cell boundary phase in a quenched sample causing the weak pinning strength of the cell boundary phase in the quenched sample. The micromagnetic simulation of the microchemistry effect showed that the small gradient of K_1 decreases the pinning strength of the cell boundary phase substantially and that the enrichment in Fe of the cell boundary phase further reduces the pinning strength of the cell boundary phase against magnetic domain wall motion.

In [2], it was not only demonstrated that the transformation from repulsive pinning to attractive pinning by alloying Cu particles in Sm-Co based permanent magnets but also provided insights into the grain boundary engineering for enhancing the intrinsic coercivity of rare earth permanent magnets. The significant increase in the Cu concentration in 1:5 cell boundary phases is not only driven by the thermal diffusion during the slow cooling process, but also the concentration gradient of residual Cu located near the grain boundaries. This results in the attractive domain pinning instead of repulsive domain wall pinning in Cu-particle-alloyed Sm(Co,M)$_z$ permanent magnets. Thus, it was demonstrated that the attractive domain wall pinning, instead of widely accepted repulsive domain wall pinning, predominates in the Cu-particle-alloyed magnet, and the exceptionally high H_{ci} is attributed to the continuity and enlarged pinning strength of the 1:5 cell boundary phases in the Cu-particle-alloyed magnet.

Thus, in [2,4,5], the micromagnetic simulation was performed using a sandwich model of permanent magnets, which comprises the 2:17R/1:5H/2:17R phases alloyed with copper and iron and assumes the repulsive or/and attractive pinning of 180°-domain walls.

In [7], micromagnetic simulation was performed in MuMax3 [8], which is a GPU-accelerated software that uses finite-difference discretization for the exchanged-coupled L10-FeNi/SmCo$_5$ composite; its geometry is based on nanorods array. In the case of exchange-coupled materials, interface exchange coupling coefficient, or the exchange strength J_{ex} should be defined, which sets the strength of coupling between phases in a composite. Approaches for the production of exchange-coupled composites based on anisotropic nanostructures were proposed.

Despite the extensive experimental [9–17] and simulation studies [2,4,5] of the magnetic properties and microstructure of (Sm, Zr)(Co, Cu, Fe)$_z$ alloys, the problem of the effect of phase constituents on magnetization reversal processes remains relevant.

In the present study we consider the repulsive depinning for two variants of the modified sandwich model of a permanent magnet, which comprises additional phase layers, which are located between the 2:17R cell and 1:5H cell boundary and differ in width. The performed simulation should make a contribution to the precise engineering of the structure of (Sm, Zr)(Co, Cu, Fe)$_z$ magnets with improved hysteretic properties.

2. Materials and Methods

2.1. Model Description

To simulate the influence of the complex composition of the cell boundary to the magnetization reversals of the Sm$_2$Co$_{17}$-based magnets, finite element micromagnetic simulations were employed. Similar simulations were performed in [2,4,5], however we used two variants of modified sandwich model; these are the 2:17R/2:7R/1:5H/2:7R/2:17R phase layers and 2:17R/5:19R/1:5H/5:19R/2:17R phase layers. To simulate the demagnetization curves of a magnet, the model 80 × 30 × 30 nm in total size was considered. Schematic diagram of the model is given in (Figure 1).

The width of the 1:5H phase layer ($t_{1:5}$) was varied from 0 to 10 nm at a step of 2 nm. The width of the 2:7R phase or the 5:19R phase ($t_{2:7}$ or $t_{5:19}$), which is the sum of the right and left parts with respect to the central 1:5R phase with the fixed width, also was increased monotonically from 0 to 10 nm at a step of 2 nm. A 180° magnetic domain wall was introduced into the right 2:17R matrix phase at a distance of 10 nm from the right edge of the model. The initial of the magnetization directions of the left 2:17R phase, 1:5R, 2:7R (or 5:19R) phases, and a portion (to the left of the domain wall) of the right 2:17R phase were set to be upwards, and the initial magnetization direction of the portion of the right matrix phase 2:17R to the right of the domain wall was set to be downward. This domain wall, as was shown in [2,4,5], moves toward the interface of phase boundary in applying an external magnetic induction (field).

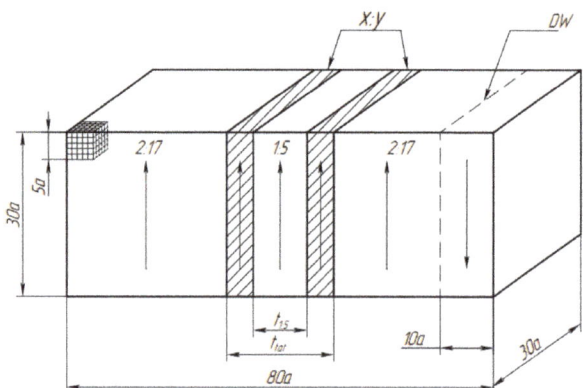

Figure 1. Schematic diagram of the initial magnetization state of the micromagnetic model containing two 2:17 phase cells and complex (x:y/1:5/x:y) cell boundary, where x:y = 2:7 or 5:19, and showing the existence of magnetic 180° domain wall inside the right 2:17 cell before applying an external magnetic induction (field); DW is the domain wall; t is the width; and a = 1 nm.

The demagnetization curves ($\mu_0 M$ vs. H) were simulated by minimizing the magnetic energy of the system in decreasing magnetic induction. The value of external magnetic induction decreases from 0 to–4 T at a step of 40 mT. The models were discretized by cubic nodes $1 \times 1 \times 1$ nm in size, the linear size of which is less than the domain wall width and the exchange length, similarly to approximations used in [2,4,5].

Calculations of the models at each node were performed by the Landau-Lifshitz-Gilbert equation [8] using software MuMax3, which is a GPU-accelerated software that uses finite-difference discretization for micromagnetic simulations. Since MuMax3 performs best with the power-of-two sizes, even values of component phase widths were used [7,8]. The exchange interaction between different material regions deserve special attention. The exchange stiffness parameter A_{ex} and magnetization saturation M_s are defined in the cell volumes, while requires a value of A_{ex}/M_s properly averaged out between the neighboring cells. For neighboring cells with different material parameters (A_{ex1}, A_{ex2} and M_{s1}, M_{s2}) MuMax3 uses approximation a harmonic mean. We used standard value of a scaling factor (S.F. = 1), which may be used to alter the exchange coupling between regions [8]. The simulation was performed under the same conditions and only the (1:5 and 2:7 or 5:19) phase thicknesses were varied.

Table 1 shows values of the magnetocrystalline anisotropy constant (K_u) and saturation magnetization (M_s) of constituent phases available in [3], which we used for the micromagnetic simulations. Table 1 also shows values of the magnetic saturation (anisotropy) field ($\mu_0 \cdot H_s = 2 \cdot K_u/M_s$) of the phases.

Table 1. Magnetic properties of constituent phases: Saturation magnetization (M_S), uniaxial anisotropy constant (K_u), and magnetic saturation field ($\mu_0 \cdot H_s$).

Compound	M_S (MA/m) [3]	K_u (MJ/m³) [3]	$\mu_0 \cdot Hs$ (T)	A_{ex} (pJ/m) [3]
Sm$_2$Co$_{17}$–R [1]	0.99	2.9	5.9	9.5
SmCo$_5$–H [1]	0.53	8.1	30.8	7.7
Sm$_2$Co$_7$–R [1]	0.55	4.0	14.6	8.0
Sm$_5$Co$_{19}$–R [1]	0.61	4.0	13.2	8.0

[1] R and H correspond to rhombohedral and hexagonal structures, respectively.

2.2. Experimental Methods

The microstructure of as-cast alloys was studied using sections prepared by traditional grinding and polishing procedures and a JXA-iSP100 (JEOL; Tokyo, Japan) scanning electron microscope equipped with an energy-dispersive analyzer. The phase compositions of the alloys were determined by X-ray diffraction (XRD) analysis performed on a Bruker D8 Advance diffractometer (Bruker, Karlsruhe, Germany) using CuK_α radiation ($\lambda = 1.54178$ Å). X-ray diffraction patterns were processed by the Rietveld method using a Bruker DIFFRAC.EVA™ (Bruker, Karlsruhe, Germany), DIFFRAC.TOPAS™ (Bruker, Karlsruhe, Germany) and ICDD PDF-2 2020 (ICDD, Newtown Square, PA, USA) software. For the quantitative phase analysis, we used Springer Materials Database for the Sm-Zr-Co system as prototypes. We improved only some atomic coordinates (particular positions) for each phase.

3. Results and Discussion

We performed micromagnetic simulations for the two variants of the modified sandwich model using the fundamental magnetic parameters estimated from experimental data in [3]. As the modulus of external induction increases, the magnetic domain wall moves leftwards, and at the first stage of the demagnetization process, was pinned at the 2:7R or 5:19R cell boundary phase. At the second stage of the demagnetization process, the magnetic domain wall was pinned at the boundary between 2:7R (or 5:19R) and 1:5H phases.

The actual position of the domain wall is determined by the ratio between the domain wall energy and magnetostatic energy for each phase [3]. The values of the domain wall energy ($\gamma_{DW} = 4 \cdot (A_{ex} \cdot K_u)^{1/2}$, mJ/m^2) and width ($\delta_{DW} = \pi \cdot (A_{ex}/K_u)^{1/2}$, nm) and the exchange length ($\delta_{ex} = (A_{ex}/\mu_0 \cdot M_s)^{1/2}$, nm) of the considered phases as well were calculated: $\gamma_{DW}^{1:5} = 31.6 > \gamma_{DW}^{2:7} = 22.6 \cong \gamma_{DW}^{5:19} = 22.5 > \gamma_{DW}^{2:17} = 21.0; \delta_{DW}^{2:17} = 5.7 > \delta_{DW}^{2:7} = 4.5 \cong \delta_{DW}^{5:19} = 4.4 > \delta_{DW}^{1:5} = 3.1; \delta_{ex}^{1:5} = 4.7 \cong \delta_{ex}^{2:7} = 4.6 > \delta_{ex}^{5:19} = 4.1 > \delta_{ex}^{2:17} = 2.1$.

As an example, Figure 2 shows the simulated demagnetization curves of the initial sandwich model (without the 2:7R or 5:19R phase layers; $t_{1:5} = 10$ nm) and our modified sandwich model (with the additional 2:7R or 5:19R phase layers; $t_{2:7}$ or $t_{5:19} = 10$ nm, $t_{1:5} = 10$ nm) of magnet, in which the first (lesser) and second (greater) plateaus in these curve results from the domain wall pinning at interfaces between 2:7R (or 5:19R) and 2:17R and between 2:17R and 1:5H, respectively.

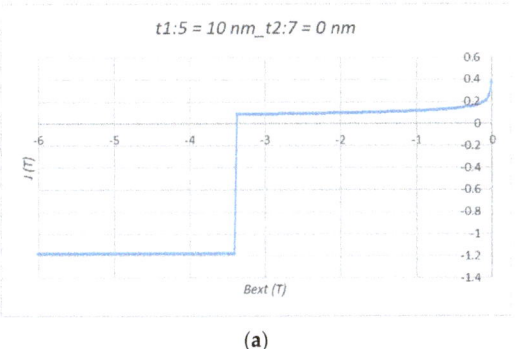
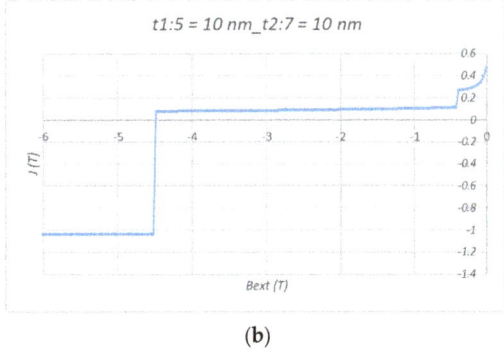

(a) (b)

Figure 2. Simulated demagnetization curves of the initial sandwich model (**a**) and modified sandwich model with the 2:7R phase layer (**b**) of (Sm, Zr)(Co, Cu, Fe)z magnet.

We quantified the effects of the total width ($t_{sum} = t_{1:5} + t_{2:7}$ or $t_{sum} = t_{1:5} + t_{5:19}$) of the 1:5H and 2:7R (or 5:19R) cell boundary phases on the coercivity ($\mu_0 \cdot_I H_c$) of the model magnet, as shown in Figure 3. An important feature of these curves ($\mu_0 \cdot_I H_c$ vs. t_{sum}) is the presence of a plateau at thicknesses of each phase of more than 4–5 nm. On the one hand, this effect can be explained by the fulfillment of the necessary geometric condition associated with the limitations of our model: $t_i > \delta_{DW} > \delta_{ex}$, where i = 1:5, 2:7, 5:19 and 2:17.

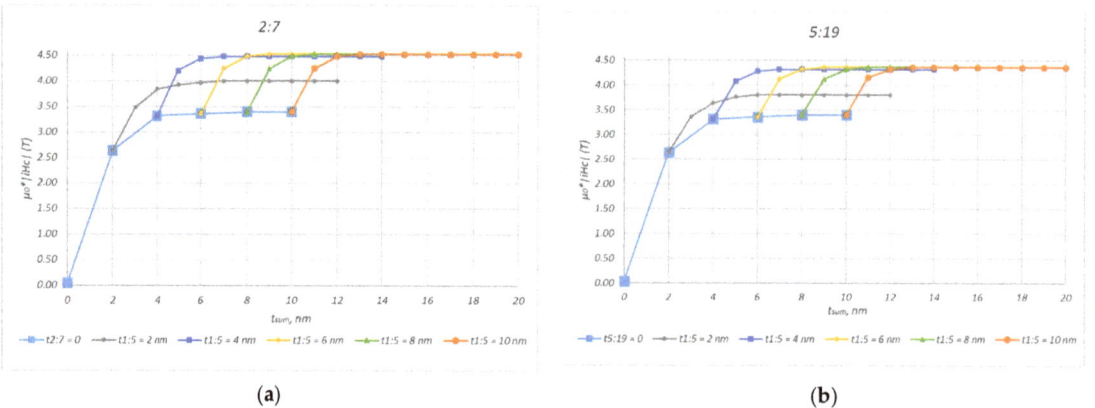

(a)   (b)

Figure 3. Calculated coercive force versus total width of cell boundary phase layer for the sandwich model 2:17R/x:yR/1:5H/x:yR/1:5H/2:17R, where x:y = 2:7R (**a**) and 5:19R (**b**).

On the other hand, similar types of curves were experimentally observed in [9]. The authors have studied the influence of temperature and annealing time of the maximum coercivity of Sm(Co, Cu, Fe, Zr)$_{7.6}$ samples and explained the observed dependence by thermally activated diffusion processes occurred during heat treatment of sintered magnets, which make a contribution to the redistribution of chemical elements and changes in the concentration profiles of the main magnetic properties in the ratio of 2:17 and 1:5 phases. We assume that, in addition to the processes of diffusive redistribution of elements between the main phases (2:17R and 1:5H) in these alloys, additional phases (2:7R or 5:19R) can be precipitated along the cell boundary. These phases are additional barriers for domain wall pinning at the cell boundary.

According to the simulation results, the total increase of the coercive force for the modified sandwich model with additional boundary phases (2:7R or 5:19R) relative to the initial magnet model (without additional phases 2:7R (or 5:19R)) was found to be 15–30%.

Obviously, it is difficult to experimentally confirm the described model of increasing coercivity of the Sm-Zr-Co-Cu-Fe alloys, which is reached at the nano-scale level. In our previous studies [10–12], we discussed different levels of heterogeneity of the Sm-Zr-Co-Cu-Fe alloys and have related the two-level heterogeneity with the formation of the high-coercivity state of them. The high-coercivity state of the alloys is formed via complex sequential transformations that occurred in the course of heat treatment (solid-solution treatment, quenching, and isothermal and stepped aging) of as-cast alloys. As a result, the periodical phase nano-structure forms and ensures the efficient coercivity mechanism, namely the domain-wall pinning at interfaces. The stable phases typical of the Sm-Co system, the 2:7 and 5:19 phases are among them, participate in the transformations. Below we demonstrate that the 2:7 and 5:19 phases, which, as was noted in [3] can be local phase variants (related to stacking faults) of the 1:5H phase, are identified in the as-cast structure of Sm-Zr-Co-Cu-Fe alloys.

Moreover, in our previous studies [10–12], it was shown that, in the case of used Sm$_{1-x}$Zr$_x$(Co$_{0.702}$Cu$_{0.088}$Fe$_{0.210}$)$_z$ samples, with x = 0.13–0.19 and z = 6.0–6.8, respectively, which are characterized by high hysteretic parameters and ultimate hysteresis loops that ensure satisfying the condition $(BH)_{MAX} \approx (4\pi J_S)^2/4$, the 2:7 and 5:19 phases are present in samples quenched after isothermal aging (during isothermal aging, as the intermediate structural constituents in cell-boundary structural constituent, it is assumed the formation of additional phases belonging to the homologous row (Sm, Zr)$_{n-1}$(Co, Cu, Fe)$_{5n-1}$–2:7 and 5:19 [14–18]) and even in samples subjected to complete heat treatment for the high-coercivity state. In this case, the high-coercivity samples based on the (Sm, Zr)(Co, Cu, Fe)z alloys are characterized by the presence of phases almost free of zirconium 2:17 [6] and X–Sm(Co$_{0.50-0.63}$Cu$_{0.22-0.40}$Fe$_{0.10-0.15}$)$_{3.5-5.0}$, which are likely to be the main structural constituents of the high-coercivity cellular structure [10–12]. It is noteworthy that our data on the presence of the 2:7 and 5:19 phases in the structure of the (Sm, Zr)(Co, Cu, Fe)z alloys at different heat-treatment stages, in part, coincide with the data of Morita et al. [18–20]. In particular, we have showed that the 2:17 and 2:7 phases are the primary-solidification phases that are maximally separated in the material structure and form the main structural constituents. This corresponds to the Liq. + 2:17 + 2:7 region in the phase diagrams of Morita et al. [19–21].

The possibility of the formation of the 2:7 and 5:19 phases, which are found in the vicinity of the 1:5 phase was observed by us experimentally using X-ray diffraction and local electron probe microanalysis. The 2:17R, 1:7H, 1:5H, 5:19R, 2:7R, 2:7H, 1:3R, and 1:3H phases were identified for the as-cast Sm$_{1-x}$Zr$_x$(Co$_{0.702}$Cu$_{0.088}$Fe$_{0.210}$)$_z$ experimental alloy with x = 0.13 and z = 6.4. The magnetic hysteretic characteristics of the alloy were studied in [8–10], respectively. Figure 4 shows the back scattered electron mode image of the structure of the alloy with x = 0.13, z = 6.4. The composition of the 5:19R phase (point 3) was found to be Sm$_{0.6}$Zr$_{0.4}$(Co$_{0.72}$Cu$_{0.09}$Fe$_{0.19}$)$_{3.9}$ and that of the 2:7R phase (points 4a and 4b) is Sm$_{0.95}$Zr$_{0.05}$(Co$_{0.66}$Cu$_{0.20}$Fe$_{0.14}$)$_{3.4}$ and Sm$_{0.4}$Zr$_{0.6}$(Co$_{0.74}$Cu$_{0.10}$Fe$_{0.16}$)$_{3.6}$, respectively. The composition of the 2:7H phase (point 4c) is Sm(Co$_{0.50}$Cu$_{0.40}$Fe$_{0.10}$)$_{3.7}$. (Points 1, 2a, 2b, and 5b correspond to 2:17R (Sm(Co$_{0.71}$Cu$_{0.05}$Fe$_{0.23}$Zr$_{0.01}$)$_{8.3}$), 1:7H (Sm$_{1-x}$Zr$_x$(Co$_{0.70}$Cu$_{0.06}$Fe$_{0.24}$)$_{5.5-7.0}$), 1:5H (Sm$_{0.94}$Zr$_{0.06}$(Co$_{0.65}$Cu$_{0.20}$Fe$_{0.16}$)$_{4.8}$), and 1:3H (Sm(Co$_{0.60}$Cu$_{0.30}$Fe$_{0.10}$)$_{2.9}$) phases, respectively; R and H denote rhombohedral and hexagonal modifications of the phases, respectively).

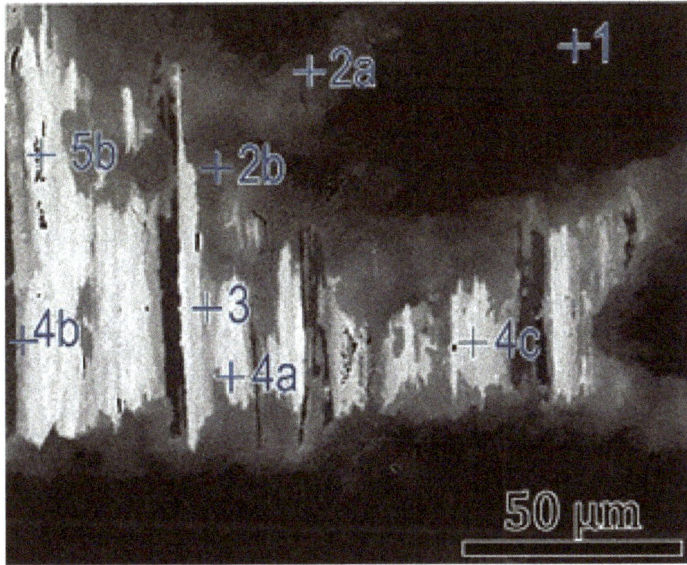

Figure 4. Microstructure (SEM images, back scattered electron mode) of the as-cast Sm$_{1-x}$Zr$_x$(Co$_{0.702}$Cu$_{0.088}$Fe$_{0.210}$)$_z$ alloy sample with *x* = 0.13 and *z* = 6.4; EMA points are shown.

Figure 5a,b demonstrate the X-ray diffraction pattern of the sample, which confirms the identification of the 2:7 and 5:19 phases present in the alloys. The lattice parameters of the phases, their percentage, and parameters of the fine structure (coherent domain size and microstresses) are shown in Table 2.

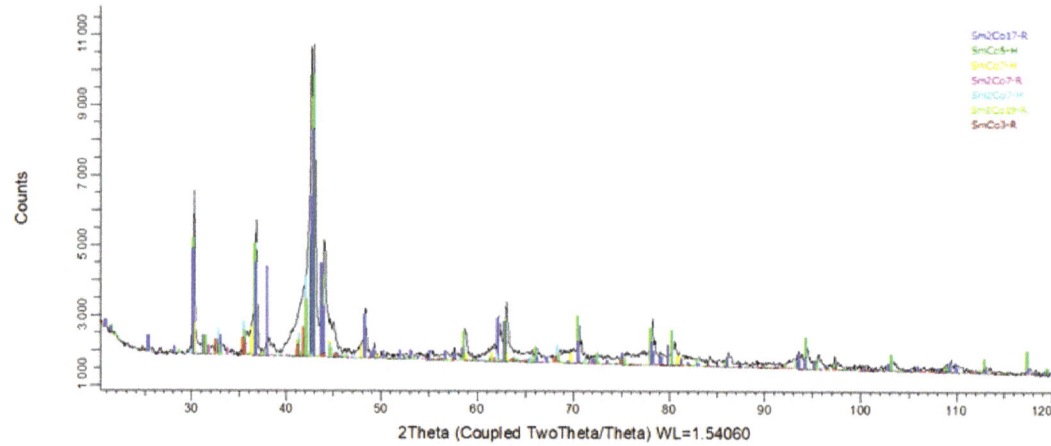

(a)

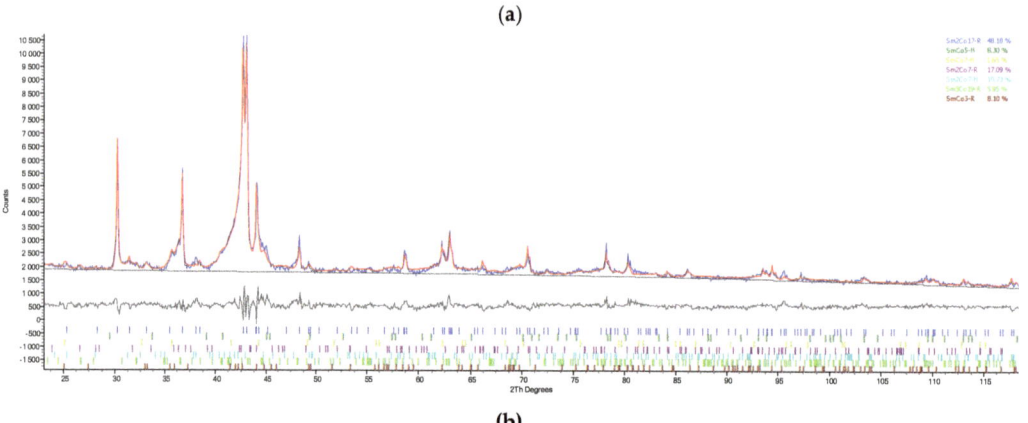

(b)

Figure 5. (a) X-ray diffraction pattern of as-cast $Sm_{1-x}Zr_x(Co_{0.702}Cu_{0.088}Fe_{0.210})_z$ alloy sample with $x = 0.13$ and $z = 6.4$; (b) Rietveld refinement of the pattern.

Table 2. Crystal lattice and fine structure parameters of the phases identified in the as-cast $Sm_{1-x}Zr_x(Co_{0.702}Cu_{0.088}Fe_{0.210})_z$ alloy sample with $x = 0.13$ and $z = 6.4$.

Phase Name	Space Group (No.)	Wt. % Rietveld	Unit Cell Volume (Å³)	Lattice Parameter a (Å)	Lattice Parameter c (Å)	Coherent Domain Size (nm)	Microstresses (%)
Sm_2Co_{17}-R	R-$3m$ (166)	48.2 ± 4.4	761.945 ± 0.100	8.452 ± 0.001	12.315 ± 0.001	96 ± 84	0.343 ± 0.010
$SmCo_5$-H	$P6/mmm$ (191)	8.3 ± 5.6	97.995 ± 0.757	5.305 ± 0.016	4.020 ± 0.020	120 ± 27	0.824 ± 0.358
$SmCo_7$-H	$P6/mmm$ (191)	1.6 ± 0.8	76.710 ± 0.176	5.013 ± 0.003	3.525 ± 0.007	207 ± 36	0.591 ± 0.061
Sm_2Co_7-R	R-$3m$ (166)	17.1 ± 6.6	789.127 ± 5.341	4.932 ± 0.006	37.456 ± 0.236	165 ± 26	0.668 ± 0.358
Sm_2Co_7-H	$P6_3/mmc$ (194)	10.7 ± 6.1	527.422 ± 13.972	5.260 ± 0.060	22.011 ± 0.299	351 ± 156	0.288 ± 0.119
Sm_5Co_{19}-R	R-$3m$ (166)	5.9 ± 4.4	1138.715 ± 4.288	5.129 ± 0.009	49.981 ± 0.073	223 ± 32	0.656 ± 0.192
$SmCo_3$-R	R-$3m$ (166)	8.1 ± 4.9	533.883 ± 22.430	5.043 ± 0.952	24.242 ± 0.029	251 ± 43	0.246 ± 0.203

Crystals **2023**, *13*, 177

The data obtained reliably confirm the presence of the above phases in the as-cast quinary alloy that is "natural" combination of them, which is reached by traditional induction melting of Sm-Co-based alloys. The as-cast state is nonequilibrium, is controlled by cooling conditions, and can likely be controllable by heat treatment conditions. This fact makes promising the results of the performed simulation.

4. Conclusions

The micromagnetic simulation was performed using the modified sandwich model of a (Sm, Zr)(Co, Cu, Fe)$_z$ magnet, which includes additional domain-wall pinning barriers in the form of 2:7R or 5:19R phase layers. These phases were not considered previously as intermediate layers that result from stacking faults formed in the 1:5 phase. As the width of the cell-boundary barrier increases, the effect of increasing coercivity was found. Moreover, the barrier width being optimum for the increase in the coercive force was determined. The increase in the width of 2:7R or 5:19R phase barriers above the optimum width does not lead to a further increase in the coercive force but can result in a decrease in the magnetization of the simulated (Sm, Zr)(Co, Cu, Fe)$_z$ magnet. The possibility of the presence of the phases was demonstrated by an example of the as-cast Sm$_{1-x}$Zr$_x$(Co$_{0.702}$Cu$_{0.088}$Fe$_{0.210}$)$_z$ alloy sample with x = 0.1 and z = 6.4, for which the EMA and X-ray diffraction analysis were performed.

The purposeful engineering of the structure of (Sm, Zr)(Co, Cu, Fe)$_z$ magnets will likely allow one to improve their hysteretic properties.

Author Contributions: Conceptualization, M.V.Z., A.G.D.; methodology, M.V.Z., V.L.K.; formal analysis, N.B.K., P.A.P.; validation, A.S.A.; data curation, N.A.D.; investigations, M.V.Z., Y.V.M., I.A.S.; writing—original draft preparation, M.V.Z.; writing—review and editing, N.B.K.; visualization, N.A.D., Y.V.M., A.S.B. All authors have read and agreed to the published version of the manuscript.

Funding: This study was carried out within the project "Development of the model of formation of high-coercivity phase and structural states of (R, Zr)(Co, Cu, Fe)$_z$ alloys and their synthesis for the production anisotropic bonded powder permanent magnets applied in electric machines.", projects No. 20-19-00689 funded by the Russian Science Foundation.

Data Availability Statement: The data presented in this study are openly available.

Conflicts of Interest: The authors declare no conflict of interest.

References

1. Buschow, K. Intermetallic compounds of rare-earth and 3d transition metals. *Rep. Prog. Phys.* **1977**, *40*, 1179. [CrossRef]
2. Chen, H.; Wang, Y.; Yao, Y.; Qu, J.; Yun, F.; Li, Y.; Ringer, S.P.; Yue, M.; Zheng, R. Attractive-domain-wall-pinning controlled Sm-Co magnets overcome the coercivity-remanence trade-off. *Acta Mater.* **2019**, *164*, 169–206. [CrossRef]
3. Goll, D.; Kronmuller, H.; Stadelmaier, H.H. Micromagnetism and the microstructure of high-temperature permanent magnets. *J. Appl. Phys.* **2004**, *96*, 6534–6545. [CrossRef]
4. Akdogan, O.; Sepehri-Amin, H.; Dempsey, N.M.; Ohkubo, T.; Hono, K.; Gutfleisch, O.; Schrefl, T.; Givord, D. Preparation, characterization, and modeling of ultrahigh coercivity Sm–Co thin films. *Adv. Electron. Mater.* **2015**, *1*, 1500009. [CrossRef]
5. Sepehri-Amin, H.; Thielsch, J.; Fischbacher, J.; Ohkubo, T.; Schrefl, T.; Gutfleisch, O.; Hono, K. Correlation of microchemistry of cell boundary phase and interface structure to the coercivity of Sm(Co$_{0.784}$Fe$_{0.100}$Cu$_{0.088}$Zr$_{0.028}$)$_{7.19}$ sintered magnets. *Acta Mater.* **2017**, *126*, 1–10. [CrossRef]
6. Xiong, X.Y.; Ohkubo, T.; Koyama, T.; Ohashi, K.; Tawara, Y.; Hono, K. The microstructure of sintered Sm(Co$_{0.72}$Fe$_{0.20}$Cu$_{0.055}$Zr$_{0.025}$)$_{7.5}$ permanent magnet studied by atom probe. *Acta Mater.* **2004**, *52*, 737–748. [CrossRef]
7. Kurichenko, V.L.; Karpenkov DYu Gostishchev, P.A. Micromagnetic modelling of nanorods array-based L10-FeNi/SmCo$_5$ exchange-coupled composites. *J. Phys. Condens. Matter.* **2020**, *32*, 405806. [CrossRef]
8. Vansteenkiste, A.; Leliaert, J.; Dvornik, M.; Helsen, M.; Garcia-Sanchez, F.; Van Waeyenberge, B. The design and verification of MuMax3. *AIP Adv.* **2014**, *4*, 107133. [CrossRef]
9. Durst, K.-D.; Kronmuller, H.; Ervens, W. Investigations of the magnetic properties and demagnetization processes of an extremely high coercive Sm(Co, Cu, Fe, Zr)$_{7.6}$ permanent magnet. II. The coercivity mechanism. *Phys. Stat. Sol.* **1988**, *108*, 705–719. [CrossRef]
10. Dormidontov, A.G.; Kolchugina, N.B.; Dormidontov, N.A.; Milov, Y.V. Structure of alloys for (Sm,Zr)(Co,Cu,Fe)$_z$ permanent magnets: First level of heterogeneity. *Materials* **2020**, *13*, 3893. [CrossRef]

11. Dormidontov, A.G.; Kolchugina, N.B.; Dormidontov, N.A.; Milov, Y.V.; Andreenko, A.S. Structure of alloys for $(Sm,Zr)(Co,Cu,Fe)_z$ permanent magnets: II. Composition, magnetization reversal, and magnetic hardening of main structural components. *Materials* **2020**, *13*, 5426. [CrossRef] [PubMed]
12. Dormidontov, A.G.; Kolchugina, N.B.; Dormidontov, N.A.; Zheleznyi, M.V.; Bakulina, A.S.; Prokofev, P.A.; Andreenko, A.S.; Milov, Y.V.; Sysoev, N.N. Structure of alloys for (Sm, Zr)(Co, Cu, Fe)z permanent magnets: III. Matrix and phases of the high-coercivity state. *Materials* **2021**, *14*, 7762. [CrossRef] [PubMed]
13. Soboleva, A.N.; Golovnia, O.A.; Popov, A.G. Embedded atom potential for Sm–Co compounds obtained by force-matching. *JMMM J. Magn. Magn. Mater.* **2019**, *490*, 165468. [CrossRef]
14. Kronmuller, H.; Goll, D. Micromagnetic analysis of pinning-hardened nanostructured, nanocrystalline Sm_2Co_{17} based alloys. *Scr. Mater.* **2022**, *47*, 545–550. [CrossRef]
15. Stadelmaier, H.H.; Goll, D.; Kronmuller, H. Permanent magnet alloys based on Sm_2Co_{17}; phase evolution in the quinary system Sm–Zr–Fe–Co–Cu. *Int. J. Mater. Res.* **2005**, *96*, 17–23. [CrossRef]
16. Goll, D.; Stadelmaier, H.H.; Kronmuller, H. Samarium–cobalt 2:17 magnets: Analysis of the coercive field of $Sm_2(CoFeCuZr)_{17}$ high-temperature permanent magnets. *Scr. Mater.* **2010**, *63*, 243–245. [CrossRef]
17. Stadelmaier, H.H.; Kronmuller, H.; Goll, D. Samarium–cobalt 2:17 magnets: Identifying $Sm_{n+1}Co_{5n-1}$ phases stabilized by Zr. *Scr. Mater.* **2010**, *63*, 843–846. [CrossRef]
18. Song, K.; Sun, W.; Chen, H.; Yu, N.; Fang, Y.; Zhu, M.; Li, W. Revealing on metallurgical behavior of iron-rich $Sm(Co_{0.65}Fe_{0.26}Cu_{0.07}Zr_{0.02})_{7.8}$ sintered magnets. *AIP Adv.* **2017**, *7*, 056238. [CrossRef]
19. Morita, Y.; Umeda, T.; Kimura, Y. Phase transformation at high temperatures and coercivity of Sm-Co-Cu-Fe magnet alloys. *J. Jpn. Inst. Met.* **1986**, *50*, 235–241. [CrossRef]
20. Morita, Y.; Umeda, T.; Kimura, Y. Phase transformation at high temperature and coercivity of $Sm(Co,Cu,Fe,Zr)_{7-9}$ magnet alloys. *IEEE Trans. Magn.* **1987**, *23*, 2702–2704. [CrossRef]
21. Morita, Y.; Umeda, T.; Kimura, Y. Effects of Zr content on phase transformation and magnetic properties of $Sm(Co,Cu,Fe,Zr)_{7-9}$ magnet alloys. *J. Jpn. Inst. Met.* **1988**, *52*, 243–250. [CrossRef] [PubMed]

crystals

Article

Burning Rate Prediction of Solid Rocket Propellant (SRP) with High-Energy Materials Genome (HEMG)

Weiqiang Pang [1], Victor Abrukov [2,*], Darya Anufrieva [2] and Dongping Chen [3]

[1]   Xi'an Modern Chemistry Research Institute, Xi'an 710065, China
[2]   Department of Applied Physics and Nanotechnology, Chuvash State University, 428015 Cheboksary, Russia
[3]   State Key Lab of Explosion Science and Technology, Beijing Institute of Technology, Beijing 100081, China
*   Correspondence: abrukov@yandex.ru

Abstract: High-energy materials genome (HEMG) is an analytical and calculation tool that contains relationships between variables of the object, which allows researchers to calculate the values of one part of the variables through others, solve direct and inverse tasks, predict the characteristics of non-experimental objects, predict parameters to obtain an object with desired characteristics and execute virtual experiments for conditions which cannot be organized or have difficulty being organized. HEMG is based on experimental data on the burning rate of various high-energy materials (HEMs) under various conditions, on the metadata on the quantum and physicochemical characteristics of HEMs components as well as on thermodynamic characteristics of HEMs as a whole. The history and current status of the emergence of HEMG are presented herein. The fundamental basis of the artificial neural networks (ANN) as a methodological HEMG base, as well as some examples of HEMG conception used to create multifactor computational models (MCM) of solid rocket propellants (SRP) combustion, is presented.

Keywords: high-energy materials genome; artificial neural networks; burning rate; multifactor computational models

Citation: Pang, W.; Abrukov, V.; Anufrieva, D.; Chen, D. Burning Rate Prediction of Solid Rocket Propellant (SRP) with High-Energy Materials Genome (HEMG). _Crystals_ **2023**, _13_, 237. https://doi.org/10.3390/cryst13020237

Academic Editor: Thomas M. Klapötke

Received: 12 January 2023
Revised: 26 January 2023
Accepted: 27 January 2023
Published: 30 January 2023

1. Introduction

When researching into the area of high-energy materials (HEMs), the goal is always to search for new HEMs with improved properties. There is currently no standard practice for researchers and developers to disclose predictive algorithms and in silico methods for developing new propellants with desirable characteristics. Therefore, generally, it was necessary to conduct many expensive and hazardous experiments to obtain a burning rate or to determine the optimal HEMs compositions. There exist a large number of experimental data about the combustion characteristics of HEMs [1–3], while there are several disadvantages for HEM investigation, such as expensive cost, safety risks, etc. With the rapid development of computer simulation science and combustion diagnosis technology, the performance prediction of solid rocket propellant (SRP) has gained much attention from researchers worldwide. The prediction of burning rate of SRPs is an important aspect of analysis, which has important theoretical significance to reveal the combustion mechanism of SRP. For example, Zhang Xiaoping et al. used genetic neural network to simulate the combustion performance of nitrate ester plasticized polyether (NEPE) propellant under high-pressure conditions and proposed 13 characterization parameters. This method has high calculation accuracy, but the model is a purely physical model that does not consider the influence of chemical composition and the structure of propellants [4]. The effect of various factors on the combustion characteristics of SRP can be predicted by using multifactor calculation model (MCM) non-linear combustion models that are based on artificial intelligence (AI) methods.

In this work, the high-energy materials genome (HEMG) method was introduced. Based on the usage of AI, this solves the problems of creation of MCM of combustion and detonation. Its history and current status, as well as examples of the results of its application, were presented. In particular, we detailed the use of artificial neural networks (ANN) to create the MCM of combustion of SRP. Additionally, the burning rate of SRP with micro-sized aluminium (mAl) or nano-sized metal (nAl and nNi) particles was calculated by means of ANN, predicted in some cases and then compared to the experimental data.

2. History and Current Status of HEMG

HEMG is based on the experimental data on the combustion and detonation characteristics of various HEMs under various conditions, being based also on the metadata on the quantum and physicochemical characteristics of HEMs components as well as the thermodynamic characteristics of HEM as a whole.

HEMG involves the principles of the Materials Genome Initiative (MGI) for Global Competitiveness that was announced through a whitepaper by the National Science and Technology Council of the USA in June 2011 [5], which is one of starting points of HEMG history.

In 2014, the US National Institute of Standards and Technology presented a strategic plan for the implementation of MGI. In the field of energetic materials (EMs), as a continuation of the MGI idea, the Energetic Materials Genome Initiative (EMGI) was launched as an idea in 2017 and it was marked that if the MGI modes were used in the development of EMs, the efficiency of EMs manufacturing would be greatly enhanced, which will benefit the society [6]. It could be noted that the research corresponding to the idea of MGI at Chuvash State University saw their proposals implemented, and the MCM for combustion and detonation obtained by means of ANN were presented [7,8]. Moreover, the MCM of various SRP combustions were obtained [9], and improved MCM for various SRP combustions and detonations were obtained [10–14].

Nowadays, in the fields of HEMs, the following works dealing with the conception of MGI are worth analysis. For instance, Wang et al. [15] depict how the MGI approach can be used to accelerate the discovery of new insensitive high-energy explosives by the identification of "genetic" features. Kang et al. [16] depict how machine learning (ML), materials informatics (MI), and thermochemical data are combined to screen potential candidates of EMs. To directly characterize the energetic performance, the heat of explosion is used as the target property. The critical descriptors of cohesive energy, averaged over all constituent elements and the oxygen balance, are found by forward stepwise selection from a large number of possible descriptors. With them and a theoretically labeled heat of explosion training data set, a satisfactory surrogate ML model is trained. The ML model is applied to the large databases NIST ICSD (NIST Inorganic Crystal Structure Database, NIST Standard Reference Database Number 3, National Institute of Standards and Technology, Gaithersburg MD, 20899, DOI: https://doi.org/10.18434/M32147, (retrieved on 28 January 2023) and PubChem to predict the heat of explosion. At the gross-level filtering by the ML model, 2732 molecular candidates based on carbon, hydrogen, nitrogen, and oxygen (CHNO) with high heat of explosion are predicted. Afterward, a fine-level thermochemical screening is carried out on the 2732 materials, resulting in 262 candidates with TNT equivalent power index Pe (TNT) greater than 1.5. Raising Pe (TNT) further to larger than 1.8 sees 29 potential candidates be found from the 2732 molecular candidates, all of which are new to the current reservoir of well-known EMs. Yuan et al. [17] remarked that the approach of taming energetic compounds via the permutation of chemical building blocks has gradually reached a crossroads. The future will leverage new tools such as AI to construct the HEMG method. Yang et al. [18] remarked that researchers have begun to apply deep learning methods to the prediction of explosive detonation performance. The deep learning method has the advantage of simple and rapid prediction of explosive detonation properties. However, some problems remain in the study of detonation properties based on deep learning. For example, there are few studies on the prediction of mixed explosives, on the prediction of the parameters of the equation for the state of explosives, or on the application of explosive properties to predict the formulation of explosives. Based on an ANN model and a one-dimensional convolutional neural network model, three improved deep learning models were established. Tian et al. [19] marked that the prediction of the properties of EMs using ML has been receiving more attention in recent years. This review summarized recent advances in predicting energetic compounds' density, detonation velocity, enthalpy of formation, sensitivity, the heat of the explosion, and decomposition temperature using ML. Moreover, it presented general steps for applying ML to the prediction of practical chemical properties from the aspects of data, molecular representation, algorithms, and general accuracy. Additionally, it raised some controversies specific to ML in EMs and its possible development directions. Important information related to MGI can be found at [20]. In June 2022, the MGI Fifth Principal Investigator Meeting took place. It was noted that techniques on well-known species of CHNO-based EMs, such as HMX, PETN, TNT, RDX, TATB and CL-20, were developed [21]. This promises rich scientific advances in the data-driven design of next-generation EMs.

In conclusion, it can be noted that the works using the MGI conception have begun to appear for EMs in recent years [9–19,21]. Much attention has been paid to the development of new molecules of insensitive high-energy explosives and the implementation of screening of potential candidates of advanced EMs from the point of view of the heat of explosion. Multifactor computational models are being created that allow researchers to solve the direct task—prediction of explosive detonation properties, in particular to approximate the dependence of the detonation velocity of explosives on various parameters, predicting energetic compounds' properties. For example, the inverse task of determining the atomic composition of the explosive molecule was solved [18,22,23]. The methodology and know-how for creating MCM of HEMs combustion by means of ANN, as well as examples of the results of the application of ANN for creating the MCM of combustion of various propellants, was outlined in [24].

3. Methodology

The application of ANN, one of the best tools for creating MCM of experimental data, is based on the Kolmogorov–Arnold theorem [25–27] and its special cases considered by Hecht-Nielsen [28]. From a computational point of view, ANN is a structure that includes a certain number of processing elements and which executes a fixed set of mathematical functions. This processing element is called an artificial neuron (AN). It consists of an input vector (X_i), synapses, a summator, a nonlinear transfer function, and an output signal value, as shown in Figure 1 [10].

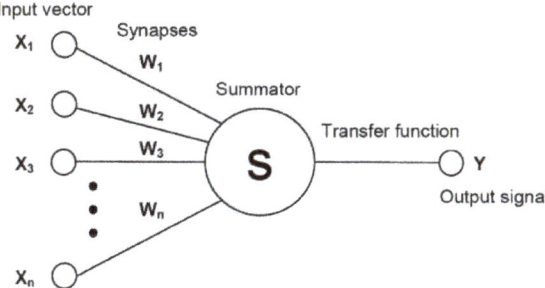

Figure 1. A scheme of an elementary processor of an artificial neuron.

The executive equation of a neuron is determined according to the following operations:

$$S = \sum_{i=1}^{n} X_i W_i \tag{1}$$

$$f(S) = \frac{1}{1 + e^{-\alpha S}} \tag{2}$$

$$Y = f(S) \tag{3}$$

The task of synapses is to multiply the input vector components, X_i, by a number characterizing the synapse strength (it is called synaptic weight, or W_i). These values obtained are summed and the sum is fed into the transfer function, Y, whose role is played by a monotonous function of one argument (usually sigmoid function $f(S)$). Thus, AN maps the vector X_i to a scalar value Y.

The simplest kind of ANN is feed-forward ANN, whose neurons are grouped into layers. The structure of feed-forward ANN is shown in Figure 2 [10].

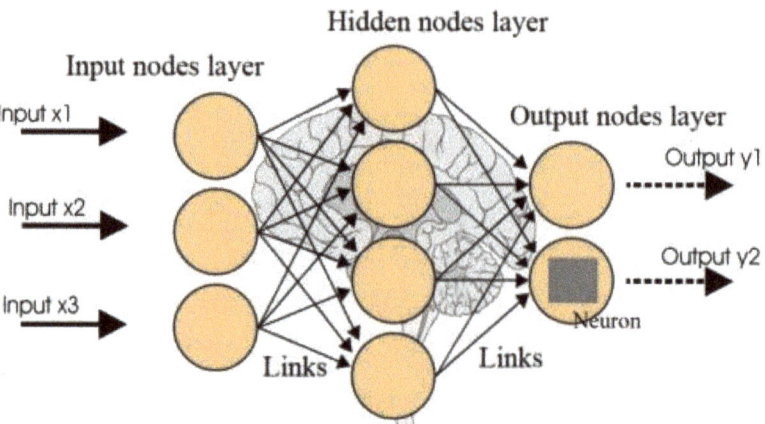

Figure 2. The structure of feed-forward ANN.

As seen in Figure 2, its structure consists of one input nodes layer (3 input nodes), one hidden layer (4 neurons) and one output layer (2 neurons), with each layer relating to its neighboring layer in "all-to-all" manner. Input nodes serve only as signal sources, while the other neurons perform the computations described above.

This computational structure can approximate the dependencies between the input variables and target (output) variables (functions) of an object after its training on a set of experimental data. The essence of training is to select the correct synaptic weights. In the process of training, weights of all synapses are determined from the requirement that ANN should map all known input vectors to the known corresponding values of the target variables with minimum errors.

This process is organized as follows. The initial synaptic weights are set using a random number generator. Then, a random input vector of real data is selected and fed into the ANN. The ANN calculates an output value, it is compared with the expected output value and the respective error is calculated. Using the "error back propagation" algorithm based on the classic gradient descent method [29–31], synaptic weights are changed by certain values. After that, a new input vector of real data is randomly selected and the whole weight update procedure is repeated. The procedure is repeated until an acceptable difference between the values computed by ANN and real values of the target variable is reached. The number of training cycles can be more than 500–1000.

The resulting ANN is able to map any input vector which is close to the vectors used during training into the respective value of the target variable i.e., it can approximate the dependence of a target variable on input factors.

The organization of real data to be used for ANN training is very important. The data for ANN training (consisting of input variable value vectors and output values corresponding to them) can be formed by means of various techniques. They can contain data measured in real experiments or data obtained from numerical simulations; they can contain data of both types when these data can complement each other. The data must be cleared, that is, contradictions, duplicates, anomalous values must be excluded. The data should be evenly distributed over the area of the input vector space, and it is necessary to avoid large differences in data density in different parts of this area (this is the requirement for data to be equally weighted).

The data should be supplemented with metadata containing additional information about the object, for example, physical or chemical constants characterizing the object under study, the parameters of the technology for creating the object, etc. The use of metadata as additional data not only increases the accuracy of the ANN model, but it also allows a deeper understanding of the physicochemical nature of the objects of research and the fine details of the mechanism of the processes under study.

Another significant circumstance is proper choice of ANN structure for which certain theoretical and empirical rules exist. For example, one of general rules (confirmed by our experience) is that the number of synapses should be 3–5 times less than the number of input vectors (examples) used in training. Use of ANN with a greater number of synapses may lead to the so-called overfitting.

The loss of the ability to generalize means that the ANN remembers training examples well and accurately reproduces the target variables for the training input vectors, but gives erroneous values of the target variables for the input vectors that it does not use in training.

To find out if the ANN has the ability to generalize the dependencies contained in the data, the following approach is used. In the process of training, the input vectors (a set of examples) are divided into two groups. A large group is used for training, and a smaller group is used only to check the ANN prediction accuracy. If the ANN accuracy in both groups is approximately the same, the ANN is not retrained and has the ability to identify and generalize the dependencies of existing data.

One more rule that has been empirically established is that it is better to use two separate ANN for each of the two "outputs" than one ANN for both "outputs" (Figure 2). The general principle for ANN structure selection is as follows. For the majority of tasks, two hidden ANN layers are sufficient to obtain an acceptable error level. Therefore, using ANN with more than two hidden layers can hardly make sense in many cases. Moreover, accuracy of networks with a single hidden layer (Figure 2) is often quite good for problems of physics and natural science where dependencies are deterministic. The final choice of the optimal ANN structure for each research task is carried out empirically by checking the exactness of different ANN (for example, with a different number of AN in the hidden layer).

It should be noted here that all questions of the methodology of ANN use for approximating experimental data have been well worked out at present, both from a theoretical and practical point of view. There exist a number of academic (free) and professional software packages which support all steps of data pre-processing, ANN training, model results visualization, model quality evaluation and validation. These make modeling experimental data simple and convenient.

Therefore, at present, it is possible to put forward the motto that experimental work cannot be considered complete until an MCM of experimental data has been created.

We believe that an autonomous executable module of the ANN model created by the authors of the article should be a mandatory supplement to any scientific article. This is explained as follows. A correctly created ANN model is, first, the most complete form of presentation of experimental results, since the ANN model contains the relationships between all the variables of the experiment. This will allow any reader of the article, having received the autonomous executable module, to independently examine in detail all the regularities contained in the ANN model and visualize in the form of graphs those regularities that the authors of the article could not cite in the article due to limitations on the volume of the article.

An additional advantage of the autonomous executable module of the ANN model is that, with its help, the reader of the article can conduct "virtual experiments", setting such combinations of factor values that were not investigated in the published article. Examples of possible scenarios for virtual experiments and the results obtained are presented in [24].

The results of the virtual introduction of copper isobutyrate catalyst into various mixtures of copper phthalate + lead catalyzed by soot greatly changes both the value of the burning rate and the dependence of the burning rate on pressure [32].

The results of of virtual simultaneously embedding two or more different types of metal powder Al, Ti, Ni, and Zr into propellants with varying composition are depicted in [12]. The results depict that the value of the burning rate and the graph of the dependence of the burning rate on the pressure vary in a complex way depending on the type of metal and the amount of simultaneously embedded metals. The results of the virtual simultaneous embedding of two additives differing in size (micro-size and nano-size) in the propellants composition are depicted in [33]. The results of virtual simultaneous use of mAl/PbO (micro-size) and nAl/PbO (nano-size) indicate that the value of the burning rate and the graph of the dependence of the burning rate on pressure change significantly.

Virtual experiments can also be carried out to execute unique experiments for such combinations of factor values that cannot be organized or are difficult to organize. The results of virtual use of only monodisperse AP particles in the propellant compositions are depicted in [34]. The two real propellant compositions have polydisperse AP particles with average sizes of AP particles 45.8 ± 30 μm and 399.6 ± 82 μm. The results of virtual use of only monodisperse AP particles 45.8 μm and 399.6 μm show that the monodispersity of AP particles does not strongly affect the combustion rate or the form of dependence of the combustion rate on pressure for both cases.

In addition to the above, one more very interesting case should be noted when the use of ANN is justified. Our experience shows that the root-mean-square (RMS) error of the ANN model is always less than the RMS error of the experimental data used to create the ANN model. This allows the ANN model to be used as a means of checking the quality of the experiment as a whole! Moreover, it can do so both from the point of view of the measurement error of the variables of the experiment, and from the point of view of the correctness of the experiment, that is, from the completeness of taking into account all the factors affecting the goal of the experiment.

In cases where the RMS error of the ANN model is too large (for example, when the RMS error of the ANN model is more than 10^{-3}), it is necessary to improve the accuracy of the experimental variables measurement and (or) change the formulation of the experimental problem, trying to take into account additional factors affecting the goal function of the experiment.

An example of the use of additional factors (metadata) that affect the goal function of the experiment, i.e., the detonation velocity, is presented in [22,23]. In contrast to [11], in which only the numbers of C, H, N, O atoms in the explosive molecules were used as input data, in [22,23] not only were the numbers of C, H, N, O atoms in the explosive molecules used, but so were combinations of the ratio of the number of atoms C, H, N, O to each other (C/H, N/O). These combinations can be considered as metadata that reflect a structure of the ratio of the explosive molecules. The evaluation of the quality of the model indicated that the RMS error is 0.00025. The maximum relative error is less than 1% over the entire range of detonation velocities, except for very low (about 1.5 km/s) and very high (more than nine km/s) detonation velocities. These errors are several times smaller than errors for the model presented in [11]. The reduction in errors and the improvement in the quality of the model occurred precisely due to the input of metadata, reflecting the structure of chemical bonds in the explosive molecule as factors into the model.

4. Results and Discussion

4.1. Experiment

The base set of burning rate data was taken from [35]. The set consists of data about combustion of RDX-CMDB and CL-20-CMDB propellants with different nanopowders and contents:

- 73.5% NG/NC + 19.5% burning rate inhibitor + 4.0% catalyst + 3.0% additives with and without nAl;
- 63.0% NG/NC + 2.3% catalyst + 2.8% additives + 26% RDX + 4.6% diethyl phthalate (DEP) + 2.6% (nAl + Al_2O_3) with and without nAlN;
- 63.4% NG/NC + 5.85% catalyst + 4.75% additives + 24% HMX with and without nDPN;
- CL-20-CMDB propellants formulation with different mass fraction of nNi;
- RDX-CMDB propellants with different mass fractions of nNi.

The base set is prepared in a special type [36] of that modeling by means of ANN demands.

4.2. Modelling

All models were obtained by using ANN that included in analytical platform Deductor (https://basegroup.ru/deductor/description (accessed on 28 January 2023).

4.2.1. Direct Task

The direct task reveals dependences of the burning rate (goal function of models) on the various factors. In our case, ANN structure (Figure 3) for solving the direct task consists of one input layer (17 neurons which correspond to 17 factors), one hidden (inner) layer (5 neurons) and one output layer (1 neuron which corresponds to goal function).

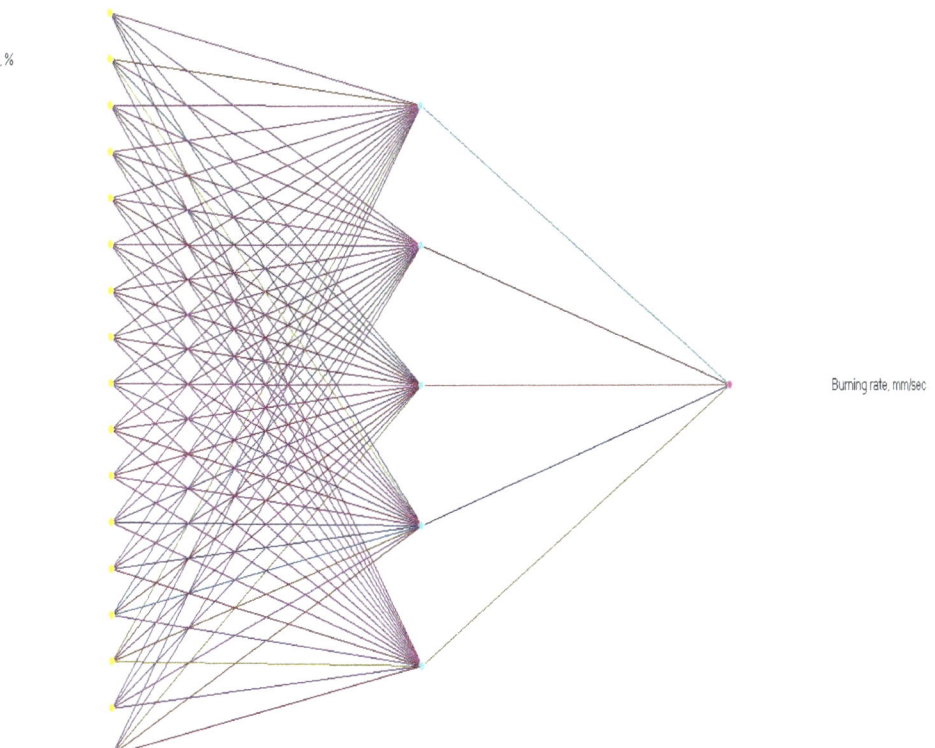

Figure 3. ANN structure for solving the direct task.

This calculation structure, after training on the experimental data, allows the determination (calculation) of the burning rate value for any set of factors values.

The advantage of the analytical platform Deductor is the automatic execution of quality assessments of the ANN model obtained. A portion (95%) of the full data set was randomly selected and used for training, and the remaining data (5%) was used for model testing (validation) only. Both the testing and training procedures are run simultaneously. The quality assessments of the ANN model obtained depicts that root-mean-square error of the ANN model training procedure equals 3.4×10^{-4} for 78% of the training data set and that the root-mean-square error of the ANN model testing equals 5.1×10^{-4} for 71% of testing data set.

It is important to note that the root-mean-square error of the testing procedure on the data that have not been used for training is about equal to the root-mean-square error of the training procedure. This observation confirms that the overfitting of ANN structure is not present.

The examples of the results of calculation of the ANN MCM that solves a direct task (two cases for various set of factors) and if two graphs of the dependence of burning rate on pressure are depicted in Table 1 and in Figures 4 and 5. In Table 1, the dependence of the burning rate on pressure value and the quantity of additives, denoted with the names "others" and nNi, it depicted.

Table 1. Examples of results of calculation of the ANN MCM that solves the direct task.

Input Factors	Values in Case 1	Values in Case 2
NC + NG, %	82.5	82.5
Burning rate inhibitor, %	0	0
Catalyst, %	5.75	5.75
nAl, %	0	0
Others, %	6.25	5.75
RDX, %	0	0
DEP, %	0	0
nAlN, %	0	0
Al_2O_3, %	0	0
HMX, %	0	0
mAl, %	0	0
nDPN, %	0	0
Al	5.5	5.5
CL-20	0	0
nNi, %	0	0.5
SUM	100	100
Pressure, MPa	15	10
Output Burning rate, mm/s	31.9	34.4

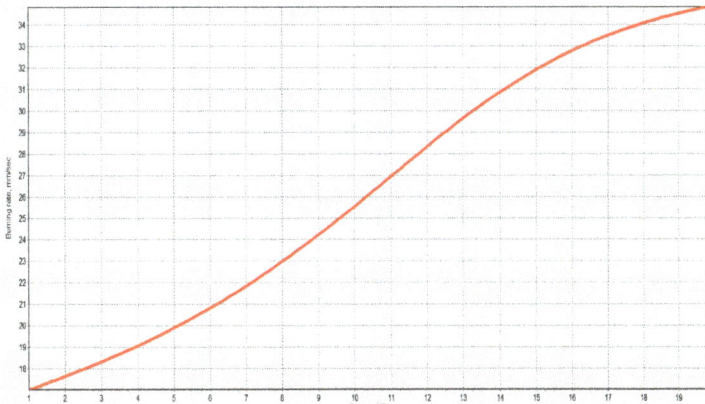

Figure 4. A representative dependence of burning rate on pressure for the case 1 in Table 1.

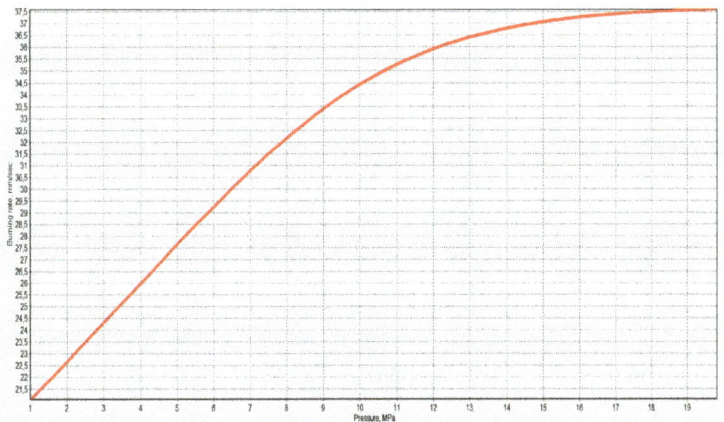

Figure 5. A representative dependence of burning rate on pressure for the case 2 in Table 1.

The graphs like Figures 4 and 5 are obtained, by means of the Deductor, in auto mode in every time the researcher calculates a burning rate and for any set of input values.

The ANN MCM obtained can be considered as a specialized calculator that solves the direct task and contains all the links between the goal and function of the model—the burning rate and 17 factors. It can instantly give the value of the burning rate for any set of factor values and present graphs of the burning rate versus any factor, not just pressure. Much more examples of such results are presented in [24].

4.2.2. Inverse Problem (Task)

The one of possible ANN structure for solving the inverse task (Figure 6) consists of one input layer (17 neurons which correspond 16 factors and 1 goal function—burning rate which we have to obtain), one hidden (inner) layer (5 neurons) and one output layer (1 neuron which corresponds 1 factor—pressure which have to help us to rich the required value of burning rate).

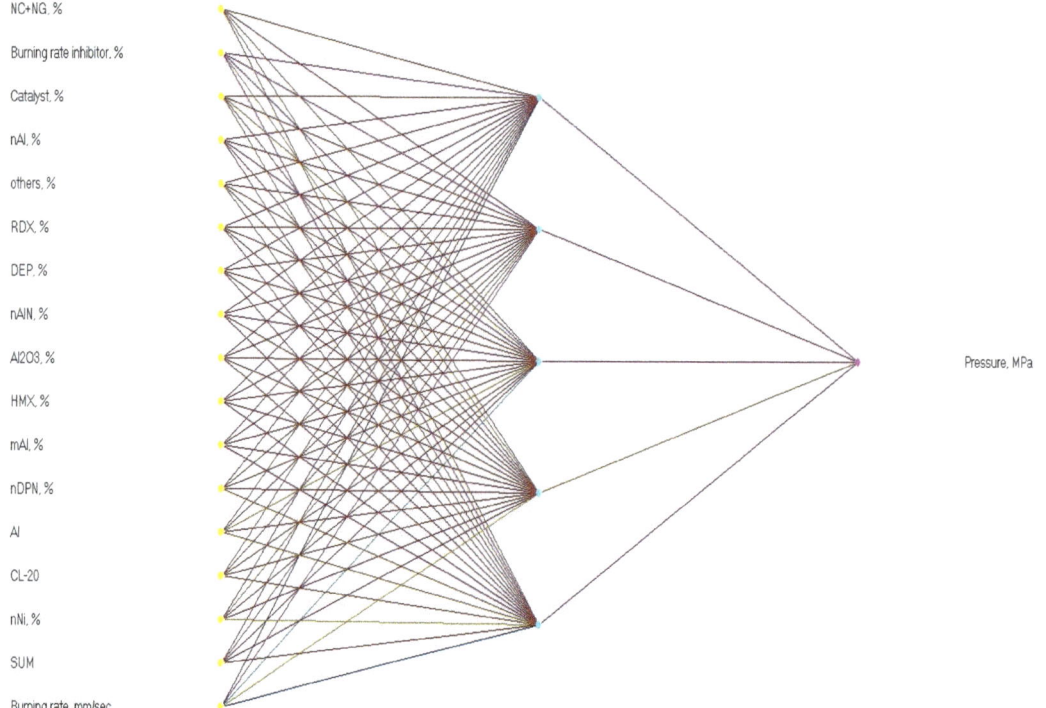

NC+NG, %

Burning rate inhibitor, %

Catalyst, %

nAl, %

others, %

RDX, %

DEP, %

nAlN, %

Al2O3, %

HMX, %

mAl, %

nDPN, %

Al

CL-20

nNi, %

SUM

Burning rate, mm/sec

Pressure, MPa

Figure 6. ANN structure for the simplest form of inverse task.

This calculation structure, after training, allows the determination (calculation) of the pressure values that can help us to enrich the required value of burning rate for any set of factors values.

The quality assessments of the ANN model obtained depicts that the root-mean-square error of the ANN model training procedure equals 1.7×10^{-2} for 86% of training data set and that the root-mean-square error of the ANN model testing equals 7.3×10^{-3} for 43% and 1.4×10^{-2} for 29% of testing data set.

The root-mean-square error of the model for solving the inverse task is greater than of the model for solving the direct task. This is quite understandable. The latter is explained by the fact that inverse tasks solved based on experimental data are incorrectly (ill-posed) set, according to Hadamard, from the point of view of pure mathematics. If the requirements for the existence of a solution and the stability of the solution to errors in the input data are satisfied in the case of using an ANN well, then the requirement for the uniqueness of the solution cannot be fully met. First of all, this is due to the significant multifactor nature of the task of determining the pressure that provides one or another burning rate, both since the same burning rate can be obtained both due to a change in pressure (with a constant propellant composition) and due to the composition of the propellant (at a constant pressure).

The examples of the results of that calculation of the ANN MCM that solves the inverse task and the graph of connection of the burning rate and pressure are depicted in Table 2 and Figure 7.

Table 2. The example of result of calculation of the ANN MCM that solves the inverse task.

Input Factors	Values
NC + NG, %	63
Burning rate inhibitor, %	0
Catalyst, %	2.3
nAl, %	0
Others, %	2.8
RDX, %	26
DEP, %	4.6
nAlN, %	1.3
Al_2O_3, %	0
HMX, %	0
mAl, %	0
nDPN, %	0
Al	0
CL-20	0
nNi, %	0
SUM	100
Burning rate, mm/s	18
OutputPressure, MPa	15.2

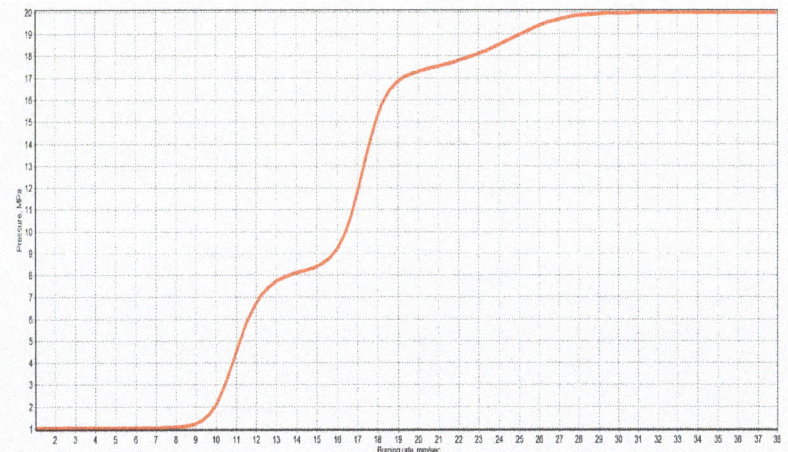

Figure 7. An example of a graph of connection of burning rate and pressure. The graph is valid only for the set of input factors values indicated in Table 2. For other sets of values, the graphs will be different.

The ANN MCM (calculator) that solves the inverse task allows us to solve various problems related to determining the composition of HEM and the level of pressure to needed obtain the required burning rate. We have presented some such results at [24,37].

4.2.3. Virtual Experiments

The ANN MCM for solving a direct task has a very interesting feature. This allows for a virtual experiment. The essence of the virtual experiment was as follows.

The virtual experiments are computational experiments carried out using ANN MCM, and during this such combinations of factor values are established that were not investigated in a real experiment. For example, a virtual experiment includes the extrapolation of the dependencies identified by the ANN model, for example, the task of predicting the values of the burning rate for pressure values for which experiments have not been carried out. Another example of a virtual experiment is a computational experiment, during which such a combination of factor values (such a set of factor values) is specified, for which the real experiment was not carried out.

The result of the virtual experiment is depicted in Table 3 and Figures 8 and 9. In Table 3, on the left is the example of the results of calculation of ANN MCM for solving the direct task for a propellant composition for which real experiments were carried out. On the right is the example of the results of calculation of ANN MCM for solving the direct task for a virtually changed propellant composition.

Table 3. The example of result of calculation of the ANN MCM that executes virtual experiments (right side).

Input Factors	Values for the Real Experiment	Values for the Virtual Experiment
NC + NG, %	63.4	63.4
Burning rate inhibitor, %	0	0
Catalyst, %	5.85	5.85
nAl, %	0	0
others, %	4.75	4.75
RDX, %	0	0
DEP, %	0	0
nAlN, %	0	0
Al_2O_3, %	0	0
HMX, %	24	24
mAl, %	2	2
nDPN, %	0	0.7
Al	0	5.5
CL-20	0	0
nNi, %	0	0
SUM	100	106.2
Pressure, MPa	15	15
Output Burning rate, mm/s	22.3	25.5

During the virtual experiment, we have included such components as nDPN and Al in the propellant composition simultaneously (this composition was not really studied in the experiment). The model instantly calculated the value of the burning rate for the new propellant composition.

4.2.4. Comparison of Predicted Burning Rate with Experimental Data of SRP

The results of experimental research of combustion performance of double-base SRP with micro- and nano-sized additives were taken, dealing with the effects of different nano-sized additives on the burning rate of double-based SRP with such micro- and nano-sized metals. It has been used for the creation of combustion multifactor computational models that solve direct and inverse tasks, and the predicted burning rate of SRP was determined as well. Figure 10 shows the comparison curves of predicted burning rate with the experimental data of SRP. It can be seen that the predicted burning rate data agree well with the experiments ones. The predicted burning rate curves are smoother than the experiment curves.

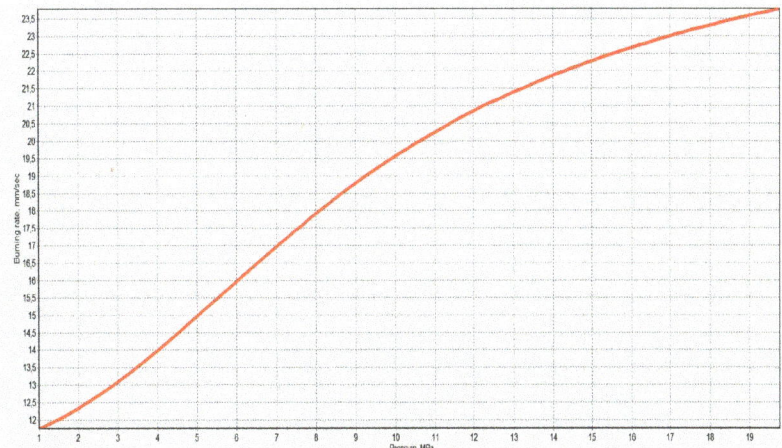

Figure 8. The graphs of the dependence of the burning rate on pressure, corresponding to the results of calculation of ANN MCM for real experiment (left side of the Table 3, direct task for the propellant for which real experiments were carried out).

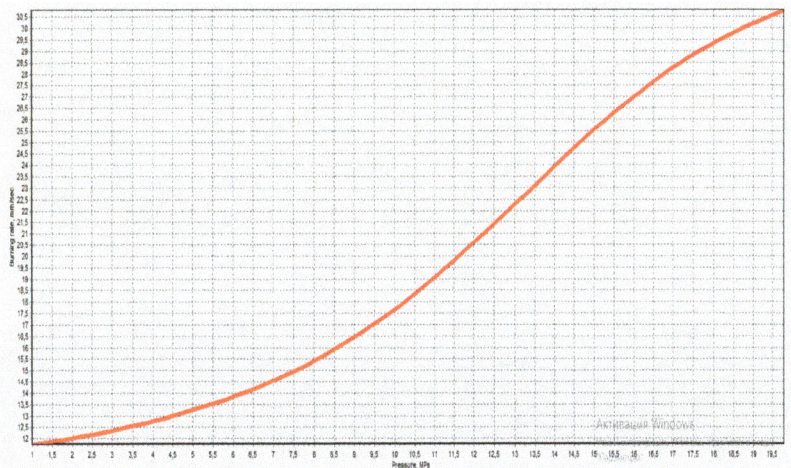

Figure 9. The graphs of the dependence of the burning rate on pressure, corresponding to the results of calculation of ANN MCM for virtual experiment (right side of the Table 3, direct task for a virtually modified propellant composition.

Pressure, MPa	Burning rate, mm/s	
	Experimental	ANN
1	1.37	1.82
2	1.81	1.93
4	2.50	2.22
8	2.93	3.05
12	3.89	4.04
15	4.59	4.70
20	5.96	5.55

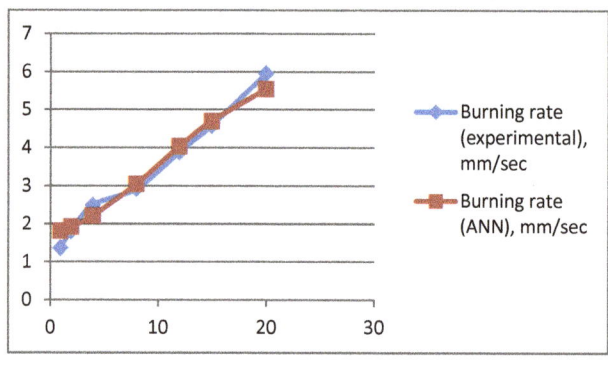

(**a**)

Pressure, MPa	Burning rate, mm/s	
	Experimental	ANN
5	11.18	11.59
8	14.21	14.15
10	15.89	15.71
12	17.08	16.97
15	18.45	18.27
18	21.10	18.97

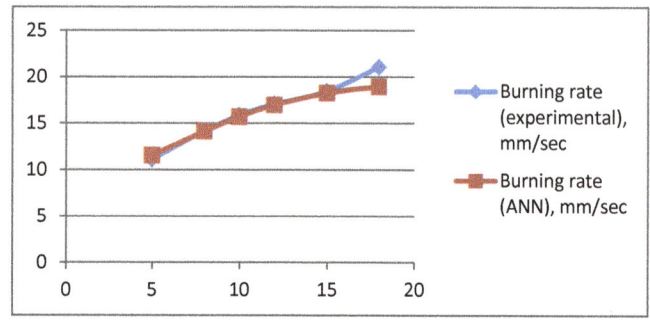

(**b**)

Pressure, MPa	Burning rate, mm/s	
	Experimental	ANN
5	14.86	14.66
8	16.35	16.62
10	18.09	17.86
12	19.38	18.93
15	19.05	20.08
18	20.00	20.8

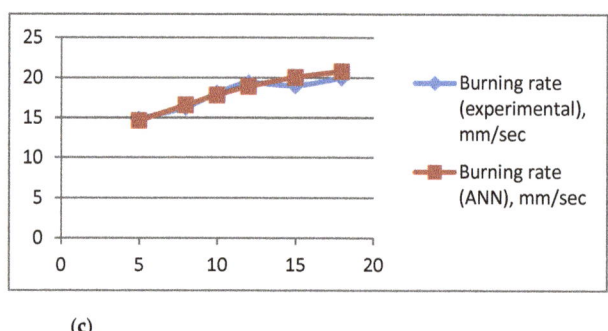

(**c**)

Figure 10. Comparison curves of predicted burning rate with experimental data of solid propellants. (**a**): 73.5% NG/NC + 19.5% burning rate inhibitor + 4.0% catalyst + 3.0% additives; (**b**): 63.0% NG/NC + 2.3% catalyst + 2.8% additives + 26% RDX + 4.6% diethyl phthalate (DEP) + 1.3% (nAlN); (**c**): 63.0% NG/NC + 2.3% catalyst + 2.8% additives + 26% RDX + 4.6% diethyl phthalate (DEP) + 1.3% (Al$_2$O$_3$).

5. Conclusions

1. The usage of ANN for the creation of new MCM of the propellants combustion and detonation, that solve the direct and inverse tasks as well execute the virtual experiments, depict that ANN have the wide possibilities for propellants combustion and detonation research and development of new kind of advanced propellants. The results presented in this article depict no more than 1% of the propellants combustion patterns contained in the obtained MCM.

2. The autonomous computer module of MCM allows reader to independently and in detail study all the regularities contained in the ANN model, visualizing in the form of hundreds of graphs those regularities that the authors of the article could not present in the article due to the limitations on the volume of the article. Instructions for using the executable ANN model are included with the module.

3. The autonomous computing module of MSM can be utilized. This allows researchers to calculate the values of the burning rate for energetic compositions at various conditions, visualize the patterns contained in the experimental data, conduct virtual experiments, and predict the burning rate of propellants at different pressures. The virtual experiments are a very promising means to develop new and advanced solid propellants in the framework of HEMG.

Author Contributions: Conceptualization, V.A. and W.P.; methodology, V.A.; software, V.A. and D.A.; investigation, V.A., W.P. and D.A.; resources, W.P.; data curation, V.A. and W.P.; writing—original draft preparation, V.A. and D.A.; writing—review and editing, W.P. and D.C. All authors have read and agreed to the published version of the manuscript.

Funding: This research received no external funding.

Data Availability Statement: https://www.researchgate.net/publication/361367833_Data_about_combustion_of_RDX-CMDB_and_CL-20-CMDB_propellants_with_different_nano_powders_and_contents, https://www.researchgate.net/publication/366078129_Abrukov-Pang-Anufrieva-Loginom (accessed on 28 January 2023) and https://www.researchgate.net/publication/367412407_Abrukov-Pang-25-01-23 (accessed on 28 January 2023).

Conflicts of Interest: The authors declare no conflict of interest.

Abbreviations

HEMG	high-energy materials genome
HEMs	high-energy materials
ANN	artificial neural networks
MCM	multifactor computational models
SRP	solid rocket propellant
NEPE	nitrate ester plasticized polyether
AI	artificial intelligence
mAl	micro-sized aluminium
nAl	nano-sized aluminium
MGI	materials genome initiative
ML	machine learning
EMGI	energetic materials genome initiative
MI	materials informatics
ICSD	inorganic crystal structure database
CHNO	carbon, hydrogen, nitrogen, and oxygen
AN	artificial neuron
RMS	root-mean-square
nNi	nano-sized nickel
RDX	hexogen
CL-20	hexanitrohexaazaisowurtzitane
CMDB	compound-modified double base
NG	nitroglycerin
NC	nitrocellulose
DEP	diethyl phthalate
Al_2O_3	aluminium trioxide
nDPN	a type of nano-sized composite
HMX	octogen
nAlN	nano-sized aluminium nitride

References

1. Yan, Q.-L.; Zhao, F.-Q.; Kuo, K.K.; Zhang, X.-H.; Zeman, S.; DeLuca, L.T. Catalytic effects of nano additives on decomposition and combustion of RDX-, HMX-, and AP-based energetic compositions. *Prog. Energy Combust. Sci.* **2016**, *57*, 75–136. [CrossRef]
2. Pang, W.-Q.; Wang, K.; Xu, H.-X.; Li, J.-Q.; Xiao, L.-Q.; Fan, X.-Z.; Li, H. Combustion features of nitrate ester plasticized polyether solid propellants with ADN and Fox-12 particles. *Int. J. Energy Mater. Chem. Prop.* **2020**, *19*, 11–23. [CrossRef]
3. Pang, W.-Q.; Wang, K.; DeLuca, L.T.; Trache, D.; Fan, X.-Z.; Li, J.; Li, H. Experiments and simulations on interactions between 2,3-bis(hydroxymethyl)-2,3-dinitro-1,4-butanediol tetranitrate (DNTN) with some energetic components and inert materials. *FirePhysChem* **2021**, *1*, 166–173. [CrossRef]
4. Zhang, X.-P.; Dai, Z.-L. Calculation for high-pressure combustion properties of high-energy solid propellant based on GA-BP neural network. *J. Solid Rocket. Technol.* **2007**, *30*, 229–232.
5. Kalil, T.; Wadia, C. Materials Genome Initiative for Global Competitiveness, A Whitepaper, Executive Office of the President National Science and Technology Council, Washington, D.C. 20502. 24 June 2011. Available online: https://www.mgi.gov/sites/default/files/documents/materials_genome_initiative-final.pdf (accessed on 28 January 2023).
6. DeLuca, L.T.; Shimada, T.; Sinditskii, V.P.; Calabro, M.; Manzara, A.P. Innovation of Energetic Materials by Materials Genome Initiative. An Introduction to Energetic Materials for Propulsion. In *Chemical Rocket Propulsion*; DeLuca, L.T., Shimada, T., Sinditskii, V., Calabro, M., Eds.; Springer Aerospace Technology: Cham, Switzerland, 2017. [CrossRef]
7. Abrukov, V.S.; Malinin, G.I.; Volkov, M.E.; Makarov, D.N.; Ivanov, P.V. Application of artificial neural networks for creation of "black box" models of energetic materials combustion. In *Advancements in Energetic Materials and Chemical Propulsion*; Kuo, K.K., Hori, K., Eds.; Begell House Inc. of Redding: Danbury, CT, USA, 2008; pp. 377–386.
8. Abrukov, V.S.; Karlovich, E.V.; Afanasyev, V.N.; Semenov, Y.V.; Abrukov, S.V. Creation of propellant combustion models by means of data mining tools. *Int. J. Energetic Mater. Chem. Propuls.* **2010**, *9*, 385–396. [CrossRef]
9. Chandrasekaran, N.; Bharath, R.S.; Oommen, C.; Abrukov, V.S.; Kiselev, M.V.; Anufrieva, D.A.; Sanal Kumar, V.R. Development of the Multifactorial Computational Models of the Solid Propellants Combustion by Means of Data Science Methods-Phase II. In Proceedings of the 2018 Joint Propulsion Conference, AIAA Propulsion and Energy Forum, (AIAA 2018–4961), Cincinnati, OH, USA, 9–11 July 2018. [CrossRef]
10. Abrukov, V.; Lukin, A.; Anufrieva, D.; Oommen, C.; Sanalkumar, V.; Chandrasekaran, N.; Bharath, R. Recent advancements in study of effects of nano micro additives on solid propellants combustion by means of the data science methods. *Def. Sci. J.* **2019**, *69*, 20–26. [CrossRef]
11. Chandrasekaran, N.; Oommen, C.; Sanalkumar, V.R.; Abrukov, V.S.; Anufrieva, D.A. Prediction of detonation velocity and N-O composition of high energy C-H-N-O explosives by means of the data science methods. *Prop. Explos. Pyrotech.* **2019**, *44*, 579–587. [CrossRef]
12. Mariappan, A.; Choi, H.; Abrukov, V.S.; Anufrieva, D.A.; Sankar, V.; Sanalkumar, V.R. The Application of Energetic Materials Genome Approach for Development of the Solid Propellants Through the Space Debris Recycling at the Space Platform. In Proceedings of the AIAA Propulsion and Energy 2020 Forum, AIAA 2020–3898, Online. 24–28 August 2020. [CrossRef]
13. Abrukov, V.S.; Chandrasekaran, N.; Oommen, C.; Thianesh, U.K.; Mariappan, A.; Sanal Kumar, V.R.; Anufrieva, D.A. Genome approach and data science methods for accelerated discovery of new solid propellants with desired properties. In Proceedings of the AIAA Propulsion and Energy 2020 Forum, AIAA 2020–3929, Online. 24–28 August 2020. [CrossRef]
14. Abrukov, V.S.; Oommen, C.; Sanal Kumar, V.R.; Chandrasekaran, N.; Sankar, V.; Kiselev, M.V.; Anufrieva, D.A. Development of the Multifactorial Computational Models of the Solid Propellants Combustion by Means of Data Science Methods-Phase III. Technology and Investment. In Proceedings of the 2019 55th AIAA/SAE/ASEE Joint Propulsion Conference 2019, AIAA Propulsion and Energy Forum AIAA 2019–3957, Indianapolis, IN, USA, 19–22 August 2019. [CrossRef]
15. Wang, Y.; Liu, Y.-J.; Song, S.-W.; Yang, Z.-J.; Qi, X.-J.; Wang, K.-C.; Liu, Y.; Zhang, Q.-H.; Tian, Y. Accelerating the discovery of insensitive high-energy-density materials by a materials genome approach. *Nat. Commun.* **2018**, *9*, 2444. [CrossRef] [PubMed]
16. Kang, P.; Liu, Z.; Abou-Rachid, H.; Guo, H. Machine-learning assisted screening of energetic materials. *J. Phys. Chem. A* **2020**, *124*, 5341–5351. [CrossRef] [PubMed]
17. Yuan, W.-L.; He, L.; Tao, G.-H.; Shreeve, J.M. Materials-genome approach to energetic materials. *Acc. Mater. Res.* **2021**, *2*, 692–696. [CrossRef]
18. Yang, Z.-H.; Rong, J.-L.; Zhao, Z.-T. Study on the prediction and inverse prediction of detonation properties based on deep learning. *Def. Technol.* **2022**, *in press*. [CrossRef]
19. Tian, X.-L.; Song, S.-W.; Chen, F.; Qi, X.-J.; Wang, Y.; Zhang, Q.-H. Machine learning-guided property prediction of energetic materials: Recent advances, challenges, and perspectives. *Energetic Mater. Front.* **2022**, *3*, 177–186. [CrossRef]
20. Materials Genome Initiative. Available online: https://www.mgi.gov/ (accessed on 10 January 2023).
21. Physics-Informed Meta-Learning for Design of Complex Materials. Available online: https://www.mgi.gov/sites/default/files/documents/MGI_PI_20220628.pdf/ (accessed on 10 January 2023).
22. Anufrieva, D.A.; Koshcheev, M.I.; Abrukov, V.S. Application of Data Mining Methods in Physics Research. Multifactor Detonation Models. In the Collection: High-Speed Hydrodynamics and Shipbuilding. Collection of Scientific Papers of the XII International Summer Scientific School-Conference Dedicated to the 155th Anniversary of the Birth of Academician A.N. Krylov. 2018. S. 221–226. Available online: https://www.researchgate.net/publication/333210244_primenenie_metodov_intellektualnogo_analiza_dannyh_v_fiziceskih_issledovaniah_mnogofaktornye_modeli_detonacii (accessed on 10 January 2023).

23. Anufrieva, D.A.; Abrukov, V.S.; Oommen, C.; Sanalkumar, V.R.; Chandrasekaran, N. Generalized Multifactor Computational Models of the Detonation of Condensed Systems. Available online: https://www.researchgate.net/publication/334126580_Generalized_multifactor_computational_models_of_the_detonation_of_condensed_systems (accessed on 10 January 2023). [CrossRef]

24. ResearchGate. Victor Abrukov. Available online: https://www.researchgate.net/profile/V-Abrukov/research/ (accessed on 10 January 2023).

25. Kolmogorov, A. On the representation of continuous functions of several variables by superpositions of continuous functions of a smaller number of variables. *Izvestiya AN SSSR* **1956**, *108*, 179–182, English translation: *Amer. Math. Soc. Transl.* **1961**, *17*, 369–373.

26. Arnold, V. On the function of three variables. *Izvestiya AN SSSR* **1957**, *114*, 679–681, English translation: *Amer. Math. Soc. Transl.* **1963**, *28*, 51–54.

27. Gorban, A.N. Generalized approximation theorem and computational capabilities of neural networks. *Sib. J. Comput. Math.* **1998**, *1*, 11–24.

28. Hecht-Nielsen, R. Kolmogorov's Mapping Neural Network Existence Theorem. In Proceedings of the IEEE First Annual International Conference on Neural Networks, San Diego, CA, USA, 1987; Volume 3, pp. 11–13.

29. Werbos, P.J. Beyond regression: New Tools for Prediction and Analysis in the Behavioral Sciences. Ph.D. Thesis, Harvard University, Cambridge, MA, USA, 1974.

30. Galushkin, A.I. *Neural Networks Theory*; Springer: Berlin/Heidelberg, Germany, 29 October 2007; eBook; ISSN 978-3-540-48125-6. [CrossRef]

31. Rumelhart, D.E.; Hinton, G.E.; Williams, R.J. Learning Internal Representations by Error Propagation. In *Parallel Distributed Processing*; MIT Press: Cambridge, MA, USA, 1986; Volume 1, pp. 318–362.

32. Abrukov, V.S.; Anufrieva, D.A.; Sanalkumar, V.R.; Amrith, M. Multifactor Computational Models of the Effect of Catalysts on the Combustion of Ballistic Powders (Experimental Results of Denisyuk Team) Direct Tasks, Virtual Experiments and Inverse Problems. 2020; pp. 1–20. Available online: https://www.researchgate.net/publication/344727996_Multifactor_Computational_Models_of_the_Effect_of_Catalysts_on_the_Combustion_of_Ballistic_Powders_experimental_results_of_Denisyuk_team_Direct_Problems_Experiments_Virtverse (accessed on 28 January 2023).

33. Abrukov, V.; Anufrieva, D.A.; Oommen, C.; Sanalkumar, V.R.; Chandrasekaran, N. Effects of Metals and Termites Adds on Combustion of Double-Based Solid Propellants. Development of the Multifactor Computational Models of the Solid Propellants Combustion by Means of Data Science Methods. Virtual Experiments and Propellant Combustion Genome. 2019. Available online: https://www.researchgate.net/publication/334172872_effects_of_metals_and_termites_adds_on_combustion_of_doble-_based_solid_propellants_development_of_the_multifactor_computational_models_of_the_solid_propellants_combustion_by_means_of_data_science_me (accessed on 28 January 2023).

34. Abrukov, V.S.; Anufrieva, D.A.; Sanalkumar, V.R.; Amrith, M. Comprehensive Study of AP Particle Size and Concentration Effects on the Burning Rate of Composite AP/HTPB Propellants by Means of Neural Networks. Development of the Multifactor Computational Models. Direct Tasks and Inverse Problems & Virtual Experiments. 2020; pp. 1–20. Available online: https://www.researchgate.net/publication/344494607_comprehensive_study_of_ap_particle_size_and_concentration_effects_on_the_burning_rate_of_composite_aphtpb_propellants_by_means_of_neural_networks_development_of_the_multifactor_computational_models_di#fullTextFileContent (accessed on 28 January 2023).

35. Pang, W.; Li, Y.; DeLuca, L.T.; Liang, D.; Zhao, Q.; Liu, X.; Xu, H.; Fan, X. Effect of metal nanopowders on the performance of solid rocket propellants: A review. *Nanomaterials* **2021**, *11*, 2749. [CrossRef] [PubMed]

36. Data about Combustion of RDX-CMDB and CL-20-CMDB Propellants with Different Nano Powders and Contents. Available online: https://www.researchgate.net/publication/361367833_data_about_combustion_of_RDX-CMDB_and_CL-20-CMDB_propellants_with_different_nano_powders_and_contents/ (accessed on 10 January 2023).

37. Abrukov-Pang-25-01-23. Available online: https://www.researchgate.net/publication/367412407_Abrukov-Pang-25-01-23/ (accessed on 25 January 2023).

Article

Experiment and Molecular Dynamic Simulation on Performance of 3,4-Bis(3-nitrofurazan-4-yl)furoxan (DNTF) Crystals Coated with Energetic Binder GAP

Yue Qin [1], Junming Yuan [1,*], Hu Sun [1], Yan Liu [1], Hanpeng Zhou [1], Ruiqiang Wu [2], Jinfang Chen [2] and Xiaoxiao Li [2]

[1] School of Environmental and Safety Engineering, North University of China, Taiyuan 030051, China
[2] Shanxi North Xing'an Chemical Industry Co., Ltd., Taiyuan 030008, China
*   Correspondence: junmyuan@163.com

Abstract: To investigate the crystallization of DNTF in modified double-base propellants, glycidyl azide polymer (GAP) was used as the coating material for the in situ coating of DNTF, and the performance of the coating was investigated to inhibit the crystallization. The results show that GAP can form a white gel on the surface of DNTF crystals and has a good coating effect which can significantly reduce the impact sensitivity and friction sensitivity of DNTF. Molecular dynamics was used to construct a bilayer interface model of GAP and DNTF with different growth crystal surfaces, and Molecular dynamics calculations of the binding energy and mechanical properties of the composite system were carried out. The results showed that GAP could effectively improve the mechanical properties of DNTF. The values of K/G, γ and ν are higher than those of DNTF crystals, and the values of C_{12}-C_{44} are positive, indicating that GAP can improve DNTF ductility while also improving toughness. Combining the experimental results with the simulation calculations, energetic binder GAP can be referred to as a better cladding layer for DNTF, which is feasible for inhibiting the DNTF crystallization problem in propellants.

Keywords: 3,4-Bis(3-nitrofurazan-4-yl) furozan (DNTF); crystallization; energetic binder GAP; mechanical sensitivity; molecular dynamics simulation

Citation: Qin, Y.; Yuan, J.; Sun, H.; Liu, Y.; Zhou, H.; Wu, R.; Chen, J.; Li, X. Experiment and Molecular Dynamic Simulation on Performance of 3,4-Bis(3-nitrofurazan-4-yl)furoxan (DNTF) Crystals Coated with Energetic Binder GAP. *Crystals* **2023**, *13*, 327. https://doi.org/10.3390/cryst13020327

Academic Editors: Thomas M. Klapötke, Rui Liu, Yushi Wen and Weiqiang Pang

Received: 30 January 2023
Revised: 11 February 2023
Accepted: 13 February 2023
Published: 15 February 2023

1. Introduction

In view of the low danger of DNTF synthesis, the simple preparation process, the high energy, high density, low melting point, and other high qualities, its comprehensive performance due to HMX is close to that of CL-20, making it a better application in the field of modified double-base propellants [1–3]. However, as more research on DNTF's application in propellants is conducted, its flaws are gradually revealed. Pang Jun et al. [4] found that the tablets prepared from DNTF-CMDB propellant by a high-temperature calendering process have a crystallization phenomenon in the process of natural cooling and storage, which not only affects the appearance but also causes adverse effects such as increased propellant sensitivity, ignition difficulties, and changes in ballistic performance. The crystallization phenomenon was first found in RDX-containing propellants [5–7], and research shows that the propellant preparation process, the species and content of added energy-containing materials, and their solubility in the double-base solid solvent will have an impact on the crystallization, as will the surface coating method, the addition of low surface energy substances, and other means to inhibit crystallization and get a better effect. The crystallization phenomenon will also occur in the propellant containing DNTF, which will occur due to the content of DNTF and the solvent in the formula. Zheng Wei et al. [8] conducted a study on the inhibition of crystallization in modified dual-base propellants containing DNTF, and the amount of crystallization can be significantly reduced by using acetone solution coated with 3% NC, or adding a small amount of polymer to the propellant formulation can completely inhibit crystallization, but there is no study on the binder as a coating layer of DNTF explosive crystals as a means of inhibition. There is very litter existing literature on the mechanism and inhibition of DNTF crystallization in modified double-base propellants at home and abroad. In this paper, we try to use an energetic binder to cover DNTF to form microsphere particles to achieve the effect of propellant crystallization inhibition.

It is critical to fully consider the properties of both the material and the composite system when selecting the cladding material to ensure that the propellant energy, ignition, density, and other characteristics are maintained while effectively improving the crystallization of DNTF and maintaining stable control of its crystallization amount. The azide binder is one of the more prominent types of energetic binder in the field of solid propellants because of its advantages in heat generation, high density, and low mechanical sensitivity [9,10]. In addition, the surface energy of GAP can be compared with that of adhesive. In addition, GAP has the advantage of surface energy and adhesion, which can be used as a DNTF cladding layer to try to investigate the crystallization inhibition method.

The energetic binder glycidyl azide polymer (GAP) was used as the coating material in this paper, and the in situ polymerization method was used to coat and granulate DNTF explosive particles. The morphological properties and mechanical sensitivity changes of the coated samples were studied, and the binding energy and mechanical properties of the composite system model of GAP and DNTF were calculated and analyzed using molecular dynamics simulations. Combining the experimental data and simulation results, we analyze the coating effect of energetic binder GAP on DNTF explosive particles and provide some experimental and theoretical support for improving the crystallization of DNTF in propellant.

2. Materials and Methods

2.1. Experiments and Apparatus

Materials: DNTF, purity > 99%, provided by the Xi'an Institute of Modern Chemistry, Xi'an, China; GAP, Li Ming Chemical Research Institute, Luoyang, China; curing agent (IPDI), BASF SE, Ludwigshafen Germany; tetrahydrofuran, Sinopharm Chemical Reagent Co., Ltd., Shanghai, China.

Instruments: PTY-A type analytical balance, Guangzhou Shangbo Electronic Technology Co., Ltd., Guangzhou, China; HH-SI single-hole thermostatic water bath, Gongyi Yuhua Instrument Co., Ltd., Gongyi, China; BGX-70L electric thermostatic blast drying oven, Suzhou Taiyou Mechanical Manufacturing Co., Ltd., Suzhou, China; SEM-30PLUS scanning electron microscope, Beijing Tianyao Technology Co., Ltd. (coxem China), Beijing, China.

2.2. GAP-Coated DNTF Experiment

The prepared DNTF, binder, curing agent, and solvent, tetrahydrofuran, were placed in a conical flask and mixed at 300 rpm/min for 1 hour at room temperature with ventilation turned on and the outer protective glass of the bench closed during the experiment. After the completion of the coating experiment, the specimens were poured into the surface dish and dried naturally to obtain the product.

2.3. Performance Characterization

SEM: The particle morphology of the coated microsphere samples was observed by scanning electron microscopy (SEM) and compared with that of the DNTF raw material. The samples were all treated with gold spraying and accelerated to 15 KV.

Mechanical sensitivity: Referring to GJ772A-97 Methods 601.1 and 602.1, the WL-1 impact sensitivity meter and the WM-1 friction sensitivity meter were used to test the impact sensitivity and friction sensitivity of DNTF and GAP-coated DNTF samples, respectively: For the impact sensitivity test (25 cm drop height, 30 mg dosage), a 2 kg drop hammer was used; for the friction sensitivity test (pressure 2.45 MPa, swing angle 66°, 20 mg).

2.4. Molecular Dynamics Simulation

With reference to the literature [11–13], the molecular structure of the composite system was modeled using the Visualizer module in Materials Studio software, and after geometric optimization, MD simulations were performed under the NVT ensemble synthesis in the Compass force field.

2.4.1. Vacuum Morphology Prediction of DNTF Crystals

The crystal structure data of DNTF was obtained based on the Cambridge Organic Crystal Database (CCDC) and imported into the Materials Studio simulation platform. The Forcite module was used to optimize molecular configuration and energy. The following optimization parameters were set: compass force field, "medium" accuracy, and "smart" method to optimize the cell parameters. The Bravais-Friedel-Donnary-Harker(BFDH) and Attachment Energy method(AE) in the Morphology Tools module were used to predict the vacuum morphology of DNTF crystals and obtain the surface distribution of DNTF crystals grown under vacuum conditions. Theoretically, the crystalline surface with the largest surface occupation area has the most important effect on the adhesive-filler interfacial bonding performance [14,15]. In this simulation, the interfacial interaction between GAP molecules and DNTF crystals was calculated by selecting the crystal surfaces that occupied more surface area. The vacuum growth morphology of DNTF crystals is shown in Figure 1, the left is the vacuum growth morphology of DNTF crystal in BFDH model and the right is the vacuum growth morphology of DNTF crystal in the AE model. In the figure, nitrogen is shown in blue, oxygen in red and carbon in gry. The first two growth surfaces and their corresponding percentages are listed in Table 1.

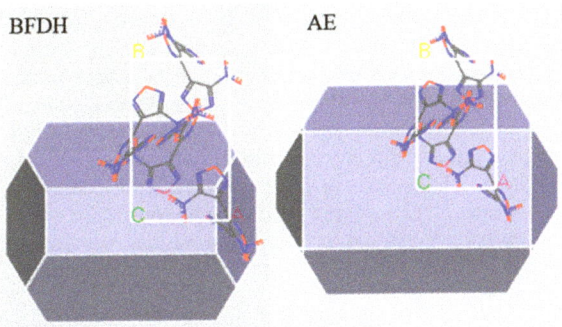

Figure 1. Vacuum growth morphology of DNTF.

Table 1. Natural growth of DNTF crystals crystal surface data.

Crystal	(h k l)	AE/(kcal·mol^{-1})	Percentage of Total Area/%
DNTF	(0 1 1)	−48.2498	53.3037
	(1 0 1)	−75.0550	24.3655

2.4.2. Model Construction of Different Crystal Surfaces of the DNTF and Polymer Interface

The modeling of the interaction between different crystal surfaces of the polymer GAP and DNTF is illustrated by the example of the GAP and DNTF (0 1 1) surfaces (see Figure 2). The supercell model of DNTF(0 1 1) surface and the AC box established by polymer GAP construct GAP-DNTF(0 1 1) interfacial crystal model.

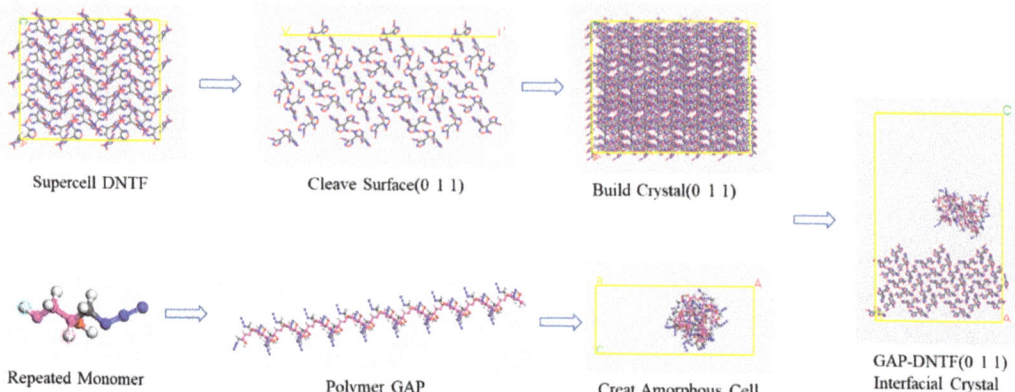

Figure 2. Establishing the interfacial crystal process.

Firstly, a homopolymer ball-and-stick structure model with a degree of polymerization of 35 was developed based on the GAP molecular formula, where the GAP binder polymer contains 35 monomers [14]. Geometry optimization of the Forcite module was selected for molecular mechanics optimization to adjust the structure and conformation of the molecules.

The polymer-bound explosive model was constructed as follows: Firstly, the $4 \times 3 \times 3$ supercell structure of DNTF was built and cut along the main growth plane (0 1 1); the vacuum-free crystal structure of DNTF (0 1 1) crystal plane was constructed; the amorphous structure of GAP molecular chain (Amorphous Cell) was constructed according to its lattice size; and their bilayer interface crystal models were obtained by the Build Layers function, where the vacuum layer Repeat the above method to obtain the bilayer interface model of GAP and other crystal surfaces of DNTF. The model construction needs to be combined with the mass ratio of energetic binder GAP to DNTF in the actual experiment, so only one GAP molecular chain is selected here for AC box construction. In addition, all steps in the model construction process need to be optimized for structure so that the system exhibits the minimum energy at each step, which is because the system loses its initial structure at each step of operation and the energy will be higher than its most stable structure.

2.4.3. Molecular Dynamics Calculation Setup

First, the geometry of the constructed interface model is optimized, and the optimization parameters are set as described in the previous section. Based on the optimized interface model, MD simulations under an isothermal isobaric (NVT) system were performed. The simulation parameters are set as follows: The temperature control method is Anderson, the time step is 1.0 fs, and the total simulation time is 200 ps, in which the first 100 ps steps are used for thermodynamic equilibrium and the last 100 ps steps are used for statistical analysis. During the simulation, van der Waals forces and electrostatic interactions are calculated using atom-based and Ewald methods, respectively, and the results are output once every 1000 steps for a total of 50 frames. The system reaches kinetic equilibrium when the fluctuation range of temperature and energy is less than 5%, and the obtained equilibrium structure is analyzed to obtain the binding energy and mechanical property parameters between components.

3. Results and Discussion

This section may be divided by subheadings. It should provide a concise and precise description of the experimental results, their interpretation, as well as the experimental conclusions that can be drawn.

3.1. Microscopic Morphological Characterization of Particles

SEM was used to characterize the microscopic morphology of DNTF explosive particles and energetic binder GAP-coated DNTF particles. In Figure 3, electron microscopic images of grain morphology characteristics of DNTF explosive are shown in a, b and c which are images of DNTF grain enlarged 50, 500 and 1000 times, respectively. The surface of DNTF without coating treatment is smooth and flat, and the grains are in the shape of regular polyhedra, with a few smaller grains attached to the surface of the grains and no other attachments. It is compatible with the crystalline shape under vacuum and can be used as a benchmark reference for the comparison of GAP binder-coated DNTF samples.

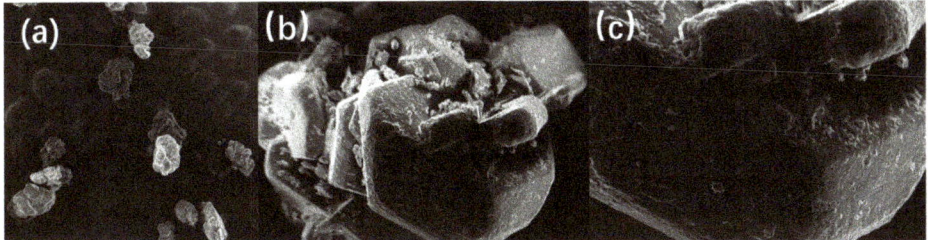

Figure 3. Electron microscopy of DNTF grain morphology: (**a**)50 times; (**b**) 500 times; (**c**) 1000 times.

In Figure 4, subfigures a, b and c are the electron microscopic morphology of the coated grains at magnifications of 500, 1000, and 5000, respectively. It can be observed that the roughness of the DNTF surface increases after GAP coating, and GAP adheres more small particles of DNTF to the surface of large particles. After magnification, it can be observed that the surface of the original DNTF grains is in a white gel state, and the surface of the grains is basically covered by a thin layer of GAP completely and uniformly. It is consistent with the growth of different crystal surfaces under MS vacuum, which indicates that the GAP-coated DNTF crystals have a good coating effect.

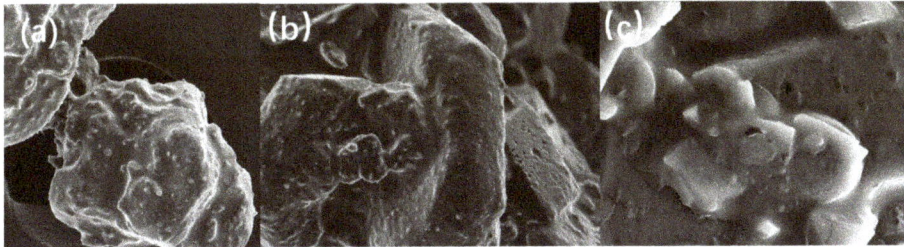

Figure 4. Electron microscopic morphology of coated grains: (**a**) 500 times; (**b**) 1000 times; (**c**) 5000 times.

3.2. Mechanical Sensitivity Analysis

The results of impact and friction sensitivity tests for DNTF and GAP/DNTF cladding are shown in Table 2. Test data of mechanical sensitivity of energetic binder GAP was obtained from the literature [15]. As can be seen from the table, the DNTF explosive grains themselves have high sensitivities, with impact and friction sensitivities of 76% and 100%, respectively; the impact and friction sensitivities of the DNTF grains after GAP cladding are reduced to 48% and 52%, respectively, indicating that the energetic binder GAP as a cladding layer for DNTF crystals can significantly reduce the mechanical sensitivities of DNTF, which will help to improve the safety of DNTF explosives.

Table 2. Impact and friction sensitivity test results.

Sample	Impact Sensitivity/%	Friction Sensitivity/%
GAP	0	0
DNTF	76	100
$M_{DNTF}:M_{GAP} = 95:5$	48	52

3.3. Binding Energy Calculation

Based on the final equilibrium structure of the NVT ensemble, Molecular dynamics simulation is used to calculate the binding energy of the energetic binder GAP and DNTF different crystal surfaces ($E_{binging}$), which is usually used to characterize the strength of the interaction between the components of the composite system, the calculation formula as (1) formula [16]:

$$E_{binding} = -E_{inter} = -(E_{total} - E_{DNTF} - E_{GAP})$$ (1)

E_{inter} is the interaction energy between the two components; E_{total} is the total energy of the composite system; E_{DNTF} is the single-point energy of the DNTF crystalline surface after removing the polymer molecular chain; and E_{GAP} is the single-point energy of the polymer after removing the DNTF crystalline surface, all in Kcal/mol. The binding energies of the polymer and the different crystal surfaces of DNTF were obtained separately according to (1). The results are listed in Table 3.

Table 3. Binding energy between GAP and DNTF (0 1 1) and DNTF (1 0 1) interfaces.

Composite System	Calculation Parameters	E_{total} (kcal·mol^{-1})	$E_{explosive}$ (kcal·mol^{-1})	E_{binder} (kcal·mol^{-1})	$E_{binding}$ (kcal·mol^{-1})
GAP/DNTF (0 1 1)	Energy	−48,551.8	−48,753.5	508.9	307.2
	vdW	−5626.3	−5552.3	23.7	97.7
	Electrostatic interaction	−8781.2	−9357.9	620.4	43.7
GAP/DNTF (1 0 1)	Energy	−47,617.5	−48,117.2	237.7	262.5
	vdW	−5369.2	−5392.3	−65.0	88.1
	Electrostatic interaction	−8117.6	−8738.1	584.8	35.7

The higher the binding energy, the stronger the interaction between the components and the more stable the system will be. Therefore, the binding energy calculation has an important influence on the compatibility of DNTF with GAP and the thermal stability of the energy-containing system. It also significantly reduces the frictional heat generation energy between DNTF and the contact surface and decreases the probability of frictional ignition. As shown in Table 3, the non-bonding energy of the bonding energy contains both vdW (van der Waals force) and electrostatic interaction, and the electrostatic interaction energy in the GAP/DNTF (0 1 1) interface system is 97.7 kJ/mol, which contributes 31.8% to the bonding energy. The electrostatic interaction energy in the GAP/DNTF (1 0 1) interface system is 199.59 kJ/mol, which contributes 33.5% to the binding energy, indicating that the interaction between GAP molecules and DNTF crystals is dominated by electrostatic forces. The binding energies of both GAP/DNTF (0 1 1) and (1 0 1) interface systems were positive, indicating that the interfaces between the binder and the explosive were both able to exist stably; the binding energy of the GAP/DNTF (0 1 1) interface system was 307.2 kcal/mol higher than that of GAP/DNTF (1 0 1), 262.5 kJ/mol, by 44.7%, indicating that the interaction strength of GAP/DNTF (0 1 1) was significantly higher than that of GAP/DNTF (1 0 1). GAP/DNTF (0 1 1) was significantly higher than that of GAP/DNTF (1 0 1), indicating that GAP was more compatible with DNTF (0 1 1) and bound more strongly. The results of the bond energy calculation for GAP/DNTF show that the higher the bond energy between the DNTF interface and the polymer GAP, the stronger they are bonded and the lower the interfacial slip, so that the probability of frictional ignition will be significantly lower. The results of the binding energy calculation of this system are in good agreement with the trend of the test results of the friction sensitivity of the GAP/DNTF cladding down to 52%, which better reflects the intrinsic factor of the influence of GAP as a cladding layer on the friction sensitivity of DNTF. In addition, it is mentioned in the literature [17] that Gibbs free energy also exists at the interface of solid and liquid. The smaller the value of the Gibbs free energy, the better the thermal stability and toughness of the system, and the less it is affected by temperature. It can be inferred that the less likely it is to generate hot spots stimulated by external unexpected energy, which is might be profit to reduce the sensibility of the GAP/DNTF system.

3.4. Mechanical Properties Calculation

The main parameters of the elastic mechanical properties include the elastic coefficient, the effective isotropic modulus, and Poisson's ratio.

The stress-strain (σ-ε) relationship for the stressed system obeys generalized Hooke's law [18,19].

$$\sigma_i = C_{ij}\varepsilon_j \tag{2}$$

The σ_i is the stress tensor, GPa; the ε_j is the strain tensor, %; C_{ij} (i, j = 1–6) is the symmetric matrix of elasticity coefficients, characterizing the stress-strain relationship; the larger the value, the greater the stress required to produce the same strain.

Theoretically, for describing the stress-strain behavior of an arbitrary material, 21 independent variables are required since $C_{ij} = C_{ji}$. This can be simplified by using the Lamé coefficients, as shown in Equation (3) [20,21].

$$[C_{ij}] = \begin{bmatrix} \lambda + 2\mu & \lambda & \lambda & 0 & 0 & 0 \\ \lambda & \lambda + 2\mu & \lambda & 0 & 0 & 0 \\ \lambda & \lambda & \lambda + 2\mu & 0 & 0 & 0 \\ 0 & 0 & 0 & \mu & 0 & 0 \\ 0 & 0 & 0 & 0 & \mu & 0 \\ 0 & 0 & 0 & 0 & 0 & \mu \end{bmatrix} \tag{3}$$

Static mechanical property analysis of the NVT-ensemble MD simulation data at equilibrium was performed to obtain the elastic coefficients (C_{ij}) and effective isotropic moduli, including the Young modulus(E), shear modulus (G) and bulk modulus (K), as well as K/G values and Cauchy pressure (C_{12}-C_{44}) for different crystalline interface models of GAP and DNTF. Based on the Lamé coefficients and the phase relationship between isotropic material moduli [21–23]:

$$E = \frac{\mu(3\lambda + 2\mu)}{\lambda + \mu} \tag{4}$$

$$K = \lambda + \frac{2}{3}\mu \tag{5}$$

$$G = \mu \tag{6}$$

$$\gamma = \frac{\lambda}{2(\lambda + \mu)} \tag{7}$$

The Poisson's ratio (ν) of the composite system can be obtained according to Equation (8), and the results of each modulus calculation are shown in Table 4.

$$E = 2G(1 + \nu) = 3K(1 - 2\nu) \tag{8}$$

The modulus parameter of the mechanical properties can be used as an indicator to evaluate the stiffness of the material but also as a measure of the material's ability to resist elastic deformation, and the plastic and fracture properties of the material are associated with the modulus. The coefficient of elasticity C reflects the different elastic effects of the material at each location. Young modulus, bulk modulus, and shear modulus can all be used to measure the strength of the material's stiffness, i.e., its ability to resist elastic deformation of the material under the action of external forces [24]. The shear modulus (G) is related to the stiffness, which indicates the ability to prevent plastic deformation of the material, and the larger its value, the higher the stiffness and yield strength of the material; the bulk modulus (K) can also be used to correlate the fracture strength of the material, and the larger its value, the greater the energy required to fracture the material, i.e., the greater the fracture strength of the material.

Table 4. Mechanical properties of different GAP and DNTF crystal surfaces.

Modules/GPa	GAP/DNTF (011)	GAP/DNTF (101)	DNTF
C_{11}	2.9624	2.0087	12.8557
C_{22}	7.2005	8.5956	10.6884
C_{33}	7.8268	8.4528	9.1765
C_{44}	2.8645	3.1927	2.1781
C_{55}	1.8319	1.8894	2.0604
C_{66}	1.0651	2.2839	8.0035
C_{12}	2.9988	3.3890	3.2970
C_{13}	1.8833	3.4845	1.1050
C_{23}	3.9747	5.5195	3.2776
E	4.8566	5.8189	10.4586
γ	0.1850	0.2644	0.1531
K	3.0786	3.4359	5.0247
G	1.9205	2.4553	4.5350
K/G	1.2539	1.7891	1.1080
ν	0.2644	0.3282	0.1531
C_{12}-C_{44}	0.1343	0.1963	1.1189

The ratio of bulk modulus to shear modulus (K/G) is used to measure the toughness of a material system, i.e., the ability of a material to withstand large deformations under impact or vibration loads without being damaged. K/G is also used to measure the ductility of a material, and the larger the value, the better the ductility of the material [24]. Like K/G, the Cauchy pressure (C_{12}-C_{44}) can also predict the ductility of the system, i.e., the ability of the material to deform without cracking; it is also used to reflect the degree of brittleness of the material; with negative values of the Cauchy pressure, the material is brittle, and the smaller the negative value, the more brittle it is. Conversely, when the Cauchy pressure is positive, it indicates that the material is more ductile and shows toughness. Poisson's ratio (ν) is defined as an elastic constant for the ratio of the transverse deformation to the longitudinal deformation when the material is deformed by tensile or compressive forces. The data on the mechanical property parameters of the energetic binder GAP/DNTF for different crystalline interface systems and pure DNTF are shown in Table 4.

With reference to the parameters of the DNTF crystals, it is easily seen that the energetic binder GAP effectively enhances the isotropic material modulus of the DNTF crystals. Compared to the pure DNTF crystals, the Young modulus (E), bulk modulus (K), and shear modulus (G) of the GAP/DNTF (0 1 1) and (1 0 1) composite systems decreased, indicating that the hardness, yield strength, and fracture strength of DNTF decreased, its stiffness decreased, its elasto-plasticity increased, and Poisson's ratio increased [21]. The K/G, and values of the GAP/DNTF (0 1 1) facet system and the GAP/DNTF (1 0 1) facet system are higher than those of pure DNTF crystals, and both systems have positive C_{12}-C_{44} values. It can be concluded that the energetic binder GAP can improve DNTF ductility while also improving toughness. According to the above analysis and the aforementioned impact sensitivity test data, it can be seen that GAP as a cladding layer of DNTF can effectively improve the mechanical properties of DNTF, especially the elasto-plasticity enhancement of the system, which can better absorb the impact energy of external stimuli and reduce the probability of impact ignition of DNTF explosives.

It is thus shown that the reduction of impact sensitivity of GAP/DNTF cladding from 76% to 48% is mainly due to the fact that GAP as a cladding layer greatly improves the elasto-plastic mechanical properties of DNTF, so the results of molecular dynamics calculations of the mechanical properties of the GAP/DNTF system better reflect the essential reasons for the significant reduction of impact sensitivity of GAP/DNTF cladding on a macroscopic scale.

The E, K, G, and C_{12}-C_{44} values of the GAP/DNTF (1 0 1) interfacial crystals are higher than those of the GAP/DNTF (0 1 1) system. The mechanical modulus of the DNTF (0 1 1) surface, as the first growth surface of the crystal, is smaller than that of the interfacial system corresponding to the DNTF (1 0 1) surface, which will help to improve the mechanical properties of [21].

4. Conclusions

(1) The SEM microscopic morphological characterization shows that the surface of pure DNTF is smooth and flat, and the surface roughness of DNTF increases after GAP coating. A white gelatinous layer is observed on the surface of the original DNTF grains after magnification, indicating that GAP has a better coating effect on DNTF crystals.

(2) The mechanical sensitivity test shows that using GAP as the cladding layer of DNTF crystal can significantly reduce the sensitivity. DNTF sensitivity, impact sensitivity reduced from 76% to 48%, and friction sensitivity reduced from 100% to 52% will help improve the safety of explosives. At the same time, the mechanical sensitivity test results also show that the coating effect of GAP on DNTF is good.

(3) The binding energy of the GAP/DNTF (0 1 1) interfacial system (307.2 kcal/mol) was higher than that of the GAP/DNTF (1 0 1) system (262.5 kcal/mol), indicating that GAP was more compatible with DNTF (0 1 1) and had a stronger and more stable binding. The high binding energy of the GAP/DNTF interfacial system is one of the main intrinsic reasons for its significantly lower friction sensitivity.

(4) Molecular dynamics calculations show that the polymer GAP effectively enhances the isotropic material modulus of DNTF crystals, resulting in less rigidity, more flexibility, and better ductility and toughness of DNTF, which helps the system absorb external impact energy and reduce the impact susceptibility of the GAP/DNTF system.

In summary, GAP can be referred to as a better cladding layer for DNTF, which is feasible for inhibiting the problem of DNTF crystallization in propellants. This study provides a potential new crystallization inhibition pathway for the DNTF modified double-base propellants. Nevertheless, this study still has some limitations. For example, consider whether the propellant calendering process at high temperature affects the crystal structure of the cladding, as well as a more comprehensive characterization of the thermal properties of the GAP-coated DNTF. Next, we will also continue to study this in depth.

Author Contributions: Conceptualization, J.Y.; methodology, J.Y.; validation, Y.Q., H.S., Y.L. and H.Z.; formal analysis, Y.Q., H.S. and H.Z.; data curation, Y.Q. and Y.L.; writing—original draft preparation, Y.Q.; writing—review and editing, J.Y.; visualization, R.W., J.C. and X.L. All authors have read and agreed to the published version of the manuscript.

Funding: This study was carried out within the project "Research on the mechanism of DNTF crystallization in propellant and inhibition technology", projects No.20210579 is funded by the Bottleneck Technology and JCJQ Foundation.

Data Availability Statement: The data presented in this study are openly available.

Conflicts of Interest: The authors declare no conflict of interest.

References

1. Zheng, W.; Wang, J.; Ren, X.; Zhang, L.; Zhou, Y. An Investigation on Thermal Decomposition of DNTF-CMDB Propellants. *Propellants Explos. Pyrotech.* **2010**, *32*, 520–524. [CrossRef]
2. Zheng, W.; Wang, J.N.; Han, F.; Tian, J.; Song, X.D.; Zhou, Y.S. Chemical Stability of CMDB Propellants Containing DNTF. *Chin. J. Explos. Propellants* **2010**, *33*, 10–13.
3. Tian, J.; Wang, B.C.; Sang, J.F.; Zhang, F.P.; Wang, J.N. Experimental research on the properties of CMDB propellant containing DNTF. *Chin. J. Explos. Propellants* **2015**, *38*, 76–79.

4. Pang, J.; Wang, J.N.; Zhang, R.E.; Xie, B. Application of CL-20, DNTF and FOX-12 in CMDB propellants. *Chin. J. Explos. Propellants* **2005**, *28*, 19–21.
5. Li, X.T. Crystallization of RDX from a composite modified double-base propellant. *Acta Armamentarii* **1979**, *1*, 23–27.
6. Jia, Z.N.; Yi, L.H. Research on a surface coating method to prevent crystallization of RDX. *Chin. J. Explos. Propellants* **1994**, *16*, 6–8.
7. Jia, Z.N. Research on the crystallization phenomenon and anti-crystallization of RDX. *Acta Armamentarii* **1995**, *16*, 37–40.
8. Zheng, W.; Xie, B.; Hu, Q.; Cao, L.; Wang, J.N.; Zhang, J. Crystal analysis of DNTF-containing modified double-based propellants. *J. Solid Rocket Technol.* **2016**, *39*, 509–512.
9. Liang, L.; Yun, N.; Geng, X.H.; Lin, S.T. Synthesis and properties of (glycidyl azide polymer) GAP. *J. North Univ. China (Soc. Sci. Ed.)* **2014**, *35*, 177–181.
10. Badgujar, D.M.; Talawar, M.B.; Zarko, V.E.; Mahulikar, P.P. New directions in the area of modern energetic polymers: An overview. *Combust. Explos. Shock Waves* **2017**, *53*, 371–387. [CrossRef]
11. Xia, L.; Xiao, J.J.; Fan, J.F.; Zhu, W.; Xiao, H.M. Molecular dynamics simulation of the mechanical properties and interfacial interactions of nitrate plasticizers. *Acta Chim. Sin.* **2008**, *66*, 874–878.
12. Wang, H.; Gao, J.; Tao, J.; Luo, Y.M.; Jiang, Q.L. Safety and molecular dynamics simulations of DNTF/HATO hybrid systems. *Chin. J. Energ. Mater.* **2019**, *27*, 897–901.
13. Meng, L.L.; Qi, X.F.; Wang, J.N.; Fan, X.Z. Molecular dynamics simulation and experimental study of the plasticization properties of NC by DNTF. *Chin. J. Explos. Propellants* **2015**, *38*, 86–89.
14. Liu, C.; Zhao, Y.; Xie, W.X.; Liu, Y.F.; Huang, H.T.; Zhang, W.; Zhang, X.H. Molecular dynamics simulation of interfacial interactions in GAP/Al composite systems. *New Chem. Mater.* **2018**, *46*, 186–189.
15. Li, Z.F.; Feng, Z.G.; Hou, Z.L. Synthesis and performance analysis of azide binder GAP. *J. Qingdao Inst. Chem. Technol.* **1997**, *18*, 155–163.
16. Xiao, J.J.; Wang, W.R.; Chen, J.; Ji, G.F.; Zhu, W.; Xiao, H.M. Study on structure, sensitivity and mechanical properties of HMX and HMX-based PBXs with molecular dynamics simulation. *Comput. Chem.* **2012**, *999*, 21–27. [CrossRef]
17. Eslami, H.; Khanjari, N.; Müller-Plathe, F. A local order parameter-based method for simulation of free energy barriers in crystal nucleation. *J. Chem. Theory Comput.* **2017**, *13*, 1307–1316. [CrossRef]
18. Ma, S.; Li, Y.J.; Li, Y.; Luo, Y.J. Research on structures, mechanical properties, and mechanical responses of TKX-50 and TKX-50 based PBX with molecular dynamics. *J. Mol. Model* **2016**, *22*, 1–11. [CrossRef]
19. Weiner, J.H. Statistical Mechanics of Elasticity. *J. Appl. Mech.* **1984**, *51*, 707–708. [CrossRef]
20. Xiao, J.J.; Huang, Y.C.; Hu, Y.J.; Xiao, H.M. Molecular dynamics simulation of mechanical properties of TATB/fluorine-polymer PBXs along different surfaces. *Sci. China Ser. B Chem.* **2005**, *48*, 504–510. [CrossRef]
21. Wang, K.; Li, H.; Li, J.Q.; Xu, H.X.; Zhang, C.; Lu, Y.Y.; Fan, X.Z.; Pang, W.Q. Molecular dynamic simulation of performance of modified BAMO/AMMO copolymers and their effects on mechanical properties of energetic materials. *Sci. Rep.* **2020**, *10*, 1–17. [CrossRef] [PubMed]
22. Xin, D.R.; Han, Q. Study on thermomechanical properties of cross-linked epoxy resin. *Mol. Simulat.* **2015**, *41*, 1081–1085. [CrossRef]
23. Hu, Y.Y.; Chen, C.Y.; Chen, K. Molecular Dynamics Simulation of Mechanical Properties of Single-base and Double-base Propellants. *J. Nanjing Xiaozhuang Univ.* **2007**, *23*, 47–50.
24. Xing, X.W.; Yuan, J.M.; Li, Y.; Sha, H.B.; Luo, Y.M.; Jiang, Q.L. Molecular dynamics calculation of the effect of GAP-ETPE on the performance of DNTF explosives. *J. Ordnance Equip. Eng.* **2022**, *43*, 293–298.

Article

Simulation Analysis of the Safety of High-Energy Hydroxyl-Terminated Polybutadiene (HTPB) Engine under the Impact of Fragments

Zheng Liu [1], Jianxin Nie [1,*], Wenqi Fan [2], Jun Tao [3], Fan Jiang [3], Tiejian Guo [4] and Kun Gao [4]

[1] State Key Laboratory of Explosion Science and Technology, Beijing Institute of Technology, Beijing 100081, China
[2] School of Aerospace Engineering, Tsinghua University, Beijing 100084, China
[3] Xi'an Modern Chemistry Research Institute, Xi'an 710065, China
[4] Xi'an Changfeng Research Institute of Mechanical-Electrical, Xi'an 710065, China
* Correspondence: niejx@bit.edu.cn

Abstract: The safety of solid rocket engine use seriously affects the survivability and combat effectiveness of weaponries. To study the engine safety against fragment in complex battlefield environments, the fragment impact safety simulation study of a high-energy four-component HTPB propellant solid engine (hereafter referred to as high-energy HTPB propellant engine) was conducted. The equation of state parameters and reaction rate equation parameters of the detonation product of high-energy HTPB propellant were calibrated by using a 50 mm diameter cylinder test and Lagrange test combined with genetic algorithm. The nonlinear dynamics software LS-DYNA was used to build a finite element model of the fragment impact engine and simulate the mechanical response of the high-energy HTPB propellant under different operating conditions. This study shows that the critical detonation velocity decreases with the increase of the number of fragments. When the number of fragments is more than 5, the influence of this factor on the critical detonation velocity is no longer obvious. Under the same loading strength conditions, the greater the metal shell strength and the greater the shell wall thickness, the more difficult it is for the high-energy HTPB propellant to be detonated by the shock. This study can provide a reference for the design and optimization analysis of solid rocket engine fragment impact safety.

Keywords: safety engineering; fragmentation impact; high-energy HTPB propellant; shock initiation; equation of state

Citation: Liu, Z.; Nie, J.; Fan, W.; Tao, J.; Jiang, F.; Guo, T.; Gao, K. Simulation Analysis of the Safety of High-Energy Hydroxyl-Terminated Polybutadiene (HTPB) Engine under the Impact of Fragments. *Crystals* **2023**, *13*, 394. https://doi.org/10.3390/cryst13030394

Academic Editor: Thomas M. Klapötke

Received: 8 February 2023
Revised: 21 February 2023
Accepted: 21 February 2023
Published: 24 February 2023

1. Introduction

Composite solid propellants have been widely used in modern rocket engines, missile engines, rocket boosters and other power devices since the 1840s. As a typical composite propellant, high-energy HTPB propellant has the advantages of excellent combustion performance and mechanical properties, low flame temperature, low molecular combustion products and low infrared radiation. However, it also has a high probability of detonation and the risk of detonation. To improve the specific impulse and other performance, the proportion of nitramine propellant continued to increase, which reduced the critical detonation diameter of the nitramine composite propellant. This leads to an increased possibility of accidental explosive accidents during the actual assembly, storage, transportation and operation of the engine, posing a great threat to personnel safety and the environment. Therefore, it is extremely important to study the critical conditions for the occurrence of hazardous reactions of solid rocket motors under the action of impact loads [1].

The detonation safety of solid propellants began to be researched early. The research contains the critical diameter of the propellant [2], impact initiation [3–6], combustion to detonation [7–9] and other aspects. The detonation parameters are mainly focused on the study of the detonation velocity. Hot spot theory was first proposed by Bowden and other scholars [10] in 1948. HTPB four-component propellant as a nonhomogeneous explosive, and the formation of hot spots is currently considered to be the cause of its initiation of shock detonation. Price et al. [11] found that AP/Al/1,3,5,7-tetranitro-1,3,5,7-tetrazocane(HMX)/Wax propellants exhibit the properties of a second type of explosive, and the critical diameter of the propellant for detonation becomes larger as the charge porosity decreases. Dick [12] conducted a wedge test to study the detonation process of different formulations of AP/Al/HMX/Wax propellants. The tests showed that propellants with HMX content less than 20% could not be shock-detonated to produce a self-sustaining burst; when the HMX content was 44%, the detonation behavior of the propellants approximated that of high-energy explosives. Baker et al. [4] conducted a drop hammer experiment to obtain the impact sensitivities and critical impact initiation energies of three HTPB-based propellants. It was shown that the propellants were detonated only when the propellant nitramine content was high and when the critical detonation energy of the propellants was high. Kohga et al. [13] showed that the detonation velocity of ammonium nitrate (AN)/nitroamine-based composite propellant increases linearly with the increase in the mass of nitramine within the unit volume of propellant, and the effect of AN on the detonation performance can be ignored. In 1991, Bai et al. [14] studied the chemical reaction process of various solid propellants, including butyl hydroxyl propellants, under different pressure shock waves for the first time in China based on the Lagrangian test. Yang et al. [15] conducted numerical simulations of the process of flat plate breakers impacting flat shells, adiabatic layers and propellants. Li et al. [16] conducted a more comprehensive combination of low susceptibility propellant studies under mechanically stimulated conditions such as bullet impact and fragment impact.

In summary, the high nitramine content of the composite propellant blast work capacity, especially the blast-driven metal acceleration capacity, is less studied. Although the main components of nitramine composite propellants and plastic-bonding explosives (PBX) are similar, they generally have a higher ammonium perchlorate (AP) content and lower nitramine content; the detonation process is nonideal. Therefore, the existing studies of PBX explosive burst-driven metal acceleration capability are not sufficient to support the characterization of the detonation performance of nitramine composite propellants. There is a lack of basic test data to support solid rocket engine detonation hazard assessment.

To deal with the safety of the engine against fragmentation impact in a complex battlefield environment, a 50 mm cylinder test and Lagrange test of high-energy HTPB propellant were designed and completed. The parameters of its equation of state were calibrated by using a genetic algorithm. The nonlinear dynamics software LS-DYNA was applied to build a fragment impact engine model and simulate the mechanical response characteristics of the high-energy HTPB engine under different operating conditions. The results of this study can provide references for the design and optimization analysis of the fragment impact safety of solid rocket engines.

2. High-Energy HTPB Propellant Equation of the State Calibration Test

2.1. Cylinder Experiment

In this paper, the high-energy four-component HTPB propellant was studied with the following components: AP/Al/cyclotrimethylenetrinitramine(RDX)/HTPB = 50/5/30/15. The density of the propellant was 1.645 g/cm^3. The propellant was tested on a copper tube with a diameter of Φ50 mm, and the test configuration is shown in Figure 1. Figure 1a shows the schematic diagram of the cylinder test, and Figure 1b shows the cylinder test configuration. The cylinder experiment device consisted of high-voltage electric detonator, detonating column, copper tube, composite propellant, electric probe, Photonic Doppler Velocimetry (PDV) and a bracket. The Φ50 mm cylinder was placed vertically on the stand, and the cylinder expansion velocity was tested by laser interference velocimetry at a height of 200 mm during the stable detonation stage of the explosive.

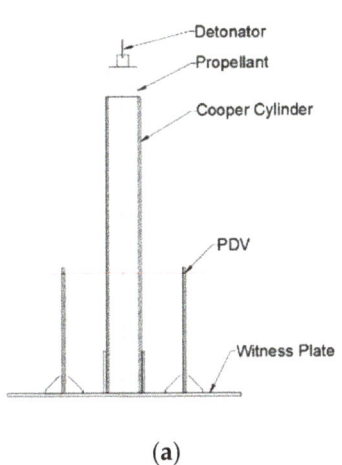

(a)   (b)

Figure 1. Schematic diagram of cylinder test. (**a**) Cylinder test design diagram. (**b**) Cylinder test configuration diagram.

The outer diameter of the copper tube was $\Phi60$ mm and the inner diameter was $\Phi50$ mm. The length of the copper tube was 495 mm, and the material was high conductivity oxygen-free copper. The detonation velocity of the composite propellant was measured using ionization probes fixed at both ends of the cylinder, and the distance between the two ionization probes was 495 mm. The recording frequency of the PDV was 24.4 MHz.

2.2. Lagrange Test

The high-energy HTPB four-component propellant used in this section remains the same as in section A. In order to ensure the uniformity of the material, firstly, the high-energy HTPB four-component propellant pillar was cut into tablets from the cast molding (the thickness of the tablets was divided into three series: 2~3 mm, 5 mm and 30 mm, which were reasonably matched according to the designed test position), and then the propellant tablets were cut into uniform diameter pillars with a $\Phi50$ mm circular cutter. The design and assembly diagram of the test device are shown in Figure 2.

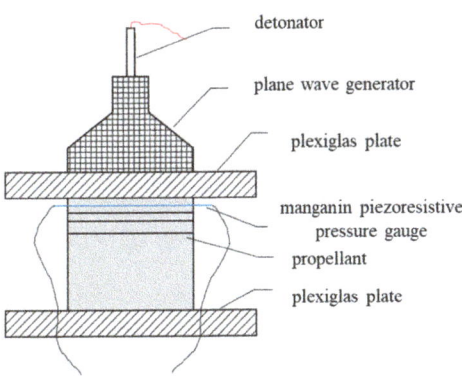

Figure 2. Design diagram and assembly diagram of Lagrange test device.

Among them, the plane wave generator was press-fitted by an 8701 explosive and a TNT explosive. The plexiglass plate was 8 cm × 10 cm square plate with thickness of 18.0 mm. The sensor adopts H-type manganin piezoresistive sensor.

2.3. Test Results and Analysis
2.3.1. Cylinder Test Results and Analysis

After the cylinder expansion experiment, the witness plate was perforated. As can be seen from Figure 3, the thickness of the 5 mm steel witness plate was perforated, and the perforation diameter reached Φ93.80 mm, indicating that the high-energy four-component HTPB propellant occurred in the form of a stable detonation reaction.

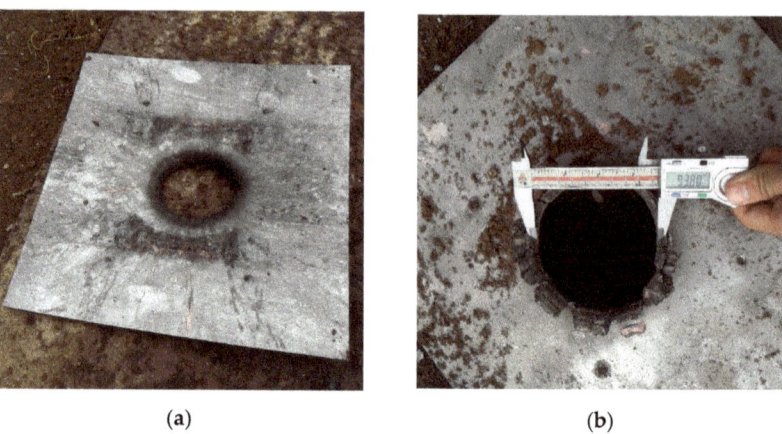

(**a**)   (**b**)

Figure 3. Shape of witness plate perforation after cylinder expansion experiment. (**a**) Witness plate front. (**b**) Back of the witness plate.

The detonation velocity of the composite propellant can be measured by the electric probes added at both ends of the cylinder. Figure 4 gives the electric probe pulse signal curve of the four-component HTPB propellant. The first sharp pulse in the curve is the upper surface of the copper tube at the pillar clamped electric probe signal; the second sharp pulse is the lower surface of the copper tube at the pillar clamped ionization probe. The time difference between the two signals is the stable detonation propagation time $\Delta \tau$ in the copper tube.

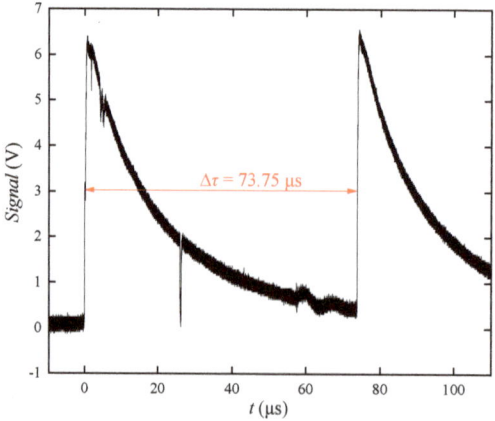

Figure 4. Velocity signal of four−component HTPB propellant.

The distance between the two ionization probes was d = 495 mm, and the time difference between the two electric probe signals was $\Delta\tau$ = 73.75 μs. This can be calculated to obtain the detonation velocity of four-component HTPB propellant V = 6712 m/s. The detonation pressure P_{CJ} of the composite propellant can be calculated with the following equation.

$$P_{CJ} = \frac{1}{\gamma + 1}\rho_0 V^2 \tag{1}$$

where ρ_0 is the charge density of the propellant; V is the detonation velocity of the propellant; and P_{CJ} is the detonation pressure of the propellant. In general, assuming γ = 3, it can be calculated to obtain the detonation pressure of HTPB propellant P_{CJ} = 18.53 GPa.

2.3.2. Lagrange Test Results and Analysis

Six Lagrange tests were conducted, and the actual Lagrange point locations are shown in Table 1.

Table 1. Actual Lagrange point location for Lagrange test.

Actual Insertion Position/x_i	x_1/mm	x_2/mm	x_3/mm	x_4/mm	x_5/mm	x_6/mm
Distance from Plexiglas plate position	0.00	3.10	5.69	8.43	14.24	20.12

The voltage data obtained at each Rasch point were subjected to data noise reduction and smoothing. The obtained pressure time course curves are shown in Figure 5.

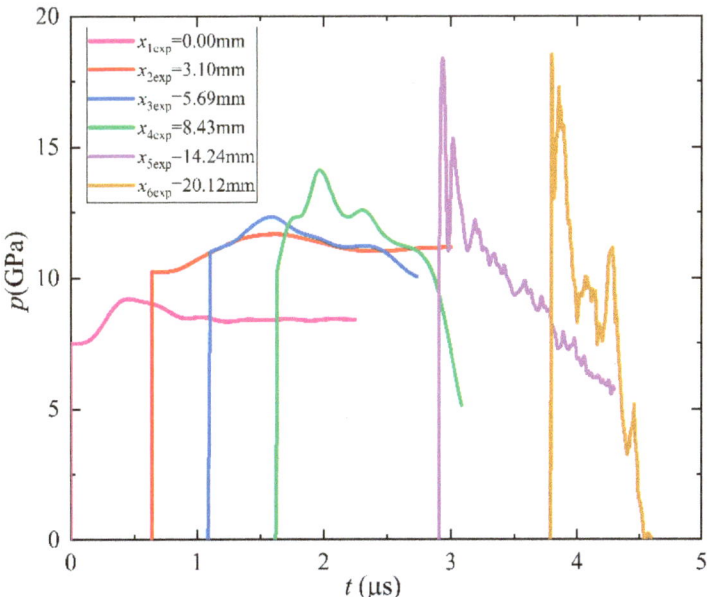

Figure 5. Pressure measurement data at each point of the Lagrange test.

2.4. Detonation Product Equation of State and Reaction Rate Equation Parameter Calibration

The Jones–Wilkson–Lee (JWL) [17] equation of state is commonly used to simulate the solid propellant detonation process, and the standard form of the equation is

$$p_s = A\left(1 - \frac{\omega}{R_1 V}\right)e^{-R_1 V} + B\left(1 - \frac{\omega}{R_2 V}\right)e^{-R_2 V} + \frac{\omega E_0}{V} \tag{2}$$

where p_s is the detonation product pressure; E_0 is the volume specific internal energy; V is the relative specific volume of the detonation product; and A, B, R_1, R_2 and ω are constants, determined by the cylinder experiment.

The propellant is a typical nonhomogeneous explosive, so its shock initiation process and shock to detonation (SDT) can be analyzed by using the classical hot-spot theory. Lee and Tarver [18] proposed the ignition growth model in 1980, which has been improved and refined, and the model was widely accepted and applied. The ignition growth reaction rate model is

$$
\begin{aligned}
\frac{\partial \lambda}{\partial t} = {} & I(1-\lambda)^b \left(\frac{\rho}{\rho_0} - 1 - a\right)^x H\left(F_{ig\max} - \lambda\right) \\
& + G_1(1-\lambda)^c \lambda^d p^y H\left(F_{G_1\max} - \lambda\right) \\
& + G_2(1-\lambda)^e \lambda^g p^z H\left(\lambda - F_{G_2\min}\right)
\end{aligned}
\tag{3}
$$

where λ is the reactivity; $H(x)$ is the step function; parameter I characterizes the number of hot-spots; parameter b is the order of combustion; parameter a is the critical compression of ignition; parameter x is the ignition term duration function; $F_{ig\max}$ controls the maximum applicable reactivity of the ignition term; parameters G_1 and d define the reaction growth rate of the hot-spot early after ignition; parameter c is the order of combustion of the growth term; parameter y is the pressure index; $F_{G_1\max}$ controls the maximum applicable reactivity of the growth term; parameters G_2 and f define the reaction growth rate of the late hot-spot after ignition; parameter e is the combustion order of the completion term; parameter z is the pressure index; and $F_{G_2\min}$ controls the minimum applicable reactivity of the completion term.

2.4.1. Calibration Process

In this paper, an adaptive genetic algorithm (AGA) was used to optimize the calibration process of the detonation product parameters. The algorithm can be used to obtain the global optimal fit parameters faster. The flow chart of the calibration equation of state parameters in this paper is shown in Figure 6.

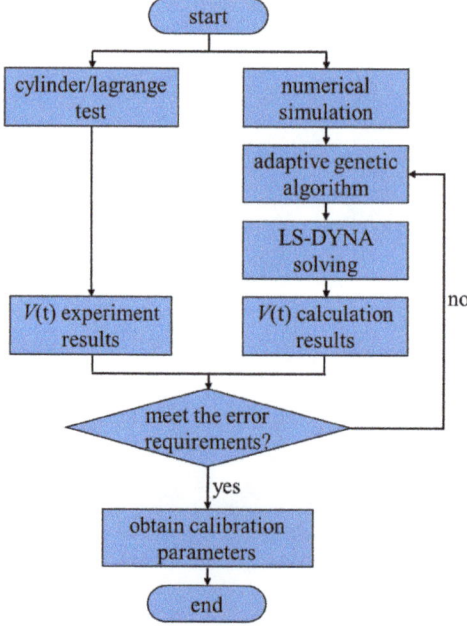

Figure 6. Flow chart of parameters calibration.

As shown in the figure, the initial parameter values are given by using an adaptive genetic algorithm. The radial expansion displacement curve is obtained by numerical simulation of nonlinear dynamics, and the curve is compared with the radial expansion displacement curve obtained experimentally to obtain the goodness of fit. Genetic operations are performed to generate a new generation population based on the fit values. Iterative calculations are performed to finally obtain the fitted parameters of the equation of state with a good degree of fit.

2.4.2. Calibration Results

The parameters are calibrated and calculated by an adaptive genetic algorithm. As shown in Figures 7 and 8, the comparison of the cylinder wall radial expansion velocity and cylinder wall radial expansion displacement with the experimental data was obtained by simulation.

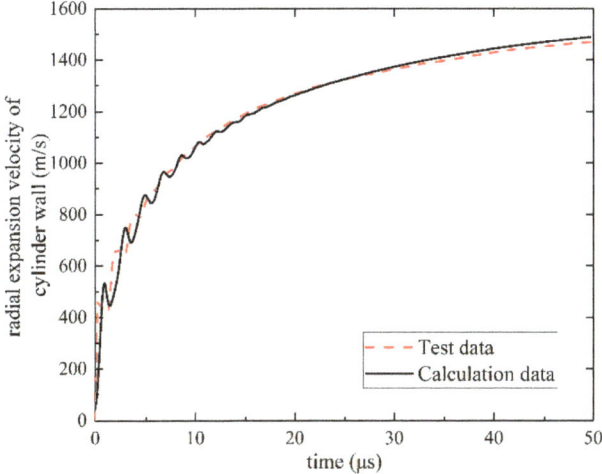

Figure 7. Wall velocity versus time curve of cylinder test.

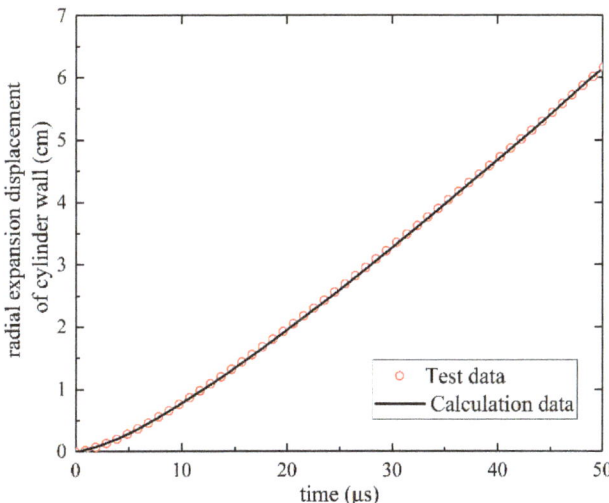

Figure 8. Wall displacement versus time curve of cylinder test.

The final obtained four-component HTPB propellant detonation product equations of state parameters are shown in Table 2.

Table 2. Propellant detonation product equations of state parameters.

ρ_0	A/GPa	B/GPa	C/GPa	R_1	R_2	ω	$E_0/(\text{kJ·cm}^{-3})$
1.645	481.34	4.5519	2.019	4.6916	1.6287	0.2796	0.0938

The pressure growth process obtained from the test and the pressure growth process obtained from the simulation is shown in Figure 9. In general, the manganin piezoresistive sensor may be destroyed after measuring the shock wave takeoff pressure within the propellant due to the action of the detonation products. After this measured pressure, data validity is poor and does not reflect the actual pressure changes during the shock initiation. Therefore, generally only the starting pressure in the Lagrange test should be used to calibrate the parameters, and the subsequent pressure is generally no longer used for calibration.

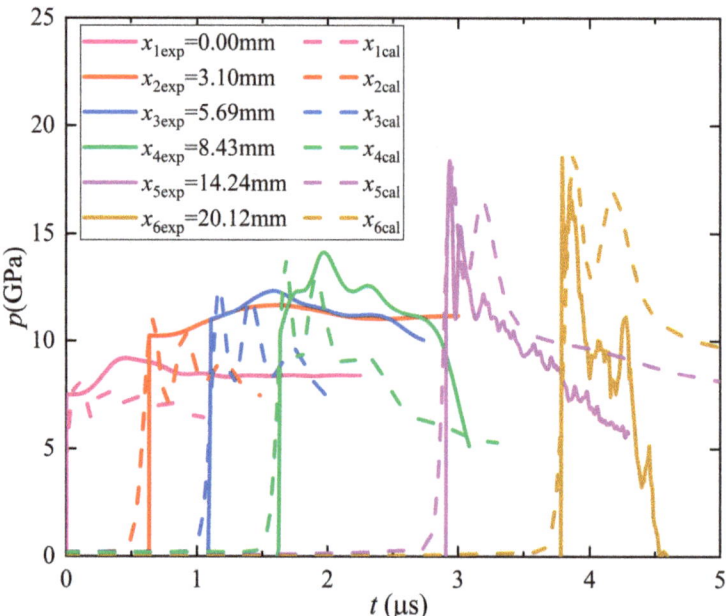

Figure 9. Comparison between the growth test and simulation calculation of the Lagrange test pressure.

The reaction rate parameters obtained for the final calibration are shown in Table 3.

Table 3. Parameters of high-energy HTPB four-component propellant reaction rate equation.

Parameter	Parameter Value	Parameter	Parameter Value	Parameter	Parameter Value
$I/\mu s^{-1}$	4427.1887	$G_1/(\text{GPa}^{-y}\cdot\mu s^{-1})$	1.8127	$G_2/(\text{GPa}^{-z}\cdot\mu s^{-1})$	261.7333
a	0.0248	c	0.6667	e	1.0
b	0.6667	d	0.1111	f	0.667
x	6.7385	y	1.0	z	2.0
$F_{ig\,max}$	0.01	F_{G1max}	1.0	F_{G2min}	0.0

3. Simulation of High-Energy HTPB Engine Fragmentation Impact Safety Experiment

3.1. Simulation Model and Parameters

3.1.1. Structural Model

A three-dimensional simulation model of the shell-loaded composite propellant loaded with different fragment impact velocities was established using TrueGrid parametric modeling software. The profile along the fragment impact direction is shown in Figure 10. The charge diameter of the high-energy HTPB propellant is $\Phi157$ mm, and the charge length is L = 400 mm. The outer diameter of the shell is $\Phi160$ mm, the wall thickness of the shell is $\delta = 1.5$ mm, and the length of the shell is L = 400 mm. Standard breakers are used, and the propellant charge is loaded by positively impacting the shell in the direction perpendicular to the outer surface of the shell. The air domain radius is 2.5 times the propellant radius.

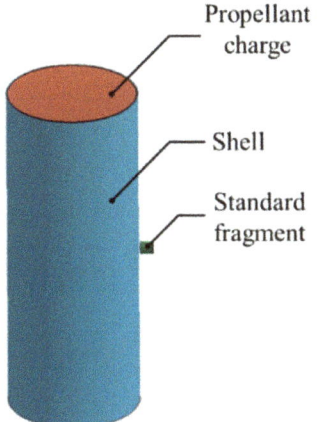

Figure 10. Schematic diagram of the standard fragmentation impact shell loading composite propellant model.

An arbitrary Lagrangian–Eulerian algorithm (ALE) was used to describe the high-energy HTPB propellant charge and the air unit. The Lagrangian algorithm was used to describe the metal shell and the fragmentation components. Among them, (a) the propellant cylindrical charge was modeled using the butterfly-type mesh modeling method, and the hexahedral cells were uniformly smoothed using the TrueGrid modeling method to make the cell mesh close to the orthogonal mesh. The accuracy of the calculation can be improved. Furthermore, (b) the hexahedral cells of the air model were also uniformly smoothed. The radius of the air domain was 2.5 times of the radius of the propellant, and the boundary was set as a reflection-free boundary to avoid the reflection of the wave at the boundary affecting the flow field calculation; (c) the metal shell adopted hexahedral cells, and five layers of cells were set in the thickness direction; (d) the standard breakers were 0.1 cm away from the outer wall of the shell. The breakers impacted the shell-mounted high-energy HTPB propellant in the vertical direction.

3.1.2. Material Model

The response of different media under different loads is different. The material model is used in numerical simulation to define the relationship between the load and response of the medium. The instantonal model in the material model mainly describes the material stress–strain relationship, and the equation of state mainly describes the medium thermodynamic state relationship. Since the material model can hardly encompass all mechanical responses of the medium, a failure model can be attached when the medium is beyond the range of applicability of the material model. The material model of each component is shown in Table 4. The mechanical behavior and damage patterns of metallic materials, such as breakers and shell materials, were described by the Johnson–Cook(J–C) material model. The equation of state of the impact process was described by the Gruneisen equation of state.

Table 4. Material model of each component.

Parts	Material Model	Equation of State	Failure Models
Propellant Charges	Fluid Elasticity Material Model	Ignition growth equation of state	/
Shell	J–C material model	Gruneisen equation of state	J–C failure model
Fragmentation	J–C material model	Gruneisen equation of state	J–C failure model
Air	Empty material model	Linear polynomial equation of state	/

The parameters of the metal material model used are shown in Table 5 [19,20].

Table 5. Metal material model parameters.

Materials	$\rho_0/(\text{g·cm}^{-3})$	G/GPa	A/GPa	B/GPa	C	n	m	T_m/K
30CrMnSiA steel	7.85	75	0.525	0.101	0.1739	0.081	1.635	1800
45# steel	7.86	200	0.790	0.510	0.015	0.27	1.05	1800

Since the failure process of the shell is accompanied by the failure process of the shell during the impact of the standard fragments, this paper uses the J–C failure model and defines the failure strain as

$$\varepsilon_f = \left(D_1 + D_2 e^{D_3\sigma^*}\right)\left(1 + D_4 \ln \dot{\varepsilon}^*_{eq}\right)\left[1 + D_5\left(1 - e^{D_6 T^*}\right)\right] \tag{4}$$

where ε_f represents the failure strain, and the unit is considered to fail when the strain of the material unit reaches this value; σ^*, $\dot{\varepsilon}^*_{eq}$ and T^* represent the stress triaxiality, equivalent effect variability and temperature, respectively; and D_1 to D_6 are the damage parameters. The parameters of the J–C failure model for metallic materials in this section are shown in Table 6.

Table 6. Parameters of the J–C failure model for metallic materials.

Materials	D_1	D_2	D_3	D_4	D_5	D_6
30CrMnSiA steel	0.0705	1.732	−0.54	−0.0123	0	0
45# steel	0.78	0	0	0	0	0

The linear polynomial equation of state used for air is

$$p = C_0 + C_1\mu + C_2\mu^2 + C_3\mu^3 + (C_4 + C_5\mu + C_6\mu^2)E \tag{5}$$

where p is the pressure; μ is the compressibility; E is the internal energy; and C_0 to C_6 are the polynomial equation coefficients.

In addition, the JWL equation of state of unreacted propellant is essentially the unreacted impact equation of state of the composite material. To simplify the calculation, the equation of state for unreacted explosives was also expressed by using the JWL equation of state form:

$$p_u = A_u e^{-R_{1u} V_u} + B_u e^{-R_{2u} V_u} + \omega_u C_{V_u} T_u V_u^{-1} \tag{6}$$

where p_u, V_u and T_u are the pressure, relative volume and temperature of the unreacted propellant, respectively; C_{V_u} is the constant volume of specific heat of the unreacted propellant; ω_u, A_u, B_u, R_{1u} and R_{2u} are the fitting constants of the equation of state.

It can be calculated by fitting the impact Hugoniot curve of the composite propellant. The fitting parameters are shown in Table 7.

Table 7. Parameters of equation of state for unreacted state high-energy HTPB propellant.

A_u/GPa	B_u/GPa	R_{1u}	R_{2u}	ω_u	C_{Vu}/(GPa·K^{-1})
36587.61	−2.789	11.0	0.4	1.69	2.5×10^{-3}

3.2. Analysis of the Factors Affecting the Safety of the Standard Fragments Impact Engine

3.2.1. Single Fragment

When v = 1830 m·s^{-1}, the impact detonation process of standard fragments impacting shell-loaded composite propellant is obtained by calculation, as shown in Figure 11.

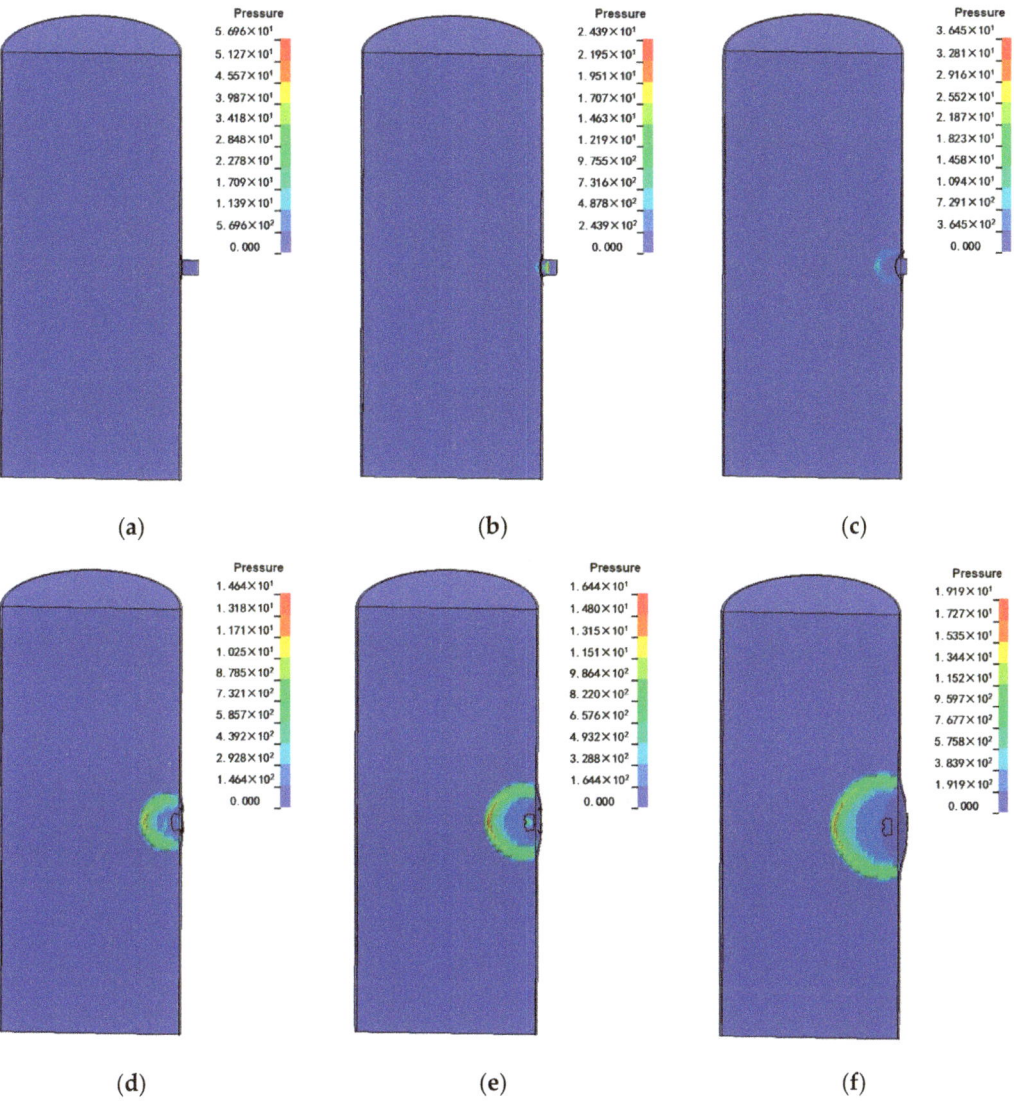

Figure 11. Internal pressure growth process when the propellant under the impact of the standard fragments of impact loading impact detonation. (**a**) $t = 0.8$ μs. (**b**) $t = 2.4$ μs. (**c**) $t = 7.0$ μs. (**d**) $t = 9.0$ μs. (**e**) $t = 11.0$ μs. (**f**) $t = 13.8$ μs.

Figure 11 shows that at $t = 0.8$ µs (a), a high-speed fragment with v = 1830 m·s^{-1} strikes the shell and generates a strong shock wave. At $t = 2.4$ µs (b), the shock wave generated by the impact propagates simultaneously in two opposite directions, radial direction of the shell and axial direction of the fragment. The metal shell at the impact is concaved and deformed under the action of the broken fragment. At $t = 7.0$ µs (c), the standard fragment penetrates the metal shell and comes into contact with the high-energy HTPB propellant charge. The shock wave within the propellant charge begins to propagate and excite the propellant reaction. At $t = 9.0$ µs (d), the fragment passes through the shell and penetrates the propellant charge. The pressure inside the propellant charge gradually increases. At $t = 11.0$ µs (e), the peak shock wave pressure within the propellant charge continues to rise, and the shell begins to expand outward driven by the explosion products at the perforation where the shell was struck. At $t = 13.8$ µs (f), a steady burst wave is generated within the propellant charge, and the burst wave continues to propagate within the propellant charge.

The critical detonation speed of the broken fragment impact is shown in Table 8.

Table 8. Fragmentation impact critical detonation velocity.

Shape of Fragment	Density $\rho_0/(\text{g·cm}^{-3})$	Characteristic Size of the Fragment $l/(\text{mm})$	Critical Detonation Velocity $V_{cr}/(\text{m·s}^{-1})$
Standard fragment	7.86	14.3	1550

3.2.2. Multiple Fragment

In the martyrdom process, solid rocket motors are often subjected to the joint action of many fragments. By establishing a simulation model of multiple fragment impact on shell-mounted composite propellant, the critical detonation velocity is obtained under different numbers of standard fragment impacting at equal intervals in the longitudinal direction. The variation of the critical impact velocity with the number of fragments is shown in Figure 12.

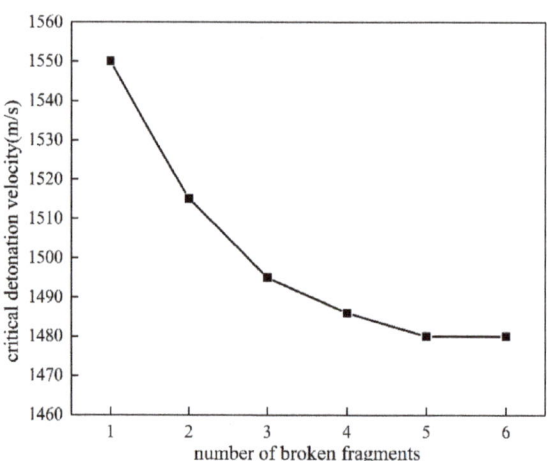

Figure 12. Multiple fragment impact critical detonation velocity curve.

As shown in Figure 12, as the number of standard fragments increases, the critical detonation speed decreases. However, when the number of fragments in the simulation model is more than five, the impact on the critical detonation velocity is no longer obvious. By analyzing the multiple fragment impact process, it can be seen that the shock wave is generated at the impact of the fragment. Shock waves generated by multiple impacts meet as they propagate through the propellant, raising the shock wave pressure, which is equivalent to raising the overall shock wave input pressure. This makes the composite propellant more prone to shock-to-burst, which reduces the critical detonation velocity of the fragmentation impact. However, when the number of standard fragments reaches a certain number, the shock wave interaction generated by the standard fragments farther apart is no longer obvious, so the impact on the critical detonation velocity is reduced.

3.2.3. Shell Material

Using the typical solid rocket engine metal shell materials 30CrMnSiA steel, D406A steel and 2024 aluminum as shell materials, the influence law of different shell materials on the composite propellant impact detonation was studied. The simulation model is consistent with Section 3.1, and a three-dimensional fluid–solid coupling model of the standard broken fragment impacting shell-mounted composite propellant is established. The model contains 677,146 mesh nodes and 638,080 hexahedral cells. The total calculation time is set to t = 30 μs, and the time step is Δt = 0.1 μs. The wave impedance of the metallic material is

$$Z = \rho_0 c \tag{7}$$

where Z is the wave impedance of the metal shell material and ρ_0 and c are the density and volume speed of sound, respectively. The shock wave impedance of each metal material is calculated and obtained as in Table 9.

Table 9. Metal material shock wave impedance.

Metal Materials	$\rho_0/(g \cdot cm^{-3})$	Speed of Sound $c/(m \cdot s^{-1})$	Shock Wave Impedance $Z/(kg \cdot m^{-2} \cdot s^{-1})$
30CrMnSiAsteel	7.85	5664	4.4×10^7
D406A steel	7.60	5918	4.5×10^7
7A04 aluminum	2.785	5330	1.5×10^7

The geometry of the shell is consistently set in the simulation model, and only the shell material is a variable. The critical detonation velocity of the standard fragment impact varies with the shell material as shown in Table 10.

Table 10. Fragmentation impact critical detonation velocity.

Materials	Density $\rho_0/(g \cdot cm^{-3})$	Critical Detonation Velocity $V_{cr}/(m \cdot s^{-1})$
30CrMnSiA steel	7.85	1550
D406A steel	7.60	1960
7A04 aluminum alloy	2.785	1125

As can be seen from Table 10, compared with aluminum alloy and steel shells, as the shell material strength increases, the standard fragments penetrating the shell will decay more energy. The weaker the composite propellant is subjected to, the initial shock wave becomes weaker, and the more difficult the occurrence of shock to blast. That is, under the same loading strength and shell wall thickness conditions, and the greater the strength of the metal shell material, the more difficult it is for the composite propellant to be shock detonated.

3.2.4. Shell Thickness

A high-energy HTPB propellant impact detonation model with different shell wall thickness protection states was established. A numerical simulation model of standard fragment impact shell-mounted composite propellant was established by using TrueGrid software, as described in Section 3.1. The general metal solid rocket motor shell wall thickness is δ = 1~3 mm, so the shell wall thickness is set as shown in Table 11.

Table 11. Shell wall thickness setting.

Shell Wall Thickness/δ_i	δ_1/mm	δ_2/mm	δ_3/mm	δ_4/mm	δ_5/mm
Shell wall thickness value	1.0	1.5	2.0	2.5	3.0

The shell inner diameter Φ157 mm and shell length L = 400 mm of the high-energy HTPB propellant are kept constant in each case. The critical impact velocity V_{cr} of the fragmentation of the high-energy HTPB propellant when the detonation occurs is calculated. The impact detonation critical impact velocity with the shell wall thickness variation law is shown in Figure 13.

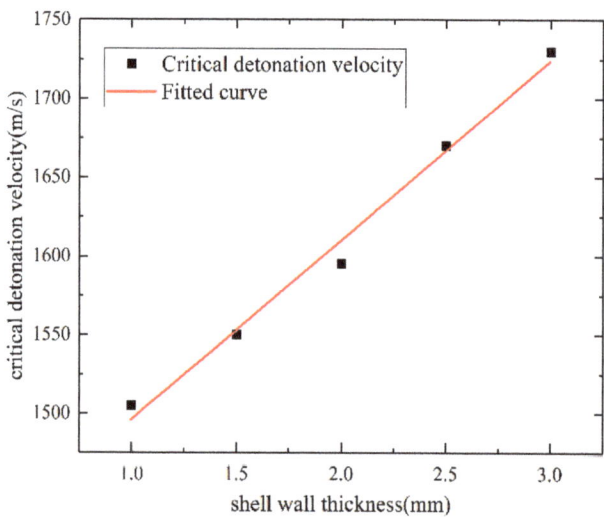

Figure 13. Impact detonation critical velocity with the shell wall thickness variation law.

Figure 13 shows that, as the shell wall thickness increases, the shock wave amplitude generated by the standard fragment on the shell surface becomes weaker. Therefore, the initial shock loading on the composite propellant becomes weaker, and it is more difficult for shock initiation to occur. The fitted equation was calculated to obtain

$$V_{cr} = 1382(1 + 0.082\delta) \tag{8}$$

where V_{cr} is a composite propellant impact detonation of the critical impact velocity of the standard fragments, unit: m·s^{-1}; δ is the shell wall thickness, unit: mm; fitting formula applicable range: 1.0 mm $\leq \delta \leq$ 3.0 mm; and fitting formula fit is R_2 = 0.98904.

4. Conclusions

In this paper, the equation of state and the parameters of the reaction rate equation of the detonation product of high-energy HTPB four-component propellant were obtained by the cylinder test and the Lagrange test, combined with the genetic algorithm. Based on these parameters, the response process of high-energy HTPB engine under the effect of fragment impact was simulated and analyzed. The following conclusions were obtained:

(1) Multiple fragment loading can increase the shock wave input pressure and reduce the critical detonation velocity of the fragment impacting high-energy four-component HTPB propellant. When the number of longitudinally distributed fragments is more than five, the critical detonation velocity no longer decreases with the increase in the number of fragments.

(2)   When the loading strength and shell wall thickness remains constant and the strength of the metal shell is greater, the more difficult it is for the composite propellant impact detonation to occur. In the case of wall thickness δ = 1.5 mm, the critical detonation velocity of 30CrMnSiA steel shell is V_{cr} = 1550 m/s; the critical detonation velocity of the D406A steel shell is V_{cr} = 1960 m/s; and the critical detonation velocity of the 7A04 aluminum alloy shell is V_{cr} = 1125 m/s.

(3)   Under the conditions of loading strength and metal materials remaining constant, the greater the shell wall thickness, the more difficult for the composite propellant impact detonation. The relationship formula between fragment critical impact velocity and shell wall thickness is V_{cr} = 1382 (1 + 0.082δ) m/s.

Author Contributions: Conceptualization, J.N. and W.F.; methodology, Z.L. and J.N.; validation, W.F. and J.T.; formal analysis, Z.L. and J.N.; investigation, Z.L., F.J., T.G. and K.G.; resources, J.N.; data curation, Z.L., W.F. and J.T.; writing—original draft preparation, Z.L. and W.F.; writing—review and editing, J.N.; supervision, J.T. and T.G.; project administration, J.N.; funding acquisition, J.N. All authors have read and agreed to the published version of the manuscript.

Funding: This research received no external funding.

Data Availability Statement: The data that support the findings of this study are available from the corresponding author upon reasonable request.

Conflicts of Interest: The authors declare no conflict of interest.

References

1.   Song, L.F.; Li, H.X.; Zheng, Z.; Li, S.W.; Wu, Z. Research progress on the effect of fragments on the safety of solid propellant charges. *Aerodyn. Missile J.* **2019**, *1*, 92.

2.   Salzman, P.K.; Irwin, O.R.; Andersen, W.H. Theoretical detonation characteristics of solid composite propellants. *AIAA J.* **1965**, *3*, 2230.

3.   Bai, C.H.; Ding, J. Study of the shock initiation and the detonation process of composite propellants. *Explos. Shock. Waves* **1989**, *9*, 199.

4.   Baker, P.J.; Coffey, C.S.; Mellor, A.M. Critical impact initiation energies for three HTPB propellants. *J. Propuls. Power* **1992**, *8*, 578. [CrossRef]

5.   Huang, F.L.; Zhang, B.P. Study on the Detonation Danger of Solid Propellants. *J. Beijing Inst. Technol.* **2004**, *13*, 341.

6.   Wu, J.Y.; Chen, L.; Lu, J.Y.; Feng, C.G.; Wang, Y.J. Research on shock initiation of the high energy solid propellants. *Acta Armamentarii* **2018**, *29*, 1315.

7.   Bernecker, R.R. The deflagration-to-detonation transition process for high-energy propellants-a review. *AIAA J.* **1986**, *24*, 82. [CrossRef]

8.   Liu, D.H.; Peng, P.G.; Wang, Z.F.; Pan, M.C. An investigation of the deflagration-to-detonation transition of the AP/HMX/HTPB composite propellant. *Acta Armamentarii* **1994**, *15*, 32.

9.   Qin, N.; Liao, L.Q.; Jin, P.G.; Xu, H.Y.; Li, J.Q.; Fan, H.J. Experimental study on deflagration-to-detonation transition of several typical solid propellants. *Chin. J. Explos. Propellants* **2010**, *33*, 86.

10.   Bowden, F.P.; Gurton, O.A. Birth and growth of the explosion in solids initiated by impact. *Nature* **1948**, *161*, 348. [CrossRef]

11.   Price, D.; Clairmont, A.R. Explosive behavior of simplified propellant models. *Combust. Flame* **1977**, *29*, 87. [CrossRef]

12.   Dick, J.J. Detonation initiation behavior of some HMX/AP/A1 propellants. *Combust. Flame* **1980**, *37*, 95. [CrossRef]

13.   Kohga, M.; Shigi, D.; Beppu, M. Detonation properties of ammonium nitrate/nitramine-based composite propellants. *J. Energetic Mater.* **2019**, *37*, 309. [CrossRef]

14.   Bai, C.H.; Ding, J. Reaction of solid propellants under shock loading. *Acta Armamentarii* **1991**, *12*, 38.

15.   Yang, K.; Xu, B.H.; Guo, Y.Q.; Wu, Q. Calculation of detonation threshold of fragments impact on solid rocket motors. *J. Solid Rocket. Technol.* **2018**, *41*, 566.

16.   Li, H.T.; Wu, Z.; Wang, Y.; Wang, Z.; Li, S.W.; Huang, Y.; Cheng, H.; Xu, S.; Song, L.F. Research Progresses of Experiment, Mechanism, and Formulas of Low Vulnerable Propellants under Mechanical Stimulations. *Equip. Environ. Eng.* **2019**, *16*, 57.

17.   Lee, E.L. *Adiabatic Expansion of High Explosive Detonation Products*; UCRL250422; Univ. of California Radiation Lab. at Livermore: Livermore, CA, USA, 1965.

18.   Lee, E.L.; Tarver, C.M. Phenomenological model of shock initiation in heterogeneous explosives. *Phys. Fluids* **1980**, *23*, 12. [CrossRef]

19.   Zhang, W.; Xiao, X.K.; Wei, G. Constitutive relation and fracture model of 7A04 aluminumalloy. *Explos. Shock. Waves* **2011**, *31*, 81.

20.   Li, Y.; Dong, M.H.S.; He, J.H.; Ye, K.C. Study on milling of 30CrMnSiA alloy steel based on abaqus. *Tool Eng.* **2016**, *50*, 35.

Article

Experiment and Numerical Simulation on Friction Ignition Response of HMX-Based Cast PBX Explosive

Junming Yuan [1,*], Yue Qin [1], Hongzheng Peng [2], Tao Xia [1], Jiayao Liu [1,2], Wei Zhao [2], Hu Sun [1] and Yan Liu [1]

[1] School of Environment and Safety Engineering, North University of China, Taiyuan 030051, China; s202114014@st.nuc.edu.cn (Y.Q.); s2014028@st.nuc.edu.cn (T.X.); s2014038@st.nuc.edu.cn (J.L.); sz202114018@st.nuc.edu.cn (H.S.); sz202114036@st.nuc.edu.cn (Y.L.)

[2] Chongqing Hongyu Precision Industry Group Co., Ltd., Chongqing 402760, China; wll@hongyu.com (H.P.); hty@hongyu.com (W.Z.)

* Correspondence: yuanjm@nuc.edu.cn

Abstract: In order to study the ignition process and response characteristics of cast polymer-bonded explosives (PBX) under the action of friction, HMX-based cast PBX explosives were used to carry out friction ignition experiments at a 90° swing angle and obtain the critical ignition loading pressure was 3.7 MPa. Combined with the morphology characterization results of HMX-based cast PBX, the friction temperature rise process was numerically simulated at the macro and micro scale, and the ignition characteristics were judged. The accuracy of the numerical simulation results was ensured based on the experiment. Based on the thermal–mechanical coupling algorithm, the mechanical–thermal response of HMX-based cast PBX tablet under friction was analyzed from the macro scale. The results show that the maximum temperature rise is 55 °C, and the temperature rise of the whole tablet is not enough to ignite the explosive. Based on the random circle and morphology characterization results of tablet, the mesoscopic model of HMX-based cast PBX was constructed, and the microcrack friction formed after interface debonding was introduced into the model. The temperature rise process at the micro scale shows that HMX crystal particles can be ignited at a temperature of 619 K under 4 MPa hydraulic pressure loaded by friction sensitivity instrument. The main reason for friction ignition of HMX-based cast PBX is the friction hot spot generated by microcracks formed after interface damage of the tablet mesoscopic model, and the external friction heat between cast PBX tablet and sliding column has little effect on ignition. External friction affects the ignition of HMX-based cast PBX by influencing the formation of internal cracks and the stress at microcracks.

Keywords: numerical simulation; friction sensitivity; ignition; cast PBX; mesoscopic model

Citation: Yuan, J.; Qin, Y.; Peng, H.; Xia, T.; Liu, J.; Zhao, W.; Sun, H.; Liu, Y. Experiment and Numerical Simulation on Friction Ignition Response of HMX-Based Cast PBX Explosive. *Crystals* 2023, 13, 671. https://doi.org/10.3390/cryst13040671

Academic Editors: Rui Liu, Yushi Wen and Weiqiang Pang

Received: 15 March 2023
Revised: 3 April 2023
Accepted: 7 April 2023
Published: 13 April 2023

1. Introduction

Cast PBX is a kind of high polymer-bonded explosive, which is widely used in high-speed penetration and damage weapons because of its good mechanical properties and low sensitivity. During the transportation, storage and use of casting PBX charge warheads, accidental ignition may occur due to external stimulation, which greatly affects the reliability and safety of weapons. Friction is one of the important stimulation sources. A lot of experiments and numerical simulations have been done on the ignition of explosives under friction. Min-cheol Gwak et al. [1] analyzed the friction ignition process of HTPB based solid propellant from the perspectives of reaction kinetics and friction heating, and built relevant models to predict the ignition time of propellant under friction action through numerical simulation. Sun Baoping et al. [2] carried out numerical simulation of PBX tablet friction-ignition experiment based on the finite element method, and analyzed the influence of pressure, velocity, and friction coefficient on ignition. Deng Chuan et al. [3] established a test method for friction sensitivity of pendulum impact-driven sand target friction explosive tablets, and tested the friction sensitivity of three PBX tablets. R. Charley et al. [4] studied the friction ignition process of solid propellant based on friction device, which can obtain the overall deformation process of propellant by high-speed photography, and then simulate the response of solid propellant under friction by discrete element method. And compare the cloud image of numerical simulation with the topography of high-speed photography. The deformation process analyzed by the discrete element method is in good agreement with the actual deformation process and can reflect the meso-ignition process of the explosive. Dai Xiaogan et al. [5] carried out the friction ignition experiment on PBX and designed a device to calculate the friction work and friction power threshold of explosive ignition under friction and analyzed the ignition mechanism of explosive in the friction sensitivity experiment. The results show that it is difficult to heat PBX as a whole and make it ignite under friction. Zeman et al. [6] provided an overview of the main developments over the past nine years in the study of the sensitivity of energetic materials (EM) to impact, shock, friction, electric spark, laser beams and heat. Luo Yi et al. [7] and others carried out slide experiment and Numerical Simulation Research on PBX-8701, and analyzed the influence of external-friction coefficient on ignition delay time based on the theory of heat transfer and diffusion and reaction kinetics.

The ignition phenomenon of explosives under external stimulation can be explained by the hot spot theory. Dienes et al. [8] analyzed and compared four hot spot mechanisms (pore collapse; hole collapse; hole collapse) in the projectile target experiment, shock heating, shear band under plastic flow. The results show that the interface friction of closed crack is the main reason of hot spot. Randolph et al. [9] developed a friction ignition model to predict the thermal decomposition of condensed phase explosives when impacted at an oblique angle on a rigid target surface. Andersen et al. [10] established a mathematical relationship between the friction coefficient of materials and the parameters that affect friction during shearing. An et al. [11] used the ReaxFF force field to study the hot spot formation mechanism of PBXN-106 explosive under the action of shock waves. The shear relaxation of the micro-convex body at the interface of the density discontinuity caused by the impact resulted in energy deposition, and the local shear in this region formed a hot spot, which was eventually accompanied by chemical reactions leading to the explosion of the system. It is concluded that reducing the density of the binder (about 1/3 of the explosive density) can inhibit the hot spot generation. Cai et al. [12] used molecular dynamics to simulate the impact response characteristics of coarse-grained explosives and pointed out that both inter-particle friction and shear deformation between particles can generate hot spots, and the direction of shear slip between particles significantly affects the frictional heat generation efficiency. Keshavarz et al. [13] presented a novel general simple model for prediction of the relationship between friction sensitivity and activation energy of thermolysis of cyclic and acyclic nitramines on the basis of their molecular structures. Richard et al. [14] studied the response to mechanical non-shock stimulation using explosive-driven deformation test and ballistic impact chamber. According to the experimental results, the shear rate threshold as a single parameter to describe the mechanical sensitivity is challenged, and preference is given to the development of an ignition criterion based on intergranular sliding friction under the action of a normal pressure. Hu et al. [15] presented a combined computational–experimental study of the mesoscale thermo-mechanical behavior of HTPB bonded AP composite energetic material subjected to dynamic loading conditions. The computational model considers the AP-HTPB interface debonding, post debonding interface friction and temperature rise due to viscoelastic dissipation as well as dissipative interfacial processes. Jafari et al. [16] introduced a reliable method to predict friction sensitivity of quaternary ammonium-based EILs, which are based on elemental composition of cation and anion of a desired ionic liquid as well as the contribution of specific cations and anions. Gruau et al. [17] simulated the behavior of the PBX by means of the elastic-plastic damage law and the ignition criterion due to localization of plastic strain in the microstructure, and the simulation results were consistent with the experimental results. Wu et al. [18] developed a micromechanical ignition model of the hot spot formation of HMX and PETN mixed powder explosives under the impact of a falling hammer in the fine structure, estimated the temperature rise caused by plasticity and frictional dissipation, and added the self-heating reaction model of explosives to predict the hot spot ignition by thermal explosion. Joel G. Bennett et al. [19] proposed visco elastic statistical crack ignition model (visco-scram) for numerical simulation of non-shock ignition of PBX explosives. In this model, the friction heat of microcracks is taken as the main source of hot spot formation, and the mechanical behavior of PBX explosives is considered. As a classical non-shock ignition model, this model is widely used. However, there are many parameters in the model, and it is difficult to calibrate. For different explosive formulations, it is necessary to re determine the material parameters, so the pretreatment process of numerical simulation is relatively complex. Xue, H.J. et al. [20] established an improved combined microcrack and microvoid model (CMM) to study the damage and ignition behaviors of polymer-bonded explosive (PBX) under coupled impact and high-temperature loading conditions. Bai, Z.X., Li, H.T., Yin, Y. and Duarte, C.A. [21–24] have studied the friction behavior and hot spot formation of HMX explosive crystals, as well as the ignition and combustion behavior under hot spot conditions.

The research on friction ignition has gradually changed from macro to micro hot spot formation. Barua et al. [25–28] studied the temperature rise process of PBX9501 from micro scale by CFEM finite element method. Based on the digital image processing technology, a mesoscopic model was established, which was consistent with the actual situation. The friction heating of microcracks caused by the failure of the interface between particles and matrix was studied. The relationship between the particle-failure mechanism and the overall temperature rise of PBX has been researched. Amirreza Keyhani et al. [29] analyzed the ignition process of PBX9501 at meso level based on CFEM method and considered the contribution of friction and viscoelastic plastic dissipation energy of microcracks formed after damage to temperature rise. The research results show that the viscoelastic plastic dissipation energy has little effect, and the energy contributed by crack friction is the main reason for hotspot formation.

At present, there are many research studies on the ignition of pressed PBX, but relatively few studies on cast PBX, and the research on the micro ignition focuses on the impact overload, while studies on the friction overload are fewer. Therefore, HMX-based cast PBX samples were prepared and the friction-ignition experiment of cast PBX at a 90° swing angle were carried out in this paper. Based on the experimental results combined with the macro numerical simulation, the friction-ignition process and response characteristics of HMX-based cast PBX tablets are studied at the micro level, and the accuracy of the numerical simulation results is judged by friction ignition experiments.

2. Materials and Methods

2.1. Experiment

2.1.1. Preparation and Characterization of HMX-Based Cast PBX Sample

According to the typical HTPB casting PBX formulation, a small amount of casting PBX explosive samples were prepared. Cast PBX is composed of Al powder, HTPB bonding system and HMX crystal particles. All raw materials should be fully dried before preparation, and their water content must be strictly controlled. Aluminum powder and binder are thoroughly mixed and added to the main explosive in batches under heated conditions. After mixing evenly, the material is poured into the mold so that the explosive is cured at a constant temperature of 60 °C for 72 h. The formulation of cast PBX is shown in Table 1 referred from [30].

Table 1. Formulation of HMX-based cast PBX explosive.

Sample	HMX	Al	HTPB	DOA	TDI
Mass fraction	55%	33%	6%	5.8%	0.2%

Figure 1 shows the cast PBX tablets and the electron microscopic morphology. After the preparation for the cast PBX is completed, slice work is carried out, and the slice size is about Φ10 mm × 1 mm. Because the formulation contains Al powder, the color of the tablet is grey, as shown in Figure 1a,b. The tablet was characterized by scanning electron microscope, and HMX crystal particles with a particle size range of 10~80 μm are shown in Figure 1c. The electron microscope morphology provides experimental support for the establishment of the mesoscopic model in the next friction ignition simulation.

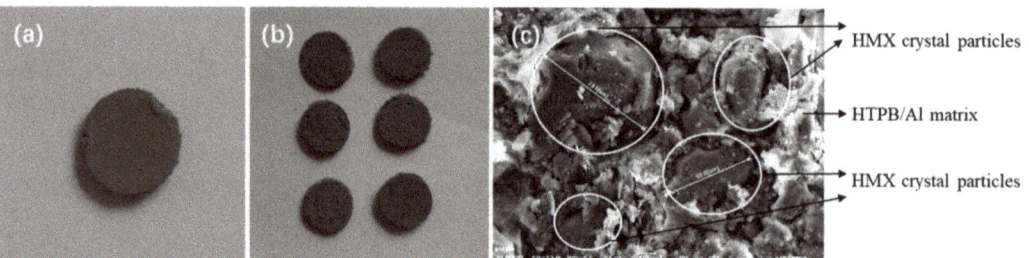

Figure 1. (**a,b**) HMX-based Cast PBX tablets; and (**c**) 1000 times electron microscope morphology of HMX-based PBX explosive sample.

The photographs in Figure 1c indicate the presence of micropores in cast PBX near the HMX particles and the binder matrix. This is also reflected in Figure 1b, where pores are visible on the surface of the poured PBX tablet. The binder curing process itself has pores, as shown in Figure 2a. Cast PBX tablets without vacuum pumping and uneven mixing can cause pores. The presence of pores weakens the cohesive strength of the interface boundaries and contributes to the development of deformation localization and local heating at the boundaries of HMX particles under the action of stresses arising from friction of the samples. Figure 2c shows the microstructure of HMX-based cast PBX tablets. The picture clearly shows that the HTPB/Al matrix composed of Al powder particles and adhesive is tightly wrapped with many HMX crystal particles of different sizes, providing theoretical support for the construction of a simplified model.

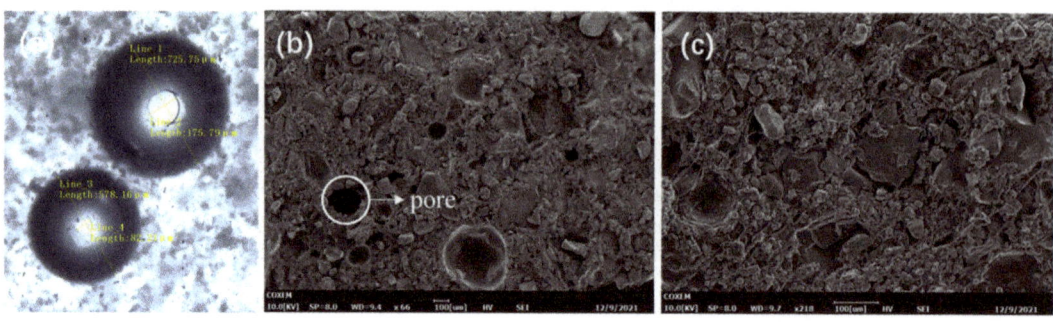

Figure 2. (**a**) HTPB binder curing pores; (**b**) Pore of HMX-based cast PBX tablets; (**c**) 200 times electron microscope morphology of HMX-based PBX explosive sample.

2.1.2. Test Process of Friction Sensitivity Experiment

The cast PBX tablet is loaded onto the sliding column end face of the friction device, and the upper sliding column is gently placed. The upper slide column is rotated for 1–2 cycles to evenly distribute the explosive sample between the end faces of the two slide columns. The loaded friction device is placed in the combustion equipment of the friction sensitivity instrument, at a 90° swing angle, and the gauge pressure raised to 4.0 MPa (or other different loading pressures). The pendulum is released, and the striking rod is struck. The striking rod causes the upper and lower sliding columns to move at high speed, causing sliding friction on the end faces of the upper and lower columns, resulting in intense friction on the test sample. After completing this operation process, the experimenter repeats the process to perform the test on the next sample.

2.1.3. Critical Ignition Loading Pressure

MGY-I friction sensitivity instrument is used to test the friction ignition of cast PBX [31]. In order to obtain the exact critical initiation conditions of friction, the Bruceton method [32] is used for reference to the high impact sensitivity characteristics. In the test of friction sensitivity, there are two variables controllable, one is the pendulum angle and the other is the loading pressure of hydraulic press. Considering the feasibility of numerical simulation, the paper uses the loading pressure as a variable. The Bruceton method is used to determine the critical ignition loading pressure of friction sensitivity. The method is calculated according to the following formula.

$$\sigma_{50}^f = \left[A + B \left(\frac{C}{D} \pm \frac{1}{2} \right) \right] \tag{1}$$

$$C = \sum i \times n_i \tag{2}$$

$$D = \sum n_i \tag{3}$$

In the formula, σ_{50}^f is the critical ignition loading pressure of explosives; A is the minimum hydraulic pressure loaded by friction sensitivity instrument for initiation; B is the set loading pressure interval, 0.5 MPa in this paper; i is the height serial number; n_i is the sum of the number of explosion times under the height of serial number i. The friction ignition experiment was carried out under a swing angle of 90° on cast PBX tablets with the same specification. In this experiment, the loading pressure range is 2 MPa~4 MPa, and the pressure gradient is 0.5 MPa; So, A = 2 MPa, B = 0.5 MPa. Finally, the critical ignition loading pressure calculated is σ_{50}^f = 3.7 MPa. The calculation parameters in Formula (1) are shown in Table 2.

Table 2. Sorting of test results.

i	n_i	C/MPa	D/MPa
1	0		
2	1		
3	3	47	12
4	4		
5	4		

2.1.4. Solution of Relative Slip Rate

In order to obtain the slip rate of friction ignition experiment, the traditional friction sensitivity instrument was improved, as shown in Figure 3. A force sensor is added at the end of the striking bar. The impact force and action time of the sliding column can be measured by the force sensor, and the displacement curve of the sliding column can be obtained by the motion equation.

When measuring the friction sensitivity, the upper sliding column starts to slide to the right under the action of pendulum. The sliding speed of the upper column is not uniform. According to the F-T curve measured by the sensor, the displacement x-t curve of the sliding column can be obtained by calculating the F-T integral. The impact force F obtained does not provide acceleration for the upper sliding column, but also overcomes friction resistance F.

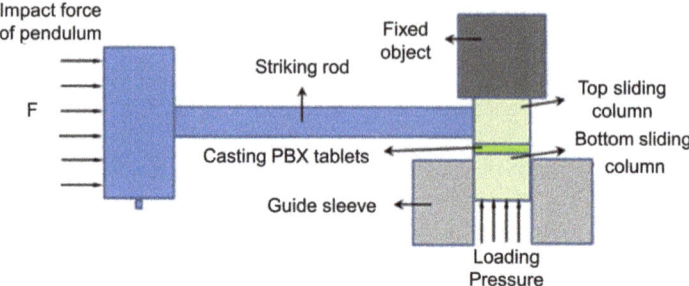

Figure 3. Pendulum impact force test device.

The velocity curve of the sliding column can be used as the boundary condition of the numerical simulation to ensure that the numerical simulation is consistent with the experiment. The slip rate curves under different loading pressures are shown in Figure 4a. Over time, under different loading pressures, the slip rate of the upper sliding column first rises and then gradually stabilizes. Under a loading pressure of 2 MPa, the slip rate curve and displacement curve are shown in Figure 4b. As can be seen from the Figure 4b, the maximum sliding speed of the sliding column is 8 m/s, and the maximum sliding displacement is about 12 mm. This is a linear system of friction, and the sliding distance increases with time. In actual friction experiments, due to the limitation of the length of the striking rod and the constraint of the guide hole, the sliding displacement of the upper sliding column generally does not exceed 2.5 mm.

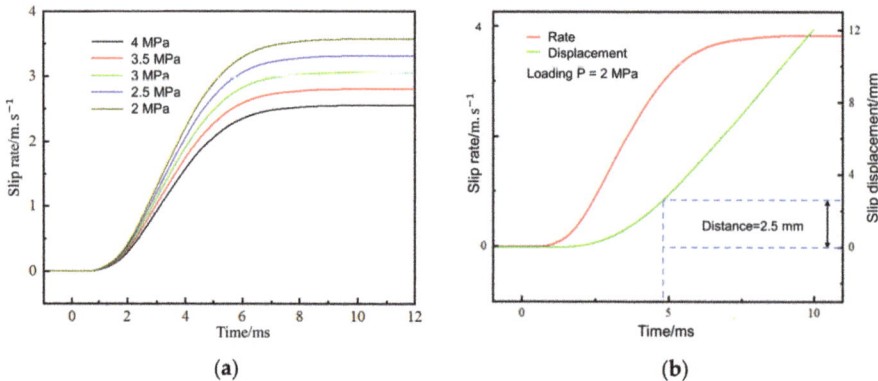

Figure 4. Slip rate and friction displacement curve of the upper sliding column.

2.2. Numerical Simulation

2.2.1. Modeling

The macro simulation physical model of friction ignition experiment is shown in Figure 5. The model is established according to the equal proportion of the size of the experimental device, in which the size of the upper and lower striking columns is the same, which is a cylinder with a diameter of 10 mm and a height of 10 mm. The diameter of HMX-based cast PBX tablets φ the thickness is 1 mm. The vertical downward pressure load is applied on the top of the sliding column on the model, and the pressure load is 2–4 MPa, and the gradient is 0.5 MPa. The friction displacement is applied horizontally, and the displacement load is applied according to the curve. The sliding column is completely fixed.

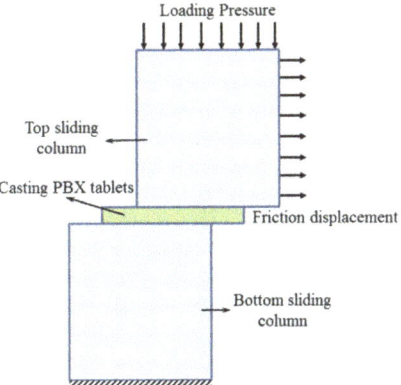

Figure 5. Macroscopic friction model of cast PBX.

The main components of the cast PBX prepared in this paper are HMX/Al powder and HTPB binder. The particle size and gradation of each component need to be considered. The mesoscopic model of cast PBX is built based on the random circle model. The proportion of each particle in the model is closely approximate with that of the cast PBX based on the above Figure 2c. Due to the large number of particles, the mesh and contact action need to be divided, so appropriate simplification is adopted. Some literatures point out a treatment method: small particles and binder matrix are regarded as homogeneous binder system, only the spatial distribution of large particles is considered, and the random spatial coordinates of large particles are given by Matlab function programming. Combined with the electron micrograph of HMX-based PBX explosive sample in Figure 2c, the simplified mesoscopic model is shown in the Figure 6.

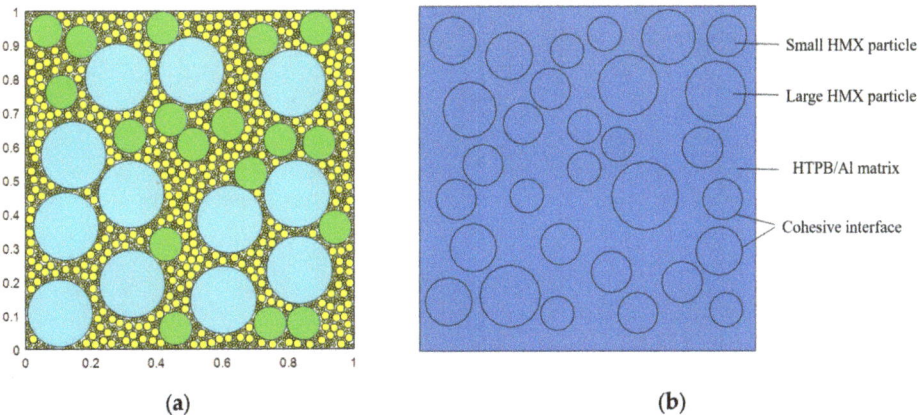

(a)   (b)

Figure 6. HMX-based cast PBX mesoscopic model: (**a**) random circular mesoscopic model; and (**b**) simplified mesoscopic model by Al and HTPB as a matrix.

A small part of the whole cast PBX tablet was taken and magnified to obtain the mesoscopic model. The friction coefficient f is 0.15. As the friction displacement of the sliding column is relatively small, it can be ignored. Pressure 1 is applied as the loading pressure as shown in Figure 7. The left boundary is fixed horizontally, and there are degrees of freedom in the vertical direction. In the model, the upper sliding column is fixed in the vertical direction, and a fixed displacement is applied to the right. The displacement and amplitude are applied according to the curve. The boundary conditions of the mesoscopic model are shown in Figure 7.

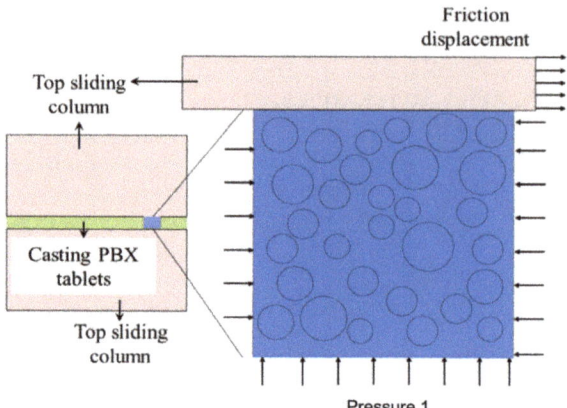

Figure 7. Boundary conditions of mesoscopic model for friction ignition.

2.2.2. Material Model

(1) Viscoelastic material model.

A viscoelastic body is set at any time τ_i have an instantaneous strain $\Delta\varepsilon_i(\tau_i)$, The instantaneous strain $\Delta\varepsilon_i(\tau_i)$, For the stress at any subsequent time t $\sigma(t)$ Impact, in order to $\Delta\sigma_i(t,\tau_i)$ The subscript of Δ indicates that the effect is caused by the i-th strain increment $\Delta\varepsilon_i(\tau_i)$, because the research is linear system, the stress change $\Delta\sigma_i(t,\tau_i)$ and instantaneous strain increment $\Delta\varepsilon_i(\tau_i)$ Satisfy relationship [33]:

$$\Delta\sigma_i(t,\tau_i) = E(t,\tau_i)\,\Delta\varepsilon_i(\tau_i) \tag{4}$$

Before time t, if the instantaneous strain increment $\Delta\varepsilon_i(\tau_i)$ is more than one (i = 1, 2, 3, $\cdots$, n), and each $\Delta\varepsilon_i$ has no influence on each other. According to the principle of Boltzmann superposition, there are:

$$\sigma(t) = \sum_{i=1}^{n}\Delta\sigma_i(t,\tau_i) = \sum_{i=1}^{n}E(t,\tau_i)\,\Delta\varepsilon_i(\tau_i) \tag{5}$$

When the instantaneous strain is continuous, it becomes $\varepsilon(\tau)$, $(n \to \infty)$, The continuous strain can be obtained $\varepsilon(\tau)$, The expression of stress response is as follows:

$$\sigma(t) = \int_{-\infty}^{t}E(t,\tau)d\varepsilon(\tau) = \int_{-\infty}^{t}E(t,\tau)\frac{\partial\varepsilon}{\partial\tau}d\tau \tag{6}$$

$$E(t,\tau) = 2(1+v)\,G(t,\tau) \tag{7}$$

In calculating the curve of relaxation modulus, the stress and strain at different relaxation times need to be measured. According to the integral formula, a method is given in document [34,35]. The exponential series fitting formula is used. The equation is as follows:

$$G(t) = G_\infty + \sum_{i=1}^{n}G_i e^{\frac{-t}{\tau_i^G}} = G_0(1 - \sum_{i=1}^{n}g_i^P e^{\frac{-t}{\tau_i^G}}) \tag{8}$$

$$g_R(t) = 1 - \sum_{i=1}^{n}g_i^P e^{\frac{-t}{\tau_i^G}} \tag{9}$$

where $G_0 = G_\infty + \sum_{i=1}^{n}G_i$ is the instantaneous relaxation shear modulus, G_∞ is the steady-state shear modulus, $g_R(t) = G_R(t)/G_0$ is the dimensionless relaxation shear modulus, $g_i^P = G_i/G_0$ is relative modulus of the i-th term. The softening effect of temperature on materials is usually described by Williams-Landel-Ferry (WFL) time-temperature equivalent equation:

$$\tau_i^G = \int_0^t \frac{d\,t'}{a(\theta(\,t'))} \tag{10}$$

$$-\lg(a(\theta(\,t'))) = \frac{\alpha(T - T_{ref})}{\beta + T - T_{ref}} \tag{11}$$

where a is the time temperature transfer function, α and β is constant, T_{ref} is the reference temperature, τ_i^G is the relaxation time. Mechanical and thermal parameters of cast PBX and HTPB/Al matrix are shown in Tables 3 and 4 respectively. Among the Tables 3 and 4, λ is the thermal conductivity coefficient. In addition, q_m is the heat released by the complete decomposition of the unit mass explosive, and E_a is the activation energy in Table 3.

(2)   Elastoplastic material model.

Table 3. Mechanical and thermal parameters of cast PBX [36,37].

τ_1/ms	τ_2/ms	τ_3/ms	τ_4/ms	G_0/MPa	ρ (kg/m³)	λ (W/m·s)	C_v (J/K·g)	q_m (J/g)
7.5×10^{-7}	7.5×10^{-6}	7.5×10^{-7}	7.5×10^{-7}	675.77	1790	0.2	1500	1155
g_1	g_2	g_3	g_4	υ	E_a (J/mol)	Z/s^{-1}	α	β/K
0.698	0.172	0.1237	0.0063	0.3	2.2×10^5	1.81×10^{19}	-10	107.54

Table 4. Mechanical and thermal parameters of HTPB/Al matrix [15,37,38].

τ_1/ms	τ_3/ms	τ_3/ms	τ_4/ms	τ_5/ms	τ_6/ms	G_0/MPa	ρ (kg/m³)	λ (W/m·s)	C_v (J/K·g)
1.04×10^{-7}	2.1×10^{-5}	1.66×10^{-3}	0.0105	0.05	0.21	109.53	1400	0.2	1419
g_1	g_2	g_3	g_4	g_5	g_6	υ	α	β/K	
33	30	25	13	8	6	0.45	-15	102	

Compared with HTPB bonding system, HMX crystal particles has higher hardness and modulus, so it is elastic-plastic. Mechanical and thermal parameters of HMX are shown in Table 5. The constitutive model is as follows:

$$\sigma = \begin{cases} E\varepsilon & \varepsilon \leq \varepsilon_0 \\ \sigma_0 + E_t(\varepsilon - \varepsilon_0) & \varepsilon > \varepsilon_0 \end{cases} \tag{12}$$

where E is Young modulus, σ_0 is the yield stress, E_t is plastic hardening rate. The material of upper sliding column and lower sliding column is special steel, the material model is same as HMX crystal particles. Material parameters of sliding column are shown in Table 6. Among Tables 5 and 6, Z is pre-exponential factor, and C_v is the specific heat capacity, respectively.

Table 5. Mechanical and thermal parameters of HMX crystal particles [39,40].

E	υ	σ_0/MPa	E_t/MPa	E_a (J/mol)	Z/s^{-1}	q_m (J/g)	ρ (kg/m³)	λ (W/m·s)	C_v (J/g·K)
13.3×10^3	0.3	260	5.7×10^3	2.2×10^5	1.81×10^{19}	2100	1865	0.456	1190

Table 6. Material parameters of sliding column [40].

E	υ	σ_0/MPa	ρ (kg/m³)	λ (W/m·s)	C_v (J/K·g)
2.1×10^5	0.3	1.83×10^3	7800	50	460

2.2.3. Interface Model

(1) Cohesive model.

In the micromechanical analysis, the interface between the binder and explosive plays an important role in the micromechanical properties and thermal response of PBX. On the one hand, the debonding of particles and binder at the interface will cause the mechanical damage of PBX. On the other hand, friction between the debonding explosive particles and the binder will form a high temperature concentration area, which may produce local hot spots. In the existing literature, a cohesive model is used to simulate the debonding phenomenon at the interface, which can better describe the debonding phenomenon between explosive particles and binder, as shown in Figure 8.

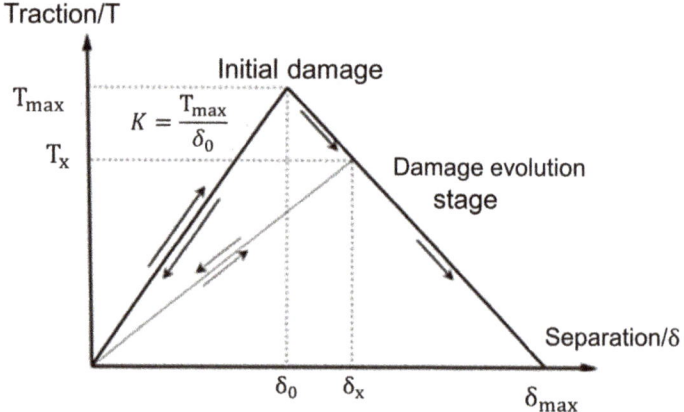

Figure 8. Bilinear cohesive zone model.

In the model, T_{max} is the maximum traction force allowed; δ_0 is the separation amount of the interface during the initial damage, K is the stiffness matrix, $K = \frac{T_{max}}{\delta_0}$. When the traction force $T > T_{max}$, the interface begins to be damaged, and the strength of the model will decrease in the damage evolution stage. When the separation amount $\delta > \delta_{max}$, the model is completely destroyed, and the bonding of the interface fails. Interface parameters of HMX and binder are shown in Table 7. Among the Table 7, μ is the coefficient of friction, and K_c is the heat distribution coefficient between two objects.

Table 7. Interface parameters of HMX crystal particle and binder [15].

K	T_{max}/MPa	δ_0/mm	μ	K_c
792.3	2.91	0.0015	0.3	1.12

The bilinear cohesive unit simulates the mechanical relationship between explosive particles and binder and can analyze the failure between explosive particles and binder. There are three ways to realize cohesive unit in ABAQUS: (a) By inserting a thick cohesive element in the interface layer; (b) Building a thick cohesive element by establishing a thin layer model; (c) Establishing a cohesive contact (Surface-based cohesive behavior) through a contact algorithm.

The first two are constructed by assigning cohesive properties to the units. Each unit is a cohesive mechanical unit. Unfortunately, the cohesive unit in ABAQUS only has mechanical properties and cannot adapt to the thermo-mechanical coupling algorithm. Therefore, the cohesive unit cannot perform thermal and temperature calculations. Calculations can only analyze the mechanical response process, so this paper adopted the last method cohesive contact method to realize the debonding of particles and binder and the friction effect after debonding.

(2)   The friction of interface.

After the failure of the cohesive interface, microcracks will be formed, and friction will occur in microcracks. The heat generated by friction is the main reason for the temperature concentration. The heat generated by friction and the distribution law of heat [41] are as follows:

$$Q_f = \mu F_N v \tag{13}$$

$$Q_f = Q_{f1} + Q_{f2} \tag{14}$$

$$K_c = \frac{Q_{f1}}{Q_{f2}} = \sqrt{\frac{\lambda_1 C_{v1} \rho_1}{\lambda_2 C_{v2} \rho_2}} \tag{15}$$

where, Q_f is the heat produced by friction, F_N is the normal pressure, v is the slip rate, μ is the coefficient of friction. Q_f can be further divided into two parts, Q_{f1} and Q_{f2} represents the friction heat of two objects; K_c is the heat distribution coefficient between two objects, λ is the heat conductivity, C_v is the specific heat capacity, ρ is density.

2.2.4. Self Heating Reaction Theory and Heat Transfer Equation of Explosives

The ignition process of explosive in low-speed friction can be explained by the hot spot theory, and the hot spot initiation can be described by the thermal initiation equation. Therefore, it is necessary to analyze the internal thermal balance process of explosive. The internal thermal balance equation of explosive is as follows [42].

$$\rho C \frac{\partial T}{\partial \tau} = \lambda \nabla^2 T + \Phi_V \tag{16}$$

where, $\nabla^2 T$ can be regarded as the inflow heat from the outside; λ is thermal conductivity coefficient and Φ_V is the intensity of internal heat source. The intensity of internal heat source Φ_V is determined by the exothermic intensity of explosive reaction. The thermal decomposition process of explosives is usually expressed as the rate by the first order Arrhenius equation. The total reaction heat per unit volume per unit time of explosives is as follows:

$$\Phi_V = Q_V = \rho q_m c^n k = \rho q_m Z \exp\left(-\frac{E_a}{RT}\right) \tag{17}$$

In this paper, the heat-generation subroutine HETVAL is written to realize the heat generation in the thermal decomposition process of explosives. In the actual decomposition process, it is necessary to consider the initial decomposition temperature of explosives. In this paper, the initial decomposition temperature is 554 K. When the temperature of the elemental particles of the explosive is greater than the initial decomposition temperature T_S (554 K), the thermal decomposition reaction of the explosive will be triggered. At lower temperatures, the thermal decomposition rate is lower, and the chemical heat generated is lower. During this process, the frictional heat of microcracks is still the main source of internal heat in explosives. If the temperature continues to rise, the reaction rate of the explosive accelerates, and the heat of chemical reaction becomes the main source of internal heat. Due to the rapid thermal decomposition process of cast PBX, the instantaneous heat flow generated will cause a sharp increase in temperature, which can be considered as the ignition response of the explosive under external stimulus.

3. Results and Discussion

3.1. Response Analysis of PBX Tablet Simulation

3.1.1. Analysis of Tablet Deformation Process

The deformation process and Mises stress nephogram of cast PBX tablet under 3 MPa loading pressure are shown in Figure 9. It can be seen that the casting PBX tablet is first pressed by the hydraulic press. With the sliding of the upper sliding column, the stress begins to concentrate. The Mises stress of the concentrated unit is much greater than that of the hydraulic press. The maximum Misses stress values at 1 ms, 1.5 ms, 2.5 ms, 4 ms, 6 ms, and 10 ms are 0.32 MPa, 8.93 MPa, 15.64 MPa, 32 MPa, 50.34 MPa, and 55.33 MPa, respectively. As time goes on, the Misses stress at the interface gradually increases. At 2.5 ms, the tablet begins to deform, and the stress-concentration area is the most obvious. Combined with the displacement curve of the upper sliding column, the sliding speed is the fastest. At 4 ms, the deformation of the tablet is intensified, but no distortion occurs. Therefore, the stress concentration area is still in the center of the tablet. With the movement of the upper sliding column, the tablet began to deform. At 6 ms, the tablet was slightly distorted, and the stress concentration area diffused due to the distortion. At 10 ms, the tablet was seriously deformed.

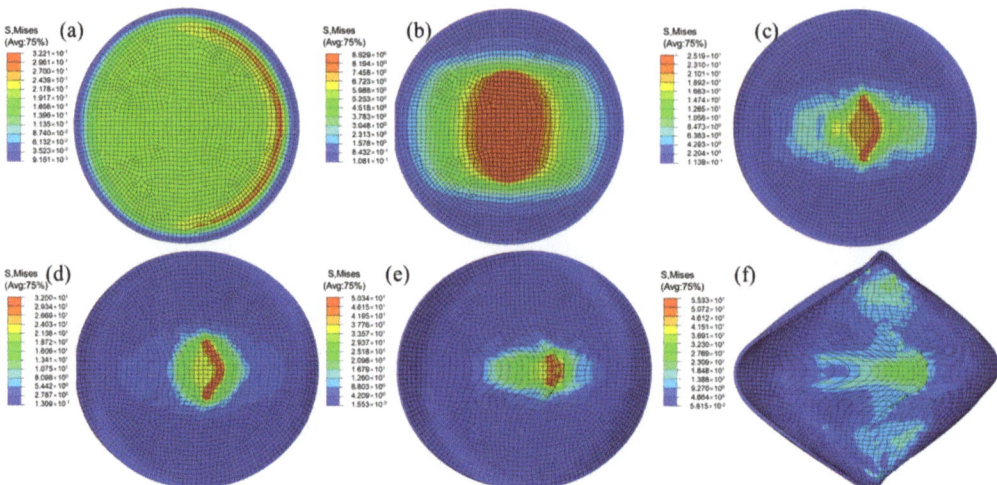

Figure 9. Nephogram of Misses stress and friction deformation of cast PBX tablet: (**a**) t = 1 ms; (**b**) t = 1.5 ms; (**c**) t = 2.5 ms; (**d**) t = 4 ms; (**e**) t = 6 ms; and (**f**) t = 10 ms.

3.1.2. Effect of Loading Pressure on Tablets

Under different hydraulic conditions, the stress and temperature rise of the tablet under friction are also different. The hydraulic pressure of 2 MPa, 2.5 MPa, 3 MPa, 3.5 MPa and 4 MPa is applied, respectively, and the displacement amplitude under the corresponding hydraulic pressure is applied. The stress component (S22) and temperature rise of the central layer unit (Unit No. 1924) of the tablet under different hydraulic pressures are analyzed. The Figure 10 shows the stress component of S22 under hydraulic pressure. With the increase of hydraulic pressure, the stress also increases. The maximum stress under different pressures is 61 MPa, 78 MPa, 87 MPa, 88 MPa and 92 MPa. The figure shows the maximum temperature rise inside the tablet under different hydraulic pressures. The maximum temperature rises under 2–4 MPa hydraulic pressure are 34 °C, 47 °C, 54 °C, 50 °C and 55 °C, respectively. It can be seen that the temperature rise gradually increases with the increase of pressure, but the overall temperature rise is not high.

Through the simulation of friction sensitivity experiment, the influence of pressure on the temperature rise of tablets was analyzed. The simulation results show that the pressure has a great influence on the temperature rise of tablets. When the pressure is low, there is no large area of temperature rise concentration area. With the increase of pressure, the temperature concentration area begins to appear. The concentration area further moves and gradually shrinks, and the maximum temperature rise increases with the increase of pressure. The external friction can make the local temperature of the tablet rise, but it can't continue. The overall temperature rises are not able to make the explosive ignite, so we need to further consider the ignition mechanism from the micro level.

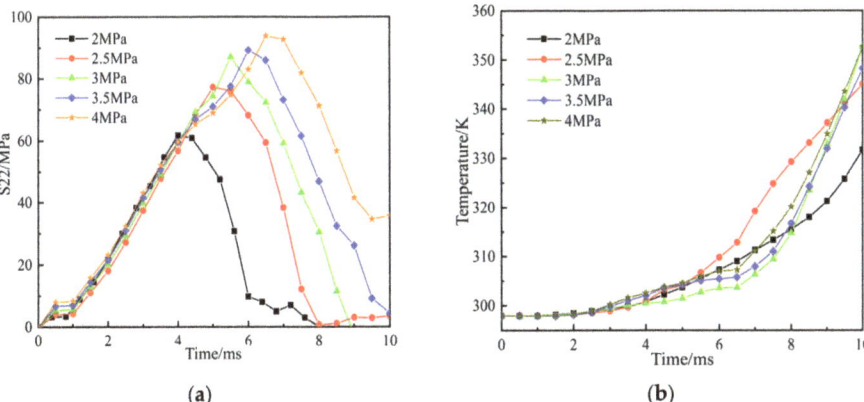

Figure 10. (**a**) Stress components of tablet unit under different hydraulic pressures; and (**b**) temperature rise of tablet unit under different hydraulic pressures.

3.2. Response Analysis of Tablet Mesoscopic Model

3.2.1. Analysis of Micro Ignition Response and Critical Loading Pressure

In order to understand the meso view of the fire process under the action of friction, take the temperature rise cloud diagram of the PBX mesoscopic model under 3.5 MPa and 4 MPa hydraulic pressure to analyze the temperature rise and ignition process are shown in Figure 11. Under 3.5 MPa hydraulic pressure, no explosive particles ignite. Under a loading pressure of 3.5 MPa, the maximum temperature values at 0.75 ms, 2.5 ms, 5 ms, 7 ms, 9 ms, and 10 ms are 298 K, 306 K, 453 K, 579 K, 425 K, and 396 K, respectively. At 0.25 ms, the friction heat at the bottom of the tablet is the main factor affecting the temperature rise. At 7 ms, it can be seen that the heat generated by the bottom friction has almost begun to diffuse into the HMX crystal particles. Due to the interface, the heat conductivity is small, and the heat diffusion is slow. The entire friction process ends at 10 ms, and the heat spreads to the entire explosive area. However, because the temperature accumulation is not obvious enough, and the friction generates little heat, the whole does not ignite.

Figure 11b shows the temperature rise under 4 MPa hydraulic pressure. Under a loading pressure of 4 MPa, the maximum temperature values at 0.75 ms, 2.5 ms, 5 ms, 5.5 ms are 299 K, 355 K, 1253 K, 1630 K. At 5.75 ms and 6 ms, the temperature rose sharply, and ignition occurred. During the sliding process, the distance increases, and the temperature hot spots gradually increase, especially the temperature at the interface between the sliding column and the tablet gradually rises. The temperature rise inside the tablet is mainly concentrated at the interface. The temperature of HTPB at the interface is higher than that of HMX crystal particles. However, as the temperature of HMX rises, the heat-generation subroutine begins. At this time, the self-heating reaction of HMX crystal particles generates a lot of heat, which breaks the original thermal equilibrium state, and the temperature rises sharply. This process can be considered as the process of ignition of HXM particles. It can be seen from the figure that the first smaller temperature rise area starts at 0.25 ms; the local temperature rise points increase with time, and the maximum temperature gradually becomes larger. At 5.5 ms, the temperature distribution cloud map before the critical ignition, the highest the temperature zone is distributed at the microcrack interface. At this time, the corresponding instantaneous temperatures of the matrix and particles are 1654 K and 619 K, respectively. At 5.75 ms, it can be seen that the color of the HMX crystal particles has deepened, indicating that the temperature has risen sharply, and ignition has occurred.

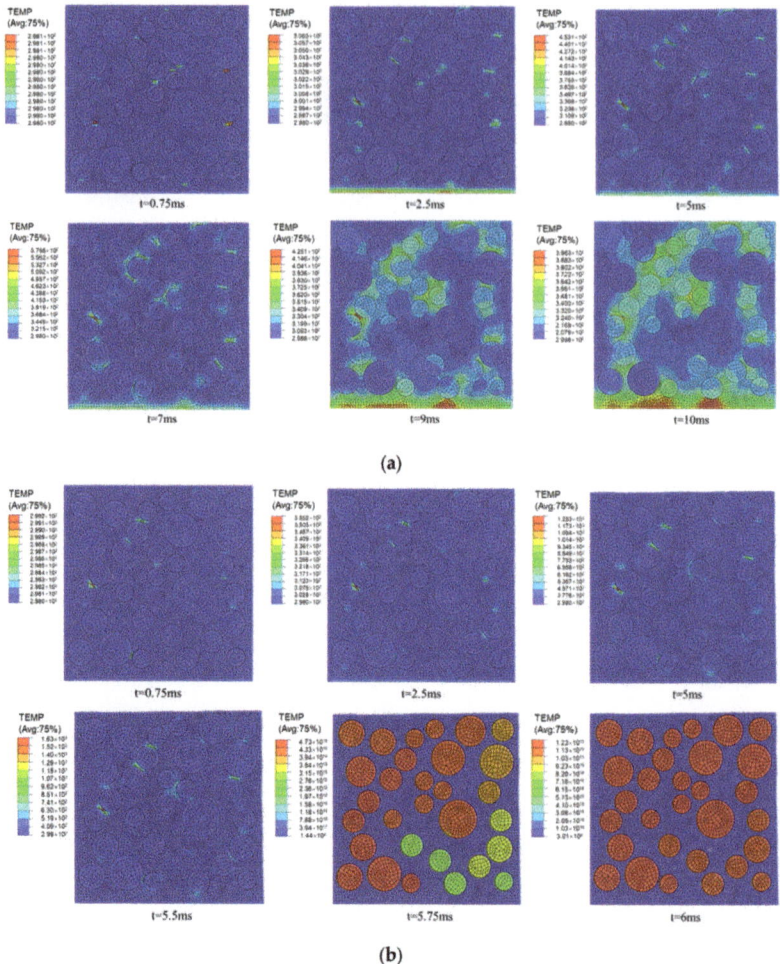

Figure 11. (**a**) Meso-level temperature rise cloud diagram at a loading pressure of 3.5 MPa; (**b**) meso-level temperature rise cloud diagram under a loading pressure of 4 MPa.

The stress distribution inside the PBX tablet is obtained through macro simulation. The tablet under different hydraulic pressures will generate different internal stresses. The stress component at the stress concentration area of the tablet is taken as the boundary load condition of the mesoscopic model, and the stress distribution and the ignition state of the mesoscopic model are analyzed. In order to control irrelevant variables, select the same unit under different pressures for analysis. Figure 12 shows the temperature rise curves of HMX crystal particles under different hydraulic pressures. It can be seen from the figure that with the increase of hydraulic pressure, the maximum temperature rise of HMX crystal particles gradually increases. When the hydraulic pressure is 4 MPa, the temperature of HMX crystal particles rises faster and the heat subroutine is generated. Therefore, it can be judged that the HMX crystal particles are ignited under the hydraulic pressure of 4 MPa. Point A (5.5 ms, 619 K) is the inflection point of temperature rise of explosive particles under 4 MPa hydraulic pressure. The abscissa reflects the time of ignition, and the ordinate is the temperature at the time of ignition. Therefore, it can be judged that the critical ignition hydraulic pressure is between 3.5 MPa and 4 MPa.

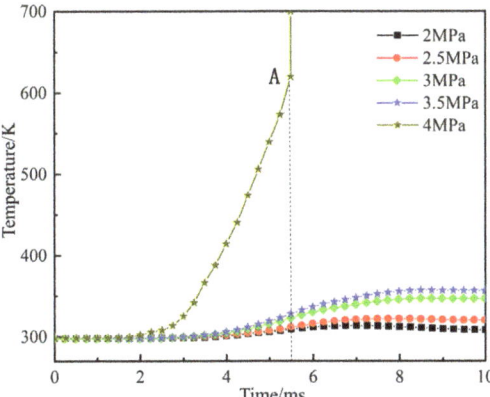

Figure 12. Temperature rises curve of mesoscopic model under different loading pressures.

3.2.2. Influence of Cohesive Interface Friction Coefficient on Ignition

The friction between the HMX crystal particles and the HTPB/Al matrix interface in the cast PBX is the main cause of ignition. At this time, the tablet is subjected to normal stress on the one hand, and frictional shear force on the other hand, under the combined action of pressure and shear force. The stress generated at the interface is relatively concentrated, by adjusting the friction coefficient μ at the cohesive interface to 0.22, 0.27, 0.37 and 0.42, respectively, and the pressure load is set to the minimum ignition pressure 4 MPa to analyze the effect of friction coefficient on ignition.

Figure 13 shows the temperature-rise curves of μ explosive particles with different interfacial friction coefficients. It can be seen from the figure that when the friction coefficient is less than 0.22, no ignition occurs. Where A, B, C, and D are the temperature inflection points when the friction coefficient is 0.27, 0.32, 0.37, 0.42, and the coordinates are A (6.5 ms, 671 K), B (5.5 ms, 620 K), C (5 ms, 704 K), D (4.5 ms, 675.5 K). The abscissa reflects the initial ignition time, and the ordinate reflects the ignition temperature. From the change of the abscissa, it can be judged that the ignition start time shortens with the increase of the friction coefficient. From the change of the ordinate, it can be seen that the friction coefficient has little effect on the initial ignition temperature.

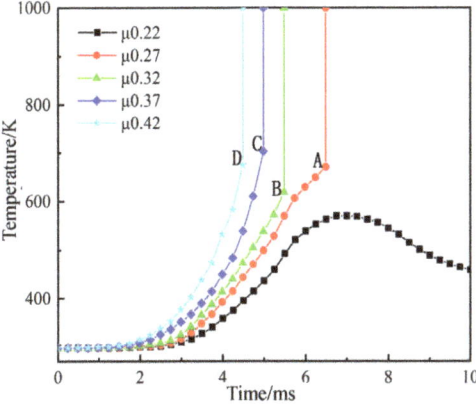

Figure 13. The influence of external friction coefficient on the temperature rise of cast PBX.

The coefficient of friction between the particle and the matrix only affects the ignition time, but has little effect on the ignition temperature. It may be due to the increase in the friction coefficient and the increase in the resistance of the interface slippage, which reduces the sliding displacement, resulting in friction at the interface. The total heat is relatively constant, and the temperature change is relatively small when the total heat is relatively constant.

3.2.3. The Influence of Friction Coefficient between Slide Column and Tablet on Ignition

In the friction-ignition experiment of the cast PBX tablets, the friction between the tablet and the spool generated higher heat. As mentioned above, the maximum temperature rises of HMX crystal particles under the pressure of 2–4 MPa are 49 °C, 62.2 °C, 72.6 °C and 75.6 °C respectively. The main reason why the cast PBX tablet is ignited by impact is the frictional heat of microcracks at the interface between the particles and the substrate. However, in addition to the internal friction of the tablet under the action of friction, the frictional heat between the tablet and the spool cannot be ignored. Based on this analysis of the influence of external friction on the ignition of the PBX, the friction coefficient f was set to 0.05, 0.25, 0.35, 0.45. The friction coefficient f was set to 0.15 in the previous paragraph. The temperature rises under different friction coefficients under 3.5 MPa and 4 MPa hydraulic pressure were analyzed, as shown in Figure 14.

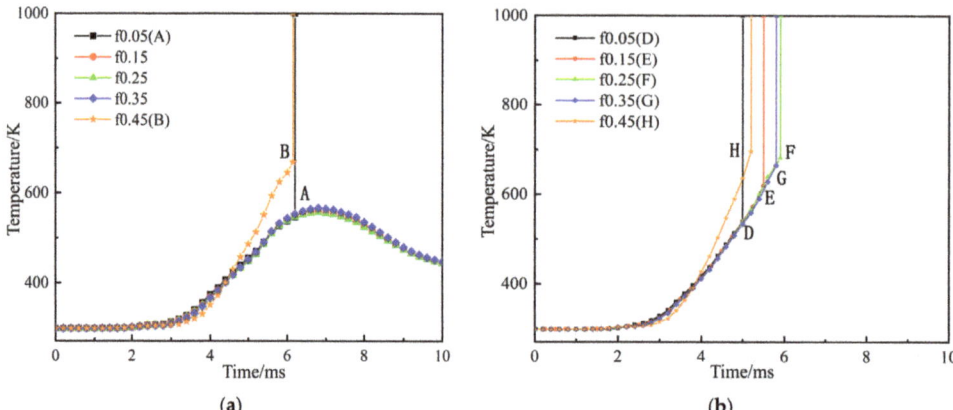

Figure 14. (**a**) The influence of external friction coefficient f on the temperature rise of cast PBX under 3.5 MPa loading pressure; and (**b**) the influence of external friction coefficient f on the temperature rise of cast PBX under 4 MPa loading pressure.

Figure 14a shows the temperature rise curve under 3.5 MPa hydraulic pressure. It can be seen that when the friction coefficient f = 0.05, 0.45, the explosive ignites. But when f = 0.15, 0.25, 0.35, no ignition occurs. The temperature rise curves in the ignition area basically coincide, and the friction coefficient has little effect on the temperature rise of explosive particles. From the analysis of the ignition area, A (6.2 ms, 545 K) and B (6.15 ms, 668 K) are f = 0.05 and 0.45, respectively. The turning point of temperature rise, point A, is suddenly ignited from the original temperature rise, and the ignition does not conform to the general temperature-rise law. It can be judged that the explosive particles of this unit are ignited due to the explosion of other explosive particles, and point B is caused by internal interface friction. The concentrated heat flow triggers the self-heating reaction of the explosive and ignites. Through the above analysis, it can be considered that the friction coefficient f has an effect on the ignition of the cast PBX. However, it does not directly affect the friction heat to affect the ignition of the explosive but affects the ignition by affecting the internal microcrack mechanical behavior.

Figure 14b shows the temperature rise curve at 4 MPa hydraulic pressure, regardless of the friction coefficient, the particles all ignite. Among them, D (5 ms, 539 K), E (5.5 ms, 620 K), F (5.8 ms, 681.5 K), G (5.8 ms, 665 K), H (5.2 ms, 696 K) are the temperature rise turning points corresponding to the friction coefficient. The temperature at the transition point of the temperature rise at point D is 539 K, which is less than the thermal decomposition starting temperature of 554 K. It can be judged that the explosive at this point was not ignited due to crack friction but was smashed. From the coordinates of points E, F, G and H, it is difficult to determine the ignition law. When f is 0.05–0.35, the temperature-rise curves basically overlap, and it can be considered that the friction coefficient has no effect on the temperature rise but has an effect on the ignition.

4. Conclusions

The critical loading pressure of cast PBX tablet was obtained by friction ignition experiment, and the device was designed to analyze the actual friction rate of the tablet. Then, based on the thermal–mechanical coupling algorithm, the friction ignition process of cast PBX was numerically simulated at the macro and micro scale. The critical conditions of the friction ignition of cast PBX were analyzed by numerical models at different scales. The results of the numerical simulations were consistent with the experimental results, which shows that the numerical simulation is feasible. The main conclusions are as follows:

(1) Friction-ignition experiments were conducted on the HMX-based cast PBX at a 90° swing angle, and the critical ignition loading pressure of 3.7 MPa was determined by the Bruceton method for frictional sensitivity test;

(2) The stress distribution and temperature distribution of the cast PBX tablet under external friction can be obtained by the macro-numerical simulation. It is difficult to draw the conclusion of ignition only by analyzing the whole friction temperature-rise process of the tablet from macro scale;

(3) Based on the morphology characterization results and mesoscopic model simulation of the cast PBX tablets, the temperature rises of HXM particles in PBX tablets under different hydraulic pressures were analyzed, and the ignition loading pressure was determined by ignition criterion. The critical ignition loading pressures determined in this paper is between 3.5 MPa and 4 MPa, which are basically consistent with the experimental critical ignition pressure of 3.7 MPa. The results of the critical ignition loading pressure by numerical simulation were in good agreement with those of the experimental test;

(4) The effects of the cohesive interface friction coefficient μ on the friction ignition of the cast PBX tablet were analyzed. The results showed that the effects of the friction coefficient μ on ignition of cast PBX tablet are obvious. When μ the larger the size of HMX crystal particles, the greater the temperature rise and the greater the possibility of ignition. At the same time, the influences of the friction coefficient f between the slide column and tablet on ignition were analyzed. The results showed that friction coefficient f has an influence on ignition response. The influence does not affect the ignition process by increasing friction heat but can adjusts the ignition result by changing the mechanical conditions inside the cast PBX tablet.

Author Contributions: Conceptualization, J.Y.; methodology, J.Y.; validation, H.P., T.X. and J.L.; formal analysis, W.Z.; data curation, J.Y.; writing—original draft preparation, J.Y. and T.X.; writing—review and editing, J.Y. and Y.Q.; visualization, H.S. and Y.L. All authors have read and agreed to the published version of the manuscript.

Funding: This research was funded by Robust Munitions Center, CAEP (No. RMC2014B03), the Bottleneck Technology and JCJQ Foundation (No. 20210579) and Special Projects of Energetic Materials (No. 20221206).

Data Availability Statement: The data presented in this study are openly available.

Conflicts of Interest: The authors declare no conflict of interest.

Nomenclature

PBX	Polymer-Bonded Explosive
HMX	Cyclotetramethylene tetranitramine
HTPB	Hydroxy terminated polybutadiene
DOA	Dioctyl adipate
TDI	Toluene diisocyanate
τ_i	Time
G_0	Instantaneous relaxation shear modulus
ρ	Density
λ	Heat conductivity
C_v	Specific heat capacity
q_m	Heat released by the complete decomposition of the unit mass explosive
g_i	Relative modulus of the i-th term
υ	Poisson's ratio
E	Young modulus
E_t	Plastic hardening rate
E_a	Activation energy
σ_0	Yield stress
A	Minimum hydraulic pressure loaded by friction sensitivity instrument for initiation
B	Set loading pressure interval
Z	Pre-exponential factor
α	Constant
β	Constant
K	Stiffness matrix
T_{max}	Maximum traction force allowed
δ_0	Separation amount of the interface during the initial damage
μ	Coefficient of friction
K_c	Heat distribution coefficient between two objects

References

1. Gwak, M.-C.; Jung, T.-Y.; Yoh, J.J.-I. Friction-induced ignition modeling of energetic materials. *J. Mech. Sci. Technol.* **2009**, *23*, 1779–1787. [CrossRef]
2. Sun, B.P.; Duan, Z.P.; Pi, A.G.; Ou, Z.C.; Huang, F.L. Approximate analysis on temperature rise in charge explosive during projectile penetration. *Explos. Shock. Waves* **2012**, *32*, 225–230.
3. Deng, C.; Shen, C.Y.; Fan, X.; Xiang, Y. Test on the Friction Sensitivity of PBX Tablet. *Chin. J. Explos. Propellants* **2012**, *35*, 22–24.
4. Charlery, R.; Saulot, A.; Daly, N.; Berthier, Y. Tribological conditions leading to ignition phenomena of energetic materials. In Proceedings of the Leeds Lyon Symposium on Tribology and Tribochemistry Forum, Lyon, France, 4–6 September 2013.
5. Dai, X.G.; Zhong, M.; Deng, C.; Zheng, X.; Wen, Y.S.; Huang, F.L. Reaction Characteristics of PBX Tablet in Friction Sensitivity Test. *Chin. J. Energ. Mater.* **2015**, *23*, 5.
6. Zeman, S.; Jungová, M. Sensitivity and performance of energetic materials. *Propellants Explos. Pyrot.* **2016**, *41*, 426–451. [CrossRef]
7. Luo, Y.; Liu, Y.; Huang, F.L. Measurement and Modification of Dynamic Friction Coefficient of Pressed Explosive. *Acta Armamentarii* **2017**, *38*, 1926–1932.
8. Dienes, J.K. Frictional hot-spots and propellant sensitivity. *MRS Online Proc. Libr.* **1983**, *24*, 373–381. [CrossRef]
9. Randolph, A.D.; Hatler, L.E.; Popolato, A. Rapid heating-to-ignition of high explosives. I. Friction heating. *Ind. Eng. Chem. Fundam.* **1976**, *15*, 1–6. [CrossRef]
10. Andersen, W.H. Role of the friction coefficient in the frictional heating ignition of explosives. *Propellants Explos. Pyrot.* **1981**, *6*, 17–23. [CrossRef]
11. An, Q.; Zybin, S.V.; Goddard, W.A., III; Jaramillo-Botero, A.; Blanco, M.; Luo, S.N. Elucidation of the dynamics for hot-spot initiation at nonuniform interfaces of highly shocked material. *Phys. Rev.* **2011**, *84*, 220101. [CrossRef]
12. Cai, Y.; Zhao, F.P.; An, Q.; Wu, H.A.; Goddard, W.A.; Luo, S.N. Shock response of single crystal and nanocrystalline pentaerythritol tetranitrate: Implications to hotspot formation in energetic materials. *J. Chem. Phys.* **2013**, *139*, 164704. [CrossRef] [PubMed]
13. Keshavarz, M.H.; Hayati, M.; Ghariban-Lavasani, S.; Zohari, N. Relationship between activation energy of thermolysis and friction sensitivity of cyclic and acyclic nitramines. *Z. Für Anorg. Und Allg. Chem.* **2016**, *642*, 182–188. [CrossRef]
14. Bouma, R.H.B.; van der Heijden, A.E.D.M. The Effect of RDX Crystal Defect Structure on Mechanical Response of a Polymer-Bonded Explosive. *Propellants Explo. Pyrot.* **2016**, *41*, 484–493. [CrossRef]
15. Hu, R.; Prakash, C.; Tomar, V.; Harr, M.; Gunduz, I.D.; Oskay, C. Experimentally-validated mesoscale modeling of the coupled mechanical–thermal response of AP–HTPB energetic material under dynamic loadin. *Int. J. Fract.* **2017**, *203*, 277–298. [CrossRef]

16. Jafari, M.; Keshavarz, M.H.; Joudaki, F.; Mousaviazar, A. A Simple Method for Predicting Friction Sensitivity of Quaternary Ammonium-Based Energetic Ionic Liquids. *Propellants Explos. Pyrot.* **2018**, *43*, 568–573. [CrossRef]

17. Gruau, C.; Picart, D.; Belmas, R.; Bouton, E.; Delmaire-Sizes, F.; Sabatier, J.; Trumel, H. Ignition of a confined high explosive under low velocity impact. *Int. J. Impact Eng.* **2009**, *36*, 537–550. [CrossRef]

18. Wu, Y.Q.; Huang, F.L. A microscopic model for predicting hot-spot ignition of granular energetic crystals in response to drop-weight impacts. *Mech. Mater.* **2011**, *43*, 835–852. [CrossRef]

19. Bennett, J.G.; Haberman, K.S.; Johnson, J.N.; Asay, B.W. A constitutive model for the non-shock ignition and mechanical response of high explosives. *J. Mech. Phys. Solids* **1998**, *46*, 2303–2322. [CrossRef]

20. Xue, H.J.; Wu, Y.Q.; Yang, K.; Wu, Y. Microcrack-and microvoid-related impact damage and ignition responses for HMX-based polymer-bonded explosives at high temperature. *Def. Technol.* **2022**, *18*, 1602–1621. [CrossRef]

21. Bai, Z.X.; Liu, Q.J.; Liu, F.S.; Jiang, C.L. Three-dimensional discrete element method to simulate the ignition and combustion of HMX explosives under hot spots. *Powder Technol.* **2022**, *412*, 118014. [CrossRef]

22. Li, H.T.; Sun, J.; Sui, H.L.; Chai, C.G.; Li, B.H.; Yu, J.X. Influencing mechanisms of a wax layer on the micro-friction behavior of the β-HMX crystal surface. *Energetic Mater. Front.* **2022**, *3*, 248–256. [CrossRef]

23. Yin, Y.; Li, H.T.; Cao, Z.H.; Li, B.H.; Li, Q.S.; He, H.T.; Yu, J.X. Crystallographic orientation dependence on nanoscale friction behavior of energetic β-HMX crystal. *Friction* **2023**, *10*, 1–14. [CrossRef]

24. Duarte, C.A.; Koslowski, M. Hot-spots in polycrystalline β-tetramethylene tetranitramine (β-HMX): The role of plasticity and friction. *J. Mech. Phys. Solids* **2023**, *171*, 105157. [CrossRef]

25. Barua, A.; Kim, S.; Horie, Y.; Zhou, M. Ignition criterion for heterogeneous energetic materials based on hotspot size-temperature threshold. *J. Appl. Phys.* **2013**, *113*, 064906.1–064906.22. [CrossRef]

26. Barua, A.; Kim, S.; Horie, Y.; Min, Z. Computational Analysis of Ignition in Heterogeneous Energetic Materials. *Mater. Sci. Forum* **2014**, *767*, 13–21. [CrossRef]

27. Kim, S.; Barua, A.; Horie, Y.; Min, Z. Ignition probability of polymer-bonded explosives accounting for multiple sources of material stochasticity. *J. Appl. Phys.* **2014**, *115*, 27–33. [CrossRef]

28. Barua, A.; Kim, S.; Horie, Y.; Min, Z. Prediction of Probabilistic Ignition behavior of polymer-bonded explosives from microstructural stochasticity. *J. Appl. Phys.* **2013**, *113*, 537–550. [CrossRef]

29. Keyhani, A.; Horie, Y.; Zhou, M. Relative importance of plasticity and fracture/friction in ignition of polymer-bonded explosives (PBXs). In Proceedings of the 20th Biennial Conference of the APS Topical Group on Shock Compression of Condensed Matter, St. Louis, MO, USA, 14 July 2017.

30. Vadhe, P.P.; Pawar, R.B.; Sinha, R.K.; Asthana, S.N.; Rao, A.S. Cast aluminized explosives. *Combust. Explos. Shock. Waves* **2008**, *44*, 461–477. [CrossRef]

31. Chen, H.T. *QJ 2913-1997. Test Method for Friction Sensitivity of Composite Solid Propellants*; Industry Standard-Aerospace: Beijing, China, 1997.

32. Che, T.Z.; Jiang, W.Q.; Zheng, Z.D.; Xu, Y.F.; Wang, B.L. *QJ 3039-1998. Test Method for Drop Weight Impact Sensitivity of Composite Solid Propellants*; Industry Standard-Aerospace: Beijing, China, 1997.

33. Chen, J.; Zeng, X.G.; Chen, H.Y. *Viscoelastic Mechanics*; Sichuan University Press: Chengdu, China, 2016; pp. 78–88.

34. Mas, E.M.; Clements, B.E.; Blumenthal, B.; Cady, C.M.; Gray, G.T.; Liu, C. A viscoelastic model for PBX binders. *AIP Conf. Proc.* **2002**, *620*, 661–664.

35. Simulia, D.S.; Fallis, A.G. ABAQUS documentation. *Mendeley* **2013**, *53*, 1689–1699.

36. Amirkhizi, A.V.; Isaacs, J.; McGee, J.; Nasser-Nemat, S. An experimentally-based viscoelastic constitutive model for polyurea, including pressure and temperature effects. *Philos. Mag.* **2006**, *86*, 5847–5866. [CrossRef]

37. Tarver, C.M.; Tran, T.D. Thermal decomposition models for HMX-based plastic bonded explosives. *Combust. Flame* **2004**, *137*, 50–62. [CrossRef]

38. Hanson-Parr, D.M.; Parr, T.P. Thermal properties measurements of solid rocket propellant oxidizers and binder materials as a function of temperature. *J. Energ. Mater.* **1999**, *17*, 1–48. [CrossRef]

39. Wu, Y.Q.; Huang, F.L. A micromechanical model for predicting combined damage of particles and interface debonding in PBX explosives. *Mech. Mater.* **2009**, *41*, 27–47.

40. Liu, Q.; Chen, L.; Wu, J.Y.; Wang, C. Two dimensional numerical simulation of shock ignition of PBX explosive mesostructure. *Chin. J. Explos. Propellants* **2011**, *34*, 10–16.

41. Kang, G.; Chen, P.W.; Zeng, Y.L.; Ning, Y.J. A Method of Generating Mesoscopic Models for PBXs with High Particle Volume Fraction. *Chin. J. Energ. Mater.* **2018**, *26*, 772–778.

42. Zhang, X.Z.; Huang, W.; Liu, Q.G. *Heat Transfer*; National Defense Industry Press: Beijing, China, 2011; pp. 9–11.

Article

Study and Design of the Mitigation Structure of a Shell PBX Charge under Thermal Stimulation

Jiahao Liang [1], Jianxin Nie [1,*], Rui Liu [1], Ming Han [2], Gangling Jiao [3], Xiaole Sun [4], Xiaoju Wang [5] and Bo Huang [5]

1   State Key Laboratory of Explosion Science and Technology, Beijing Institute of Technology, Beijing 100081, China
2   The Eighth Military Representative Office of Air Force Equipment Ministry, Xi'an 710000, China
3   Naval Institute, Beijing 100161, China
4   Chongqing Hongyu Precision Industry Co., Ltd., Chongqing 402760, China
5   Xi'an Changfeng Research Institute of Mechanical-Electrical, Xi'an 710065, China
*   Correspondence: niejx@bit.edu.cn

Abstract: To study the design method and pressure relief effect of the mitigation structure of a shell under the action of thermal stimulation, a systematic research method of theoretical calculation-simulation-experimental verification of the mitigation structure was established. Taking the shelled PBX charge as the test material, the pressure relief area that can effectively reduce the reaction intensity of the charge is obtained by theoretical calculation. The influence of the pressure relief hole area, distribution mode, and other factors on the pressure relief effect is calculated by simulation. The pressure relief effect of the mitigation structure was verified by the low-melting alloy plug with refined crystal structure for sealing the pressure relief hole and the cook-off test. The research results show that the critical pressure relief area is when the ratio of the area of the pressure relief hole to the surface area of the charge is $A_V/S_B = 0.0189$. When the number of openings increases to 6, the required pressure relief coefficient decreases to $A_V/S_B = 0.0110$; When the length/diameter ratio is greater than 5, the opening at one end cannot satisfy the reliable pressure relief of the shell. The designed low-melting-point alloy mitigation structure can form an effective pressure relief channel. With the increase in A_V/S_B from 0.0045 to 0.0180, the reaction intensity of the cook-off bomb is significantly reduced in both fast and slow cook-off, which improves the safety of the charge when subjected to unexpected thermal stimulation.

Keywords: mitigation structure; pressure relief area; pressure relief effect; cook-off; low-melting crystal

Citation: Liang, J.; Nie, J.; Liu, R.; Han, M.; Jiao, G.; Sun, X.; Wang, X.; Huang, B. Study and Design of the Mitigation Structure of a Shell PBX Charge under Thermal Stimulation. *Crystals* **2023**, *13*, 914. https://doi.org/10.3390/cryst13060914

Academic Editor: Pavel Lukáč

Received: 5 May 2023
Revised: 30 May 2023
Accepted: 2 June 2023
Published: 5 June 2023

1. Introduction

During the process of production, transportation, storage, and usage, ammunition could be stimulated by unexpected sources, such as fire, which will lead to violent reactions and bring about major safety accidents. Accidental thermal stimulation is one of the most common accidental excitable sources encountered in the whole life cycle of ammunition. The charge ignites and burns under thermal stimulation. Huge personnel and economic losses would occur when the shell is sealed because the temperature and pressure will rise rapidly in the confined space of the charge, resulting in a chain reaction from combustion to deflation to detonation. Aiming at the thermal safety of ammunition, the improvement methods mainly include insensitive explosive and charging technology, shell pressure relief technology, thermal shock buffer technology, and other aspects [1]. The principle of shell pressure relief technology is that when ammunition is under a certain thermal stimulation condition, a pressure relief channel is formed through the mitigated structure on the shell to relieve the internal pressure of the ammunition, the self-heating reaction rate of the charge is suddenly reduced, so the severity of charge response is reduced. The decrease in the internal temperature, the decrease in the natural reaction rate, and the convection driven by the product bubbles collectively lead to a delay in the ignition time [2]; these factors reduce the reaction intensity of the charge, achieving the purpose of improving the safety performance of the assembly.

Research on the design of mitigation structures had already begun when the United States and European countries developed insensitive ammunition in the last century. After considerable technical accumulation, some mitigation structure design technologies have been successfully applied to model ammunition [3–7]. Therefore, the pressure relief technology of designing vent holes on the shell is an effective control method to improve the thermal safety of ammunition. William et al. [8] studied the response characteristics of explosives through a fast cook-off test and measured the internal pressure of ammunition through a pressure sensor. The results show that a vent hole can effectively reduce the response intensity of the ammunition under the condition of fast cooking. Glascoe et al. [9] conducted a slow cook-off test equipped with a pressure relief hole, compared a molten Composition-B explosive with an HMX-based agglomerated explosive, and found that the size of the molten cast explosive pressure relief hole was too small, which may improve the response intensity of the ammunition. The gas pressure inside the condensed explosive cannot be discharged in the form of bubbles, and the pressure relief hole has almost no effect. Kinney [10] calculated the critical area of the pressure relief channel according to the dynamic equilibrium relationship between the pressure increase rate when the charge facilitated the combustion reaction and the pressure decrease rate when the pressure relief channel was relieved. Hakan et al. [11] calculated the critical pressure relief area of the pressure relief channel according to the dynamic transport equilibrium relationship between the mass of the gas generated during the combustion of the charge and the mass of the gas discharged from the channel. Wardel et al. [12] set up two different venting methods to leak the gas inside the explosive from the top center and the top edge and found that the cook-off bomb with the vent set at the top center had a more severe response. Niu Gongjie [13] studied the influence of different distribution modes of pressure relief holes on the dose-response intensity and pointed out that the pressure relief hole should be set near the ignition position of the charge. Madsen et al. [14] studied the cook-off characteristics of four explosives, including molten Composition-B explosive, PAX-28, PBXN-109, and PBXN-9, under different vent hole sizes by scale testing and analyzed the selection of a low-melting-point material for plugging the pressure relief hole.

However, the abovementioned studies were only verified through theoretical calculations or experiments. The area of the pressure relief channel, the size of the pressure relief hole, or the feasibility of the low-melting-point material as a mitigation structure were not considered comprehensively, and the actual shell installation was not considered. It is difficult to provide a complete basis for the design of the mitigation structure of a shell without forming a systematic research method of the theoretical calculation-simulation-experimental verification of the mitigation structure.

Therefore, in this study, the pressure relief area and the distribution of pressure relief holes through the systematic design method of the mitigation structure was investigated, the appropriate filling material was selected, and the pressure relief effect under different pressure relief conditions and shell failure strength thresholds were analyzed by means of simulation and experimentation. Taking a polymer bonded explosive (PBX) charge as the research object, slow cook-off, and fast cook-off tests were used for verification, which can provide a reference for the design of insensitive munitions mitigation structures.

2. Shell Mitigation Structure Design

The key to the design of the mitigation structure is determining the area of the pressure relief channel, the arrangement of the pressure relief holes, and the selection of the filling material for the pressure relief holes. The material filling the pressure hole should not affect the use under normal working conditions so that it cannot only perform a sufficient role in exhaust and pressure relief but also meet the requirements of the structural strength of the shell.

Taking a standard cook-off bomb as an example, the shell mitigation structure was designed. The materials of the shell and end cover are #45 steel. The shell is 240 mm long, 60 mm in inner diameter, 66 mm in outer diameter, and 3 mm in thickness, and both ends of the body were machined with 45 mm external threads. The thickness of the end cover is 5 mm, the diameter is 73 mm, the internal thread is processed, the pitch is 1.5 mm, the charge is Φ60 mm × 240 mm, the length-diameter ratio is 4:1, the PBX pressed charge is filled, its content component is RDX/Al/Viton F2602 is 65.5/30/4.5, and the charge density is 1927 kg m^{-3}. Among them, the RDX used is Class II, 0.075~0.300 mm. Al powder is 1–2 μm.

2.1. Pressure Relief Area Design

The pressure relief area of the mitigation structure needs to be designed according to different conditions to ensure the effect of the mitigation structure. The determination of the pressure relief area mainly considers the relationship between the pressure increase rate in the shell and the pressure release rate of the pressure relief passage. According to the research of Kinney [10], the pressure growth rate of the charge in the shell when burning is:

$$\frac{dp}{dt} = \frac{RT_B}{V}\frac{dn}{dt} = \frac{RT_B}{V}\frac{\rho}{M}\frac{\alpha}{(A-BT_0)}S_B P \tag{1}$$

where T_B (°C) is the flame temperature when the charge burns; R is the universal gas constant, which is 8.314 J·mol^{-1}·°C^{-1}; V is the volume (m^3); ρ is the density of the charge (kg m^{-3}); M is the average molar mass of the gas molecules during combustion (kg mol^{-1}); T_0 is the temperature of the charge at ignition (°C); S_B is the surface area of the charge (m^2); P is the absolute charge pressure (bar); and α, A, and B are constants for the charge burning rate and temperature.

The pressure release rate of the pressure relief channel is calculated by the following formula:

$$-\frac{dp}{dt} = \frac{A_V C_D}{V}a'P \tag{2}$$

where A_V is the area of the pressure relief hole (m^2); C_D is the exhaust coefficient, which is taken as 0.82 [15]; and a' is the speed of the air passing through the air hole (m/s).

Combining Equations (1) and (2), when the pressure increase rate in the shell and the pressure release rate of the pressure relief channel are equal, the calculation formula of the critical area of the pressure relief channel can be obtained as:

$$\frac{A_V}{S_B} = \frac{\alpha\rho RT_B}{MC_D a'(A-BT_0)} \tag{3}$$

Equation (3) shows that the ratio of the area of the exhaust passage to the combustion area of the charge can be directly obtained from the relevant physical and chemical parameters of the charge, and the critical pressure relief area can be determined.

The molar mass of the gaseous product of the elemental explosive RDX ($C_3H_6N_6O_6$) in the PBX charge is [16]:

$$M = \frac{56c + 88d - 8b}{2c + 2d + b} = \frac{27.2g}{mol} \tag{4}$$

According to ref. [17] of the burning rate parameter:

$$\frac{1}{r} = A - BT_0 \tag{5}$$

In the formula, r is the burning speed of the charge. The other parameters were calibrated from previous studies [18] and experiments and the size of the pressure relief hole required for the ignition of PBX charges at different temperatures was calculated, as shown in Table 1.

Table 1. The size of the pressure relief holes required for the ignition of PBX explosives at different temperatures.

Temperature/°C	A_V/S_B
160 [19]	0.0100
177 [19]	0.0128
237 *	0.0189

Note: * is the ignition temperature obtained from the preliminary test.

Therefore, for a PBX charge with a charge size of $\Phi60 \times 240$ mm, the minimum A_V/S_B is 0.0189 when ignited at a temperature of 237 °C.

2.2. Design of the Pressure Relief Hole Distribution

At present, most of the pressure relief holes are designed to be distributed in the tail of the projectile, the wall of the projectile, etc. The distribution mode has an important influence on the charge reaction level, and the design of the pressure relief channel near the ignition position has a significant effect.

A numerical calculation model was established based on the cook-off experiment and the data in the literature [20], and the simulation model was a standard cook-off bomb of $\Phi60 \times 240$ mm. CFD fluent software was used to calculate the response position of the PBX charge at heating rates of 3.3 °C/h (0.055 °C/min), 0.1 °C/min, 0.5 °C/min, 1 °C/min, and 3.3 °C/min. The position distribution of the pressing holes provides the design basis. Figure 1 shows the position distribution of the ignition time. Figure 2 shows the changes in ignition position under different heating rates. When the charge material and the charge structure size are determined, the ignition position is mainly affected by the heating rate. With the acceleration of the heating rate, the reaction position first moved from the center of the charge to the two ends along the axis, then moved to the corner of the cylindrical section and the end cap.

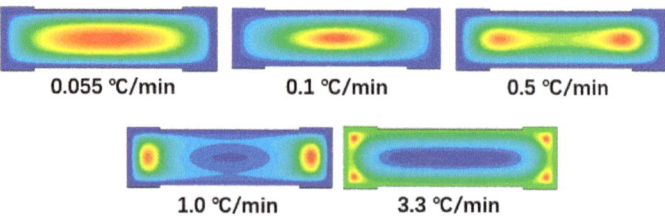

Figure 1. Ignition positions at different heating rates.

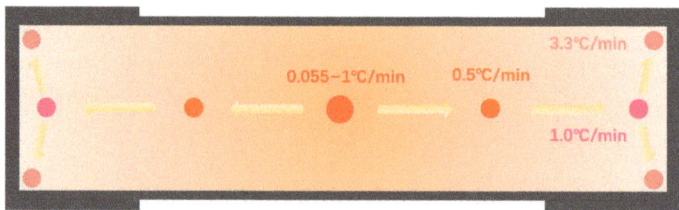

Figure 2. Schematic diagram of the ignition position change.

Since the thermal decomposition temperature of the charge is constant when the charge reaches the decomposition temperature, the charge begins to decompose, and the released heat is transferred to the low-temperature charge and the outside of the shell. When the heating rate is low, there is more time for slow thermal decomposition to occur. At the same time, because the heat of charge decomposition is greater than the heat provided by the heating of the shell, the inward transfer of heat will lead to the decomposition of the internal charge. In contrast, the outward transfer is relatively smooth, and the temperature increase process in the shell is mainly caused by the heat of the charge decomposition control, resulting in the continuous movement of the reaction center to the charge center. When the temperature of the reaction center reaches the ignition temperature, ignition occurs, and the reaction center at this moment becomes the ignition position. When the heating rate is fast, ignition occurs before the high-temperature point is transferred to the center of the charge. The pressure relief channel is designed near the ignition position, which is conducive to the timely dissipation of the heat generated by the decomposition and reduces the temperature of the charge.

Therefore, for the PBX charge with a charge size of $\Phi 60 \times 240$ mm, the pressure relief hole design can be considered at one or both ends close to the ignition position.

2.3. Design of Pressure Relief Hole Filling Material Design

Low-melting alloys are heat-sensitive, and their mechanical strength decreases at high temperatures. Under a certain ambient temperature and internal pressure, the mitigation structure of low-melting alloys are disabled and destroyed. These alloys can be used in the starting device of mitigation structures. Design requirements for mitigation structures of low-melting alloy plugs are as follows [21]:

(1) These structures have sufficient strength under storage and transportation conditions and do not affect the normal use requirements of ammunition.

(2) When the ammunition is subjected to unexpected thermal stimulation, such as high temperature, before reaching the lowest ignition point or explosion point of the ammunition, the mitigation structure must lose most of the strength, and the exhaust channel must be opened to reduce the reaction strength of the ammunition and improve the safety.

The slow cook-off response temperature of the PBX charge (RDX/Al/binder mass fraction of 65.5/30/4.5) used in this paper was obtained from the preliminary test of approximately 237 °C. Strickland et al. [22] pointed out that to effectively suppress the deflagration to detonation transition of energetic materials, the opening temperature of the pressure relief channel should be more than 60 °C lower than the slow cook-off response temperature; that is, the opening temperature of the pressure relief channel is approximately 177 °C. Therefore, the melting point of the low-melting-point alloy for the mitigation structures of the shell should be below the slow cook-off response temperature of the charge, and the mechanical strength of the low-melting alloy material is significantly reduced at a temperature of 177 °C. The structure is destroyed in this situation of internal pressure to open the pressure relief channel.

The melting point of the Sn-Zn binary alloy system is 198.5 °C, which meets the design requirements of mitigation structures. At the same time, the low-melting-point alloy must meet the requirements of temperature environment adaptability; that is, it must have good mechanical properties from −50~70 °C. Therefore, the Sn9Zn-3Al0.2La low-melting-point alloy with good mechanical properties was selected as the filling material of the pressure relief channel [23], and its mechanical properties are shown in Table 2. The addition of La element is to refine its crystal structure and improve its mechanical properties.

Table 2. Mechanical properties of Sn9Zn-3Al0.2La.

Alloy	High and Low Temperature Mechanical Properties/MPa					Melting Point/°C
	−50 °C	25 °C	70 °C	125 °C	175 °C	
Sn9Zn-3Al0.2La	133.61	67.18	60.4	31.59	9	205.6

3. Analysis of Factors Affecting the Pressure Relief Effect

3.1. Simulation Model

The pressure relief process of the shell mitigation structure is a competitive process of unburned charge combustion and gas release, and different mitigation structures (such as the length/diameter ratio of the shell, the number of pressure relief holes, and the location) have a great influence on the pressure relief of confined spaces. The explosion process time of the charge is very short, the internal pressure changes greatly, and the explosion process is dangerous. The experimental research will be limited by the conditions of the site, testing methods, and safety. The simulation calculation can easily change the conditions, such as different mitigation structures, and a comprehensive analysis of the internal pressure of the shell can be conducted. The commonly used commercial software ANSYS Fluent was used to simulate the pressure relief process of the shelled charge under three-dimensional conditions. The purpose is to obtain the internal pressure changes in the shell during the pressure relief process in different sustained-release structural conditions and provide data information support for the design of mitigation structures.

The inlet mass source term can be defined as the product of the total reaction volume A_{burn}, the explosive burning velocity r_{burn}, and the combustion product density ρ_{burn} in the shell [11]:

$$\dot{m}_{\text{inlet}} = A_{burn}r_{burn}\rho_{burn} \tag{6}$$

The simple expression for the mass outflow after opening the pressure relief channel is:

$$\dot{m}_{\text{outlet}} = A_{vent}u^*\rho^*C_D \tag{7}$$

In the formula, C_D is the exhaust coefficient, which is taken as 0.82 [15], and the other terms are solved by the following isentropic equations:

$$\rho^* = \frac{P_{vent}}{RT_{vent}} \tag{8}$$

$$u^* = \sqrt{kRT_{vent}M} \tag{9}$$

$$P^* = \frac{P_{chamber}}{\left[1 + \frac{k-1}{2}M^2\right]^{\frac{k}{k-1}}} \tag{10}$$

$$T^* = \frac{T_{chamber}}{\left[1 + \frac{k-1}{2}M^2\right]^{\frac{k}{k-1}}} \tag{11}$$

A schematic diagram of the description of the exhaust gas pressure relief process of the projectile is shown in Figure 3. The pressure-rising stage of the charge ignition stage is used as the input condition, and the UDF is loaded into the software to simulate the pressure-rising process. The pressure rise process caused by the combustion of the charge is calibrated with reference to the test results [24] (the pressure curve is shown in Figure 4). Test method: During the cook-off test, a piezoelectric pressure sensor is used to measure the pressure time-history curve inside the cook-off bombshell.

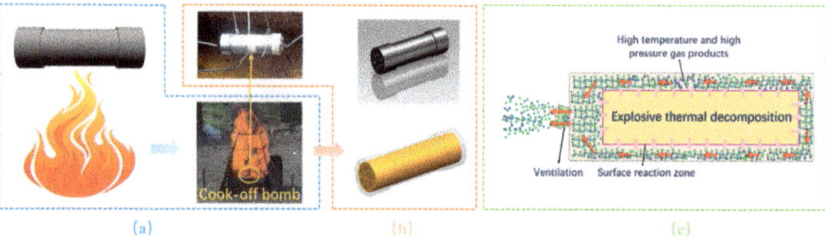

(a)   (b)   (c)

Figure 3. Schematic diagram of the exhaust pressure relief process of the projectile. (**a**) suffer from thermal stimulation; (**b**) cook-off bomb; (**c**) thermal decomposition diagram.

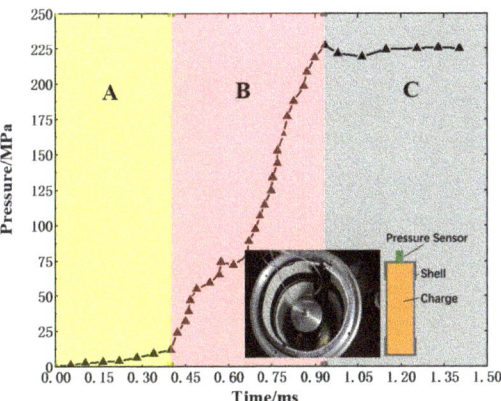

Figure 4. Pressure rise curve during charge ignition. (A—thermal decomposition; B—explosive combustion; C—shell rupture).

3.2. Simulation Results

3.2.1. Influence of the Number of Pressure Relief Holes

Figure 5 shows the variation in pressure with time when there is no mitigation structure and the critical dimension $A_V/S_B = 0.0189$. The different colored lines represent the pressure curves of different monitoring points, and the yellow stars represent the positions of the monitoring points. The curve with a sharp rise in pressure is the pressure curve designed without the mitigation structure, and monitoring points 1–7 are the critical pressure relief area pressure curve when the pressure relief channel is opened. The pressure rises when the charge starts to respond quickly is recorded as 0 times. It can be seen from the pressure curve without the mitigation structure that when the pressure relief channel is not opened, the pressure rises exponentially according to the set pressure rise rate. At 0.45 ms, the pressure reached 20.8 MPa, and finally exceeded the shell burst pressure and exploded. When the pressure relief channel is opened, when the breaking pressure of the low-melting-point alloy plug is 20.8 MPa, the pressures of monitoring points 1 and 2 near the pressure relief channel fluctuate rapidly, then fluctuate and rise and finally stabilize at 7.71 MPa and 17.08 MPa at 6 ms, respectively. Due to the pressure hysteresis effect, monitoring point 7 away from the pressure relief channel will first rise and reach a maximum value of 28.05 MPa at 1.07 ms, then oscillate and decrease. Then, the pressure in the shell is gradually decreased, then fluctuates. At approximately 4 ms, the pressure remains basically unchanged and stabilizes at 19.79 MPa. The pressure at other monitoring points has the same trend as monitoring point 7. The pressure first rises to a maximum value, then oscillates down due to the hysteresis effect over time, and the pressure at the last stable point is below 20 MPa, at this time, all areas in the shell exhibit equilibrium pressure relief. The timely opening of the pressure relief channel can effectively reduce the pressure in the shell, make the charge undergo a relatively stable combustion reaction, and reduce the probability of a more severe reaction. Since the pressure relief channel is a type of pressure opening, and the pressure at monitoring point 7, far from the pressure relief channel, is the innermost pressure point. If the equilibrium pressure of monitoring point 7 meets the design requirements, then other positions in the shell will also meet the design requirements. In the subsequent research and analysis of the pressure relief effect, only the pressure curve at monitoring point 7 away from the pressure relief channel was analyzed.

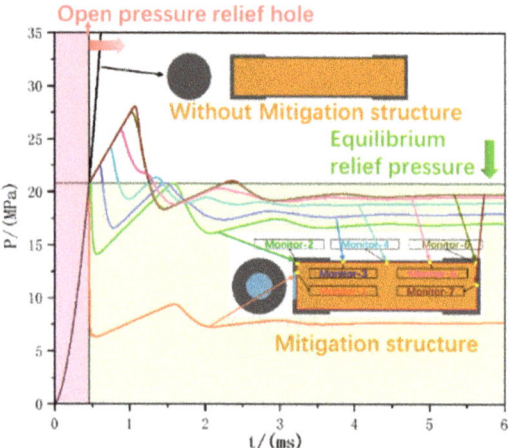

Figure 5. Pressure graph of a single pressure relief hole.

The influence of different numbers of relief holes (the number of relief holes is 1 to 6) on the pressure relief effect was studied, and the relief area at the same equilibrium pressure is shown in Figure 6. When the number of relief holes is 2, $A_V/S_B = 0.0144$ can achieve equilibrium pressure relief, and when the number of relief holes is 3~6, A_V/S_B is smaller, and only 0.0110 can achieve equilibrium pressure relief. This is because when the number of pressure relief holes increases, the air convection will be accelerated, the pressure will drop faster, and the degree of weakening of the shell will be reduced when the opening area is small. Therefore, the number of wells is 6, and $A_V/S_B = 0.0110$ is the control data.

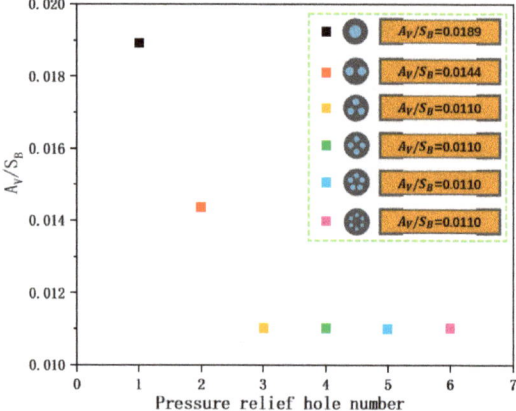

Figure 6. Pressure relief area required to reach equilibrium pressure with different numbers of relief holes.

3.2.2. Influence of Pressure Relief Area

Taking the number of openings as 6 and $A_V/S_B = 0.0110$ as the control, the influence of the area of the pressure relief holes (A_V/S_B are 0.0015, 0.0045, 0.0110, 0.0180, and 0.0268, respectively) on the pressure relief effect was investigated. The curve is shown in Figure 7. It can be seen that when the pressure relief channel is opened at 20.8 MPa, the pressure rise rate in the shell changes. However, because the monitoring point is located in the inner position, the pressure drop is delayed, so the pressure will continue to rise, and the pressure reaches a maximum value of 28 MPa at 1.06 ms. Then, according to the size of the pressure relief area, the subsequent pressure will oscillate down or up. When the aperture is 11 mm, the equilibrium pressure is maintained at 19.74 MPa at approximately 4 ms. It can be seen that the area of the pressure relief channel at this time is the equilibrium pressure threshold in this condition. When the apertures are 14 mm and 17 mm, the area is in an equilibrium pressure relief state, and the equilibrium pressures at 6 ms are 18.24 MPa and 17.00 MPa, respectively. When the aperture is 4 mm and 7 mm, disequilibrium pressure relief occurs at this time, and the pressures at 6 ms are 36.93 MPa and 22.72 MPa, respectively. Then, the pressure inside the shell will continue to rise, eventually reaching the burst pressure of the shell. It can be seen that the timely opening of the pressure relief channel can effectively reduce the pressure in the shell, and with the reduction in the pressure relief area, the pressure in the shell will change from equilibrium pressure relief to disequilibrium pressure relief.

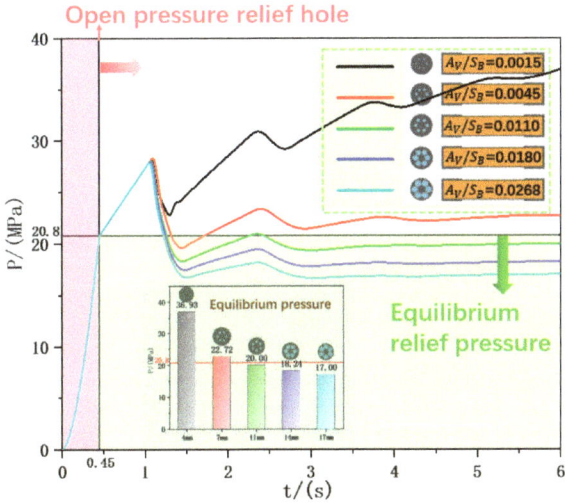

Figure 7. Pressure curves in different pressure relief areas.

3.2.3. Influence of the Length/Diameter Ratio

The pressure curves in different length/diameter ratios (we kept the hole size unchanged at this time, and the hole size under different length/diameter ratios was 11 mm) are shown in Figure 8. It can be seen that with an increasing length/diameter ratio, the pressure hysteresis effect is more serious after the pressure relief channel is opened. When the length/diameter ratio is 4, the equilibrium pressure relief state will finally be formed, and the pressure is 19.74 MPa at 6 ms. When the length/diameter ratio is less than 4, the equilibrium pressure relief is satisfied, and the pressure is all less than 20.8 Mpa. As the length/diameter ratio decreases, the pressure in the shell at equilibrium is also smaller. When the length/diameter ratio is 5 and 8, the pressure first oscillates and drops to a certain value, then continues to rise while the pressure relief channel is opened. The rising rate is related to the length/diameter ratio. The larger the length/diameter ratio, the faster the rising rate. At 6 ms, the pressure in the shell is 23.97 MPa and 35.01 MPa, which cannot provide a good pressure relief effect, resulting in unequal pressure relief. When the length/diameter ratio is 4, it is the pressure relief threshold in this condition, and when the length/diameter ratio is greater than 4, equilibrium pressure relief cannot be formed.

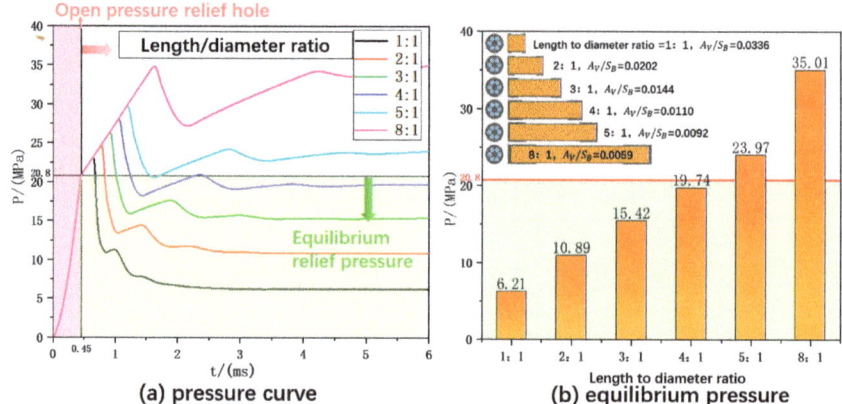

Figure 8. Pressure curves under different length/diameter ratios.

The equilibrium pressure in different length/diameter ratios was studied, the A_V/S_B required to achieve equilibrium pressure relief in different length/diameter ratios was calculated, and the results are shown in Figure 9. It can be seen that with an increasing length/diameter ratio, a larger pressure relief area is needed, and the increase is approximately exponential. When the length/diameter ratio is 5, the pressure relief area has reached the maximum critical area at this time (one end can no longer open a larger aperture). When the aspect ratio is 8, the equilibrium pressure relief cannot be performed in the condition that one end is fully open, indicating that the opening of one end cannot meet the pressure relief conditions at this time, and it needs to be considered in combination with other mitigation structure designs.

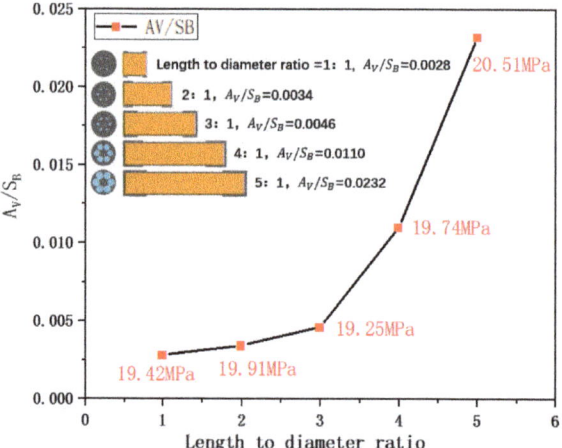

Figure 9. Balance factor at different length/diameter ratios.

3.2.4. Influence of the Pressure Relief Hole Location

Taking the hole diameter as 11 mm and comparing the influence of the position of the pressure relief hole on the pressure relief effect in different length/diameter ratios, the pressure curve is shown in Figure 10. When the length/diameter ratio is 4, an equilibrium pressure relief will be formed, and at 6 ms, the equilibrium pressure of the openings at both ends is 11.71 MPa, which is lower than the equilibrium pressure of the openings at one end of 19.74 MPa. When the ratio is 8, one end of the hole cannot meet the pressure relief requirements, but when the two ends are opened, the pressure is 18.95 MPa at 6 ms, which meets the equilibrium pressure relief requirements. When the length/diameter ratio is too large, the pressure relief method can be adopted at both ends to relieve pressure. The reason for the analysis is that when the holes are opened at both ends, the pressure at the central position can be released from both ends at the same time, which actually reduces the length/diameter ratio of the shell. Therefore, the pressure relief effect of the holes at both ends is better than that of the holes at one end. Therefore, when the holes are opened at both ends, the disequilibrium pressure relief can be transformed into an equilibrium pressure relief in certain conditions, and the pressure originally in the equilibrium pressure relief state can be lower so that the pressure can be released faster and more effectively.

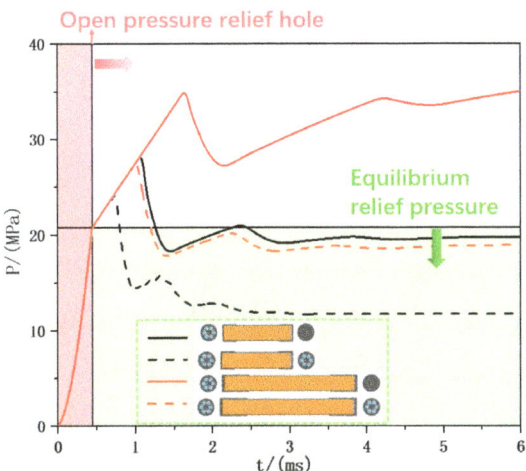

Figure 10. Pressure curves under different pressure relief hole locations.

4. Test of Pressure Relief Hole Plugging Structure Strength

4.1. Hydrostatic Pressure in the High-Temperature Test

Figure 11 shows the metallographic structure of six Sn-Zn-Al La alloys with different Al contents. It is found that as the Al content increases, the gray matrix is an Sn-Zn eutectic phase, while the dotted black is a rich Zn phase. With the addition of Al, the microstructure becomes coarser and coarser. Black dendritic tissue gradually increases and the microstructure distribution of each alloy phase in the eutectic alloy Sn9Zn is relatively uniform. As the Al content continues to increase, needle-like or dendritic Al phases gradually appear at grain boundaries or interdendritic boundaries. When the Al content increases to 10%, the black dendritic structure gradually decreases, and increasingly circular silver phases are formed in the alloy phase. The finer the crystal, the stronger the mechanical properties. Therefore, we chose Sn9Zn-3Al0.2La for the design of the sustained-release structure.

According to the self-developed hydrostatic pressure of the high-temperature test system [25], a comparative experiment was carried out to test the actual pressure-bearing capacity and pressure relief effect of low-melting-point alloys in a set high-temperature environment. The hydrostatic pressure in the high-temperature test system is shown in Figure 12. Three experiments were carried out, namely, the high-temperature blasting pressure of the shell without a mitigation structure and the normal-temperature/high-temperature blasting pressure of the Sn9Zn-3Al0.2La mitigation structure. The test methods are shown in Table 3.

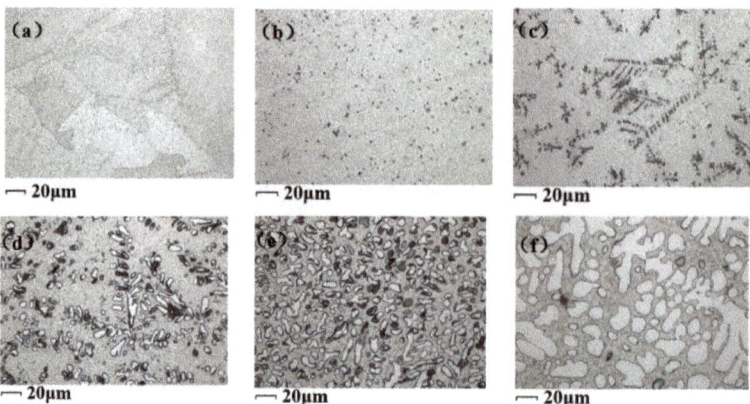

Figure 11. Microstructure of Sn-Zn-Al-La with Different Al and La Content. (**a**) Sn9Zn; (**b**) Sn9Zn-0.8Al0.2La; (**c**) Sn9Zn-3Al0.2La; (**d**) Sn9Zn-10Al0.3La; (**e**) Sn9Zn-20Al0.3La; (**f**) Sn9Zn-20Al0.3La.

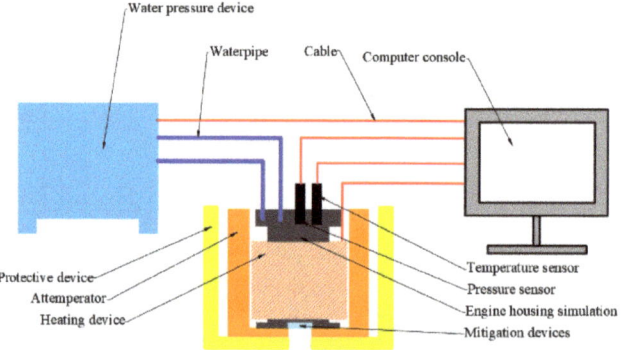

Figure 12. Schematic diagram of the hydrostatic pressure in the high-temperature test.

Table 3. Test method for hydrostatic pressure in a high-temperature test system.

Label	Mitigation Structure	Experimental Conditions	Experimental Method
1	-	high temperature	① Pressurize the shell to 20 MPa and maintain the pressure. ② Heat the shell; set the temperature to approximately 150 °C. ③ Pressurize; the maximum pressure is set to 60 MPa, until the shell bursts.
2	Sn9Zn-3Al0.2La	normal temperature	① Pressurize the shell to 2 MPa and keep the pressure for 0.5 min. ② Pressure; the maximum pressure is set to 31.5 MPa.
3	Sn9Zn-3Al0.2La	high temperature	① Pressurize the shell to 2 MPa and keep the pressure for 0.5 min. ② Heat the shell to 175 °C. ③ Pressurize; the maximum pressure is set to 31.5 MPa.

4.2. Results of the Low-Melting-Point Alloy Plug Strength Test

The pressure time-history curve of the hydraulic experiment of the shell-simulated sample is shown in Figure 13. It can be seen from the comparison that the mitigation structure has a complete structure at normal temperature. In the temperature environment of 175 °C, through the design of the pressure relief diaphragm mitigation structure, the restraint strength of the simulation shell at 175 °C is reduced by nearly 50%, from 40 MPa to 21.21 MPa, which basically meets the start-up requirements of the pressure relief channel. It can be seen from the photos of the wreckage of the pressure relief diaphragm after the experiment that the mitigation structure loses most of its strength under the action of high temperature, fails under the action of internal pressure, and forms a pressure relief channel, which can be applied to the design of the mitigation structure of the shell. This experiment verifies that the low-melting-point alloy mitigation structure can open the pressure relief exhaust channel in advance under the action of a particular temperature environment and internal pressure, which meets the design requirements of the low-melting-point alloy plug mitigation structure.

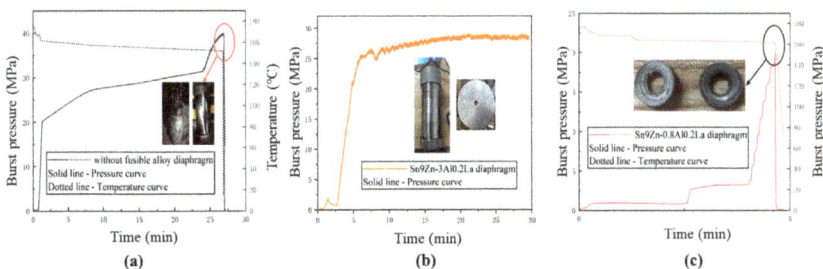

Figure 13. Pressure curve of the hydrostatic pressure in the high-temperature test of the simulation shell. (**a**) simulation shell 1, without mitigation Structure; (**b**) simulation shell 2, Sn9Zn-3Al0.2La; (**c**) simulation shell 3, Sn9Zn-3Al0.2La.

5. Mitigation Design and Cook-Off Test of the Cook-Off Bomb

To verify the pressure relief effect of the designed mitigation structure, a standard cook-off bomb was designed and sealed with low-melting-point alloy plugs. The test site was arranged to conduct fast cook-off and slow cook-off tests to study the mitigation effect of insensitive munitions for reference.

5.1. Design of the Mitigation Structure of the Shelled PBX Charge

A schematic diagram of the structure of the cook-off bomb is shown in Figure 14. Its structure is mainly composed of a shell, front and rear covers, a charge column, and a low-melting-point alloy plug. Six circular holes were opened along the periphery of one end face, and four pressure relief areas were chosen for the circular holes (without the mitigation structure, A_V/S_B were 0.0045, 0.0110, and 0.0180, respectively). The mitigation structure design of the shell was carried out under preset working conditions. The low melting point alloy material used to seal the pressure relief channel must meet the structural strength requirements. Therefore, the pressure relief mitigation structure adopted a threaded mechanical connection. The material of the low-melting-point alloy plug is Sn9Zn-3Al0.2La, which was processed according to the size of different pressure relief channels.

5.2. Test Conditions

The layout of the test site is shown in Figure 15, and the design test is shown in Table 4. In the slow cook-off test, the bomb was heated by remote control, and the temperature rose at the rate of 1 °C/min until the cook-off bomb experienced combustion and an explosion reaction or the temperature reached 400 °C and no reaction occurred. After the test was completed, the cook-off bomb was destroyed. The prepared cook-off bomb was hung horizontally on the support frame. The center was 300 mm directly above the combustion source. Aviation kerosene and an appropriate amount of gasoline were injected into the oil tank to the specified scale line, and the electric ignition device was placed into the base and energized. At the same time, a ground field overpressure sensor was placed 5 m before and after the cook-off bomb. Place the Revealer High-Speed Camera X113, the maximum shooting rate is 25,000 FPS, the maximum memory is 64 GB, the full resolution is 1280 × 1024, and the minimum exposure time is 1 μs, as shown in Figure 16. The response characteristics of the bomb under the state of cook-off were evaluated by the state of the cook-off bomb after the test, the deformation of the shell, and other effective verification methods. Refer to the US military standard MIL-STD-2105D "Nonnuclear ammunitions of risk assessment" [26] to determine the response level.

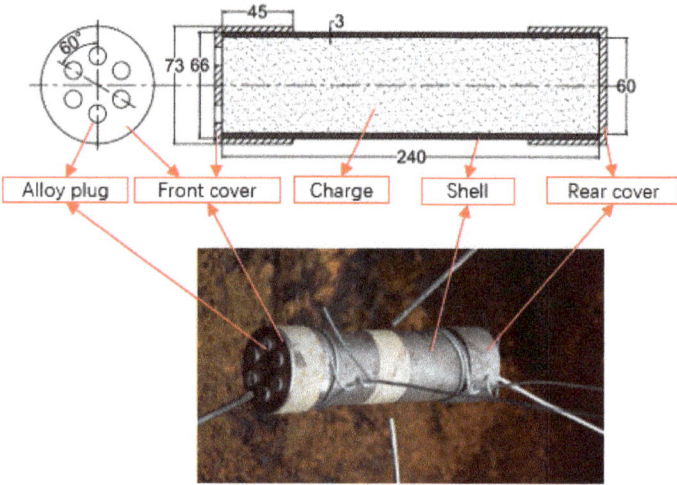

Figure 14. Standard cook-off bomb structure diagram.

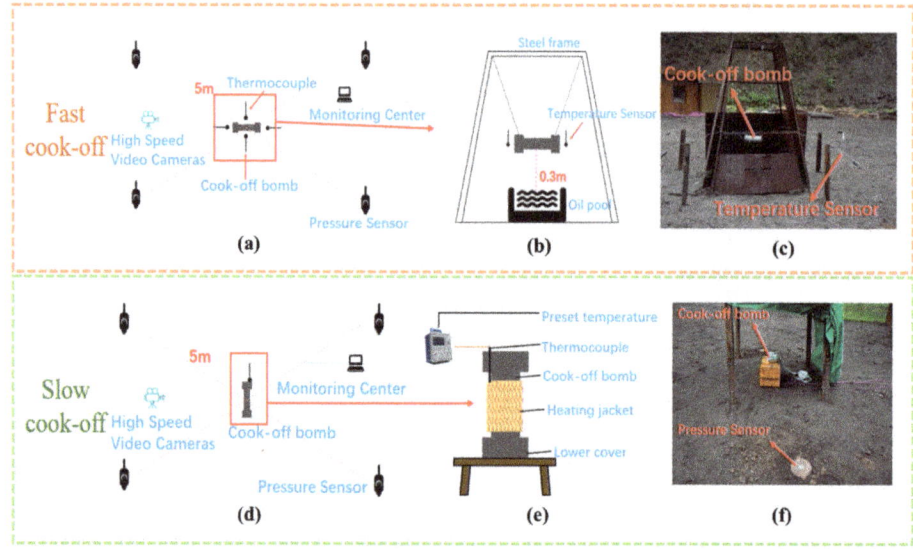

Figure 15. The layout of the cook-off site. (**a**) Layout of fast cook-off Test; (**b**) Oil pool layout plan; (**c**) Physical image of fast cook-off test; (**d**) Layout of slow cook-off Test; (**e**) cook-off bomb layout plan; (**f**) Physical image of slow cook-off test.

Table 4. Cook-off bomb design.

Test Name	Test Condition Number	(A_V/S_B)	Diameter/mm
Slow cook-off	S-1	0	-
	S-2	0.0110	11
Fast cook-off	F-1	0.0045	7
	F-2	0.0110	11
	F-3	0.0180	14

Figure 16. Revealer High-Speed Camera X113.

5.3. Results and Analysis

5.3.1. Slow Cook-Off Test

To study the pressure relief effects of the mitigation structure in the condition of slow cook-off, the cook-off bomb S-1 without mitigation structure and the cook-off bomb S-2 with $A_V/S_B = 1.10\%$ and the mitigation structure were tested. Relevant experiments were carried out, and the reaction characteristics and the effectiveness of the mitigation structure were studied.

During the test, a large amount of gas was first observed in the mitigation structure cook-off bomb, as shown in Figure 17. It can be seen that the exhaust effect of the mitigation structure is obviously. Due to the burning rate of the charge in the later stage is too fast, a thrust is formed, which makes the cook-off bomb break away from the shackles of the heating belt, which stops the heating environment, as shown in Figure 17d. The charge was kept away from the fire environment, and it failed to react completely. The residual charge is shown in Figure 17e. It was further confirmed that the mitigation structure could open well and form a pressure relief channel.

(a) (b) (c) (d) (e)

Figure 17. Effect diagram of the mitigation structure during the slow cook-off. (**a–d**) Slow cook-off test process; (**e**) Residual charge.

The slow cook-off results of the cook-off bomb are shown in Table 5, and the shell fragments are shown in Figure 18. The end cover at one end of the shell was punched open after the reaction of the S-1 cook-off bomb, and the end cover at the other end was still connected to the shell. The side wall of the shell was severely torn and broken into several large fragments, there was almost no residual explosive inside, no shock wave overpressure was detected in the test, and the reaction level was deflagration. When the temperature of the S-2 cook-off bomb is approximately 193 °C, the shell end cap screws are punched out. During the initial reaction of the charge, the gas generated by thermal decomposition forms a thrust, which keeps the bomb shell away from the fire environment but does not react completely. The end caps at both ends of the shell are slightly deflected, but the shell is intact and not damaged, and there is residual explosive inside that has not reacted completely. No shock wave overpressure is detected, and the reaction level is combustion and below. Therefore, it can be seen that the mitigation structure can effectively reduce the internal pressure of the shell during the reaction of the charge, keep the shell away from the fire environment, effectively reducing the intensity of the thermal reaction of the PBX charge, and improve the safety of the slow cook-off of the charge

Table 5. Slow cook-off test results.

Cook-Off Bomb	(A_V/S_B)	Reaction Temperature/$^\circ$C	Shock Wave Overpressure at 5 m/MPa	Response Level
S-1	0	237.0	Not detected	Deflagration
S-2	1.10	193.6	Not detected	Combustion and below

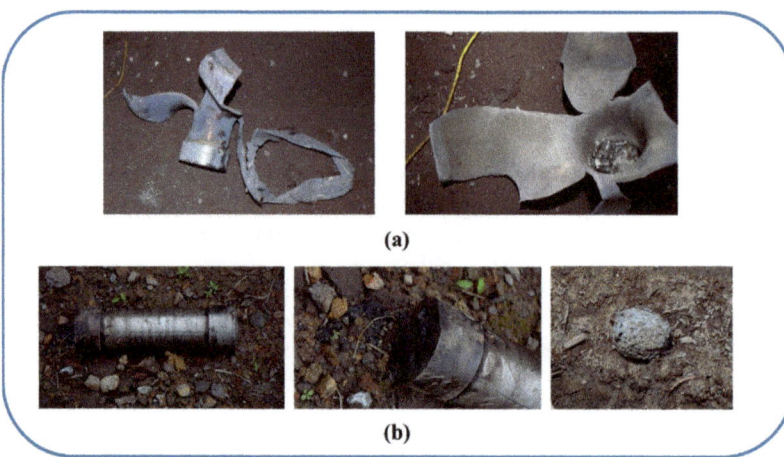

Figure 18. Slow cook-off test wreckage. (**a**) S-1 (without mitigation structure); (**b**) S-2 (mitigation structure).

5.3.2. Fast Cook-Off Test

To study the pressure relief reaction of the mitigation structure in the condition of fast cook-off, relevant experiments were carried out on three kinds of mitigation structure cook-off bombs with pressure relief areas (A_V/S_B are 0.45% (F-1), 1.10% (F-2) and 1.80% (F-3)), and the reaction characteristics and the pressure relief effect of the mitigation structure were studied. During the test, a large amount of gas can be observed from the mitigation structure cook-off bomb, as shown in Figure 19. For the cook-off bomb with the mitigation structure, there is a clear gas flow discharged from the pressure relief channel before the reaction, which confirms that the pressure relief channel can perform a good role in pressure relief.

Video screenshots of each cook-off bomb at different times are shown in Figure 20. It can be seen that it experienced two explosions at 58 s and 71 s, respectively, while F-3 experienced two explosions at 52 s and 66 s, respectively. This is because the response of the cook-off bomb is divided into two stages. In the first stage, after the internal pressure of the charge reaches the pressure threshold of the alloy plug, the alloy plug is destroyed, the gas product breaks through the mitigation structure, the pressure relief channel is opened, and a large-scale fireball is formed, which lasts for approximately 13~14 s. In the second stage, the area of the pressure relief channel formed by the mitigation structure is small, and the pressure release rate of the pressure relief channel is less than the increased rate of the pressure in the shell, resulting in a secondary explosion. The upper-end cover is slightly deformed to create a deflection until the charge in the shell is completely burned. The F-1 cook-off bomb only exploded once at 65 s because the pressure relief area is small, so the pressure release rate is much smaller than the pressure growth rate in the shell, which causes direct damage to the shell, as shown in Figure 20a at 70 s. The shell fragments impacted and destroyed the oil sump. Therefore, when A_V/S_B is more than 1.10%, it can be clearly observed that the pressure relief channel is open, while when A_V/S_B is 0.45%, the pressure relief area is small, and the pressure relief cannot be fully discharged.

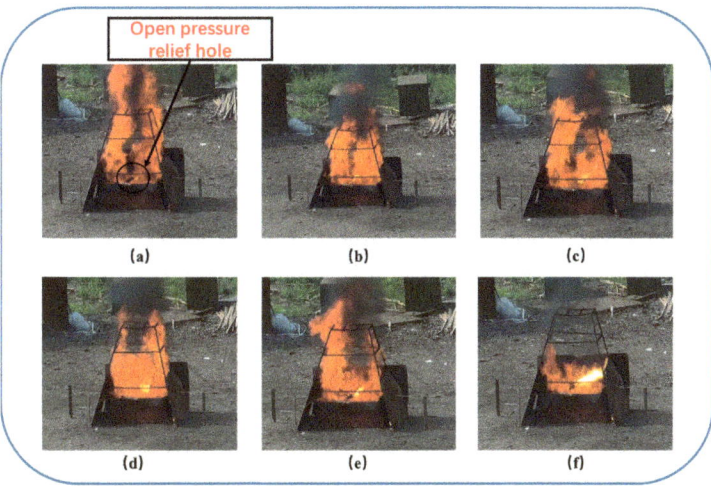

Figure 19. Effect diagram of the effect of the mitigation structure in the process of fast cook-off. (**a–f**) Fast cook-off test process.

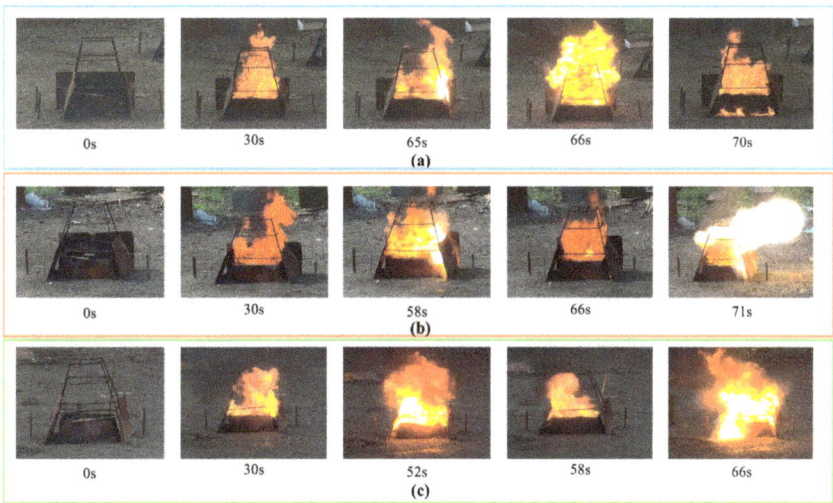

Figure 20. Video screenshot of the response process of the cook-off bomb. (**a**) F-1 video screenshot of fast cook-off response process; (**b**) F-2 video screenshot of fast cook-off response process; (**c**) F-3 video screenshot of fast cook-off response process.

The fast cook-off results of the cook-off bomb are shown in Table 6, and the shell fragments are shown in Figure 21. The difference in the reaction time of different mitigation structures is very small, and the reaction levels are also combustion, but the intensity of combustion is different. The shell of the F-1 cook-off bomb is punched open, the two end caps are washed away, the shell is torn and deformed, and the reaction level is deflagration. The end cap screws of the F-2 cook-off bomb are punched out, the shell is completely without tearing occurs, the end caps at both ends are deformed, and the reaction level is combustion. The end cap screws of the F-3 cook-off bomb are punched open, the shell is intact, and no tearing occurs. The end cap at the end with the pressure relief channel is slightly deformed, and the reaction level is burning. Deflagration and the following reactions all occur for cook-off bombs with mitigation structures. When the pressure relief area increases, although the combustion reaction occurs, the severity of the reaction gradually decreases from the rupture of the product on site and the video. This shows that the increase in the pressure relief area has a certain effect on reducing the reaction level of the ammunition, which can improve the thermal safety of the ammunition.

Table 6. Fast cook-off test results.

Cook-Off Bomb	(A_V/S_B)	Time of the Pressure Relief Channel Is Open/s	Time of Charge Reaction/s	Shock Wave Overpressure at 5 m/MPa	Response Level
F-1	0.45	—	65	Not detected	Deflagration
F-2	1.10	58	71	Not detected	Combustion
F-3	1.80	52	66	Not detected	Combustion

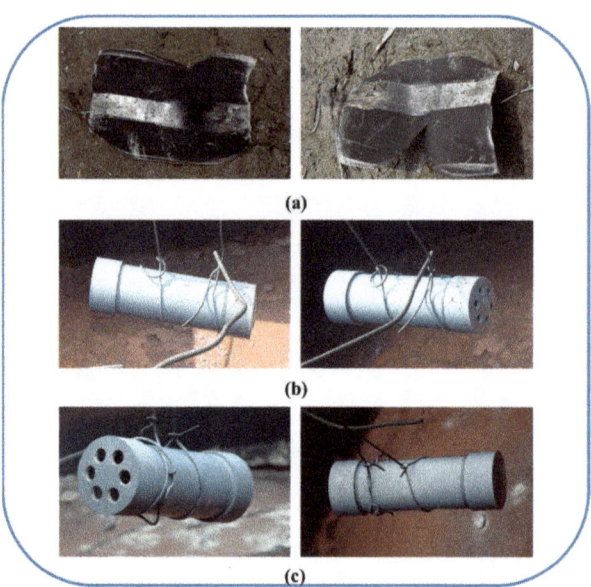

Figure 21. Cook-off bomb wreckage. (**a**) F-1 Cook-off bomb wreckage; (**b**) F-2 Cook-off bomb wreckage; (**c**) F-3 Cook-off bomb wreckage.

Figure 22 shows the wreckages of the low-melting-point alloy plug with a diameter of 14 mm in the F-3 cook-off bomb after the test. The alloy plugs of the F-1 and F-2 cook-off bombs were not found. The overall structure of the alloy plug is basically intact. The reason why the screw is ejected is because the high temperature softens the alloy plug and reduces the mechanical strength. During the action of the internal pressure of the shell, the thread fails and is damaged, and the pressure relief channel is opened to achieve effective pressure relief inside the shell. The damage to the screw cap is caused by spraying out during the first explosion and hitting the witness board or other objects, and experiencing damage. In summary, the above phenomena show that the designed mitigation structure can reliably relieve pressure during the cook-off process and improve the thermal safety of the charge.

Figure 22. Wreckages of 14 mm alloy plug after fast cook-off.

6. Conclusions

In this paper, the design method of the mitigation structure is established, the influence of different mitigation structure designs on the pressure relief effect is simulated and calculated, and the mitigation structure is designed and verified by the cook-off experiment. The study found:

(1) For PBX explosives, when the length/diameter ratio is 4 and $A_V/S_B = 0.0189$, the reaction intensity can be effectively reduced.

(2) As the number of relief holes increases, the required relief area will decrease. When the number of relief holes is 6, $A_V/S_B = 0.0110$ is the equilibrium pressure relief size.

(3) When $A_V/S_B \leq 0.011$, the balanced pressure relief of the shell pressure relief process can be effectively realized, and as the pressure relief area of the mitigation structure is larger, the reaction intensity is smaller.

The work of this paper provides a reference research method for the design of the mitigation structure of a shelled charge under the action of thermal stimulation and the mitigation structure of the shell.

Author Contributions: Conceptualization, J.N. and R.L.; methodology and formal analysis, J.L.; investigation, J.L. and M.H.; resources, G.J.; data curation, X.S.; writing—original draft preparation, J.L.; writing—review and editing, J.N.; supervision, X.W. and B.H.; funding acquisition, J.N. All authors have read and agreed to the published version of the manuscript.

Funding: This work was supported by the National Natural Science Foundation of China [grant numbers 11772058] (Jianxin Nie).

Institutional Review Board Statement: Not applicable.

Data Availability Statement: The raw/processed data required to reproduce these findings cannot be shared at this time as the data also forms part of an ongoing study.

Conflicts of Interest: The authors declare no conflict of interest.

References

1. Powell, I.J. Insensitive Munitions—Design Principles and Technology Developments. *Propellants Explos. Pyrotech.* **2016**, *41*, 409–413. [CrossRef]
2. Wang, Q.; Zhi, X.; Xiao, Y. Analysis of the effect of a venting structure on slow cookoff of Comp-B based on a universal cookoff mode. *Explos. Shock Waves* **2022**, *42*, 61–69. [CrossRef]
3. Robert, H.; Zachary, A. Spears Matthew Sanford. Advanced Precision Kill Weapon System (APKWS) IM Solutions. In Proceedings of the Insensitive Munitions & Energetic Materials Technology Symposium (2006), Bristol, UK, 24–28 April 2006.
4. Stephen, K.; Fred, B. Joint General Purpose Bomb Insensitive Munitions Program. In Proceedings of the Insensitive Munitions & Energetic Materials Technology Symposium (2006), Bristol, UK, 24–28 April 2006.
5. Tony, W. General Purpose Bomb Fast Cook-off Mitigation Techniques. In Proceedings of the Insensitive Munitions & Energetic Materials Technology Symposium (2010), Munich, Germany, 11–14 October 2010.
6. Stephen, K. Venting Techniques for Penetrator Warheads. In Proceedings of the Insensitive Munitions & Energetic Materials Technology Symposium (2010), Munich, Germany, 11–14 October 2010.
7. Stephen, K. Success in Venting Penetrator Warheads. In Proceedings of the Insensitive Munitions & Energetic Materials Technology Symposium (2012), Las Vegas, NV, USA, 14–17 May 2012.
8. William, C.; Enic, E.; Adcl, S. *Fast Cook-off Tests Report*; Center for the Simulation of Accidental Fires & Explosions: Salt Lake City, UT, USA, 2003.
9. Glascoe, E.; Dehaven, M.; McClelland, M.; Greenwood, D.; Springer, H.; Maienschein, J. Mechanisms of Comp-B Thermal Explosions. In Proceedings of the 15th International Detonation Symposium (2014), San Francisco, CA, USA, 13–18 July 2014.
10. Kinney, G.; Sewell, R. *Venting of Explosions*; NWC TM 2448; Naval Weapons Center: China Lake, CA, USA, 1974.
11. Sahin, H.; Narin, B.; Kurtulus, F.D. Development of a design methodology against fast cook-off threat for insensitive munitions. *Propellants Explos. Pyrotech.* **2016**, *41*, 580–587. [CrossRef]
12. Wardell, J.; Maienschein, J. The Scaled Thermal Explosion Experiment. In Proceedings of the 12th International Detonation Symposium (2002), San Diego, CA, USA, 11–16 August 2002.
13. Niu, G. Experimental study on the fast cook-off of ammunition with pressure mitigation structure at head. In Proceedings of the Symposium on Energetic Materials and Insensitive Munitions (2016), Nashville, TN, USA, 12–15 September 2016.
14. Madsen, T.; Fisher, S.; Baker, E.; Suarez, D.; Fuchs, B. *Explosive Venting Technology for Cook-off Response Mitigation*; Technical Report ARMET-TR-10003; Army Armament Research Development and Engineering Center Picatinny Arsenal nj Energetics Warheads and Manufacturing Technology Directorate: Picatinny Arsenal, NJ, USA, 2010.
15. Mcchristian, L.; Cistano, J.; Foxx, C.; Hartmann, W.; Murphy, W.; Stauner, R.; Takata, A. *Vulnerability of Nuclear Weapon Systems to Fire Studies of Burning Explosives*; Report RTD-TDR-63-3086; IIT Research Institute: Chicago, IL, USA, 1963.
16. Hu, R.; Zhao, F.; Gao, H.; Yao, E. Estimation of detonation performances of explosives using M, ρ, $\Delta_f H_{m\theta}$, C_p, $T_{ig\ or\ b}$ of C-H-O-N explosives and $\Delta_f H_{m\theta}$ of detonation products. *Chin. J. Explos. Propellants* **2013**, *36*, 20–23. [CrossRef]
17. Graham, J. Mitigation of Fuel Fire Threat to Large Rocket Motors by Venting. In Proceedings of the 2010 Insensitive Munitions & Energetic Materials Technology Symposium (2010), Munich, Germany, 11–14 October 2010.
18. Sinditskii, V.; Levshenkov, I.; Egorshev, Y.; Serushkin, V. Study on Combustion and Thermal Decomposition of 1,1-Diamino-2,2-dinitroethylene (FOX-7). In Proceedings of the Inter Seminar Europyro & Inter Pyrotechnics Seminar, Saint-Malo (2003), Saint-Malo, France, 23–27 June 2003.
19. Chen, K.; Huang, H.; Lu, Z.; Nie, S.; Jiang, Z. Experimental Study on Cook-Off Test for Melt-Cast and Cast-Cured Explosive at Strong Constraint. *J. Sichuan Ordnance* **2015**, *36*, 133–136. [CrossRef]
20. Kou, Y.; Chen, L.; Ma, X.; Zhao, P.; Lu, J.; Wu, J. Cook-off experimental and numerical simulation of RDX-based aluminized explosives. *Acta Armamentarii* **2019**, *40*, 978–989. [CrossRef]
21. Fisher, M. Shape Memory Polymer (SMP) Venting Mechanism for Munitions. In Proceedings of the NDIA/MSIAC Insensitive Munitions & Energetic Materials Technology Symposium (2010), Munich, Germany, 11–14 October 2010.
22. Strickland, A. Development of IM Brimstone rocket motor; An IM, minimum smoke, air-launched system. In Proceedings of the 2015 Insensitive Munitions & Energetic Materials Technology Symposium (2015), Rome, Italy, 18–21 May 2015.
23. Wei, T.; Li, N.; Nie, J.; Liang, J.; Guo, Z.; Yan, S.; Zhang, T.; Jiao, Q. The properties of Sn–Zn–Al–La fusible alloy for mitigation devices of solid propellant rocket motors. *Def. Technol.* **2022**, *18*, 1688–1696. [CrossRef]
24. Nie, J.; Li, X.; Guo, X.; Zhang, H.; Yan, S.; Zhang, T. Study on Response Characteristics of Insensitive PBX Explosive in Shell under Slow Cook-off. *J. Saf. Environ.* **2022**, *23*, 1085–1092. [CrossRef]
25. Nie, J.; Wei, Z.; Guo, X.; Yan, S.; Jiao, Q.; Fan, W.; Liang, X. A Temperature Controllable Engine Casing Hydraulic Blasting Experiment System and Method. CN112748013A, 10 June 2022.
26. *MIL-STD-2105D*; Hazard Assessment Tests for Non-Nuclear Munitions. The Department of National Defense of USA: Washington, DC, USA, 2011.

MDPI

St. Alban-Anlage 66

4052 Basel

Switzerland

www.mdpi.com

Crystals Editorial Office

E-mail: crystals@mdpi.com

www.mdpi.com/journal/crystals